A Guide to the TCP/IP Protocol Suite

BOOK 6

A Guide to the TCP/IP Protocol Suite

Second Edition

A Guide to the TCP/IP Protocol Suite

Second Edition

Floyd Wilder

Artech House
Boston • London

Library of Congress Cataloging-in-Publication Data
Wilder, Floyd
    A guide to the TCP/IP protocol suite / Floyd Wilder. — 2nd ed.
      p.   cm. — (Artech House telecommunications library)
    Includes bibliographical references and index.
    ISBN 0-89006-976-X (alk. paper)
    1. TCP/IP (Computer network protocol). I. Title. II. Series.
TK5105.585.W55   1998
004.6'2—dc21   98-25739
                                                               CIP

British Library Cataloguing in Publication Data
Wilder, Floyd
    A guide to the TCP/IP Protocol Suite. – 2nd ed.
    (Artech House telecommunications library)
    1. TCP/IP (Computer network protocol)
    I. Title
    004.6'2

    ISBN 0-89006-976-X

Cover design by Lynda Fishbourne

© 1998 ARTECH HOUSE, INC.
685 Canton Street
Norwood, MA 02062

International Standard Book Number: 0-89006-976-X
Library of Congress Catalog Card Number: 98-25739

10 9 8 7 6 5 4 3

To Marie,
Matt, Mike, Joseph and Pat

Contents

Preface

General

The TCP/IP Internet architecture and protocol specifications are maintained by the Internet Society, an international body with responsibility for Internet standards. The specifications of the TCP/IP protocols are contained in various request for comments (RFCs). The RFCs (one per topic) contain meeting notes, political statements and gripes, excellent analysis of several technologies, obsolete specifications, and precise, well-written current specifications for every protocol used in the Internet.

The problems facing a person trying to learn about the set of TCP/IP protocols (protocol suite or protocol stack) are finding the one in 2,300 (or more) RFCs that covers the particular protocol of interest and determines whether that RFC is the latest, whether it is a standard, and which RFC explains how that protocol fits in the total Internet architecture. About 1,000 RFCs were written in just the past five years, and the same amount or more will probably be written during the next five years. Although the RFCs have always been accessible, anyone with Internet access can now easily obtain published RFCs and draft specifications via the Internet. Moreover, unlike other published standards (e.g., IEEE, ANSI, ISO, and ITU-T standards), the Internet specifications are free! (An example is provided in Section 6.2.)

So why buy this book if the specifications are free? Although the Internet RFCs are easy to read compared to other standards, they are still formal documents written at a level to satisfy the needs of the software engineers, programmers, and hardware engineers who implement the standards. Furthermore, it is often necessary to read multiple RFCs and possibly other standards (e.g., IEEE and ITU-T standards) to understand some Internet protocols. This second edition, updated to reflect the constant change that characterizes the Internet RFCs, conveys a clear understanding of the major TCP/IP protocols without the formal elegance of specifications (e.g., pseudo code). It does so with graphic illustrations, easy to understand language, and summarized descriptions. The book descriptions are concise, yet substantial, resulting in a more than 10-to-1 compression ratio compared to the actual specifications. It is recommended for anyone who needs to know how the TCP/IP protocols work, including software engineers, programmers, and hardware engineers. The few readers that actually write code or design hardware components can study the book's illustrations and detailed functional descriptions before tackling the referenced RFC specifications for the latest changes and other fine details. Many others may never need to read a formal specification, RFC or other.

As mentioned previously, RFC specifications (and others) are changing daily. Readers who are writing code, designing hardware, or doing network analysis at the bit level (e.g., using an analyzer) should always check the latest RFC specifications.

As an independent data communications consultant in the late 1980s, I was involved with the design, development, and certification of X.25, DSP, QLLC, IEEE 802.x, and T1 interfaces. (Prior to that, I worked with other protocols—including BSC, HASP, IBM 2780, ACP127, 83B, and SOTUS—dating back to torn-tape centers in the early 1960s.) In 1988, I was given an opportunity to add a TCP/IP connection to a WAN node (a popular IP router). I was excited about it and scrambled feverishly through all the RFCs, just over 1,000 at the time. I read a lot and kept notes and, in the process, became a serious student of the TCP/IP protocol suite. Subsequently, I wrote the first edition of this book in 1993 and added the TCP/IP protocol suite to my training syllabus. The revisions contained in this second edition reflect my continued study of the TCP/IP protocol suite. I believe you will enjoy learning about TCP/IP also.

Purpose and Objectives

The purpose of this second edition is to provide the reader with a substantial yet concise description of all the major protocols of a TCP/IP-based network. It should be a handy technical guide at the level required by students, data communications analysts, and network software engineers. Meanwhile, for the programmer who requires the actual specifications, the book aims to provide an understanding of how it works and the precise RFC number of the specification to minimize the search time. Relevant endnotes with reference to RFC numbers are contained throughout the book. Readers should keep in mind, however, that since these RFCs may become obsolete in the future, they will need to check the periodic IAB Official Protocol Standards RFC for the latest numbers. (Currently, RFC 2200 contains the latest IAB Official Protocol Standards.) The book's secondary objective is to provide an organized description of the TCP/IP protocol suite, with many illustrations suitable for training.

Audience

This book is about how the Internet and TCP/IP protocol suite work; it is not about how to use the Internet (i.e., surf the Web). Accordingly, the reader who will benefit most from reading this book is interested in what makes the Internet tick. In an analogy to a car, it is the mechanic or automotive engineer who will benefit most—not the taxi driver or casual commuter.

The intended audience for Chapter 1 is anyone having a desire to learn the fundamentals of the TCP/IP protocol suite. Readers do not require a data communications background to understand the high-level overview of the TCP/IP protocol suite provided by Chapter 1. This chapter describes the control and data flow with illustrations and uses examples to step through a typical Internet transaction.

The intended audience for the remainder of the book is students of data communications and data communications professionals with a desire to understand the concepts, structure, and elements of a TCP/IP-based network. Many descriptions are at a level that would be seen on an analyzer, which makes it ideal for designers, implementers, and troubleshooters of TCP/IP-based networks. One designing a TCP/IP protocol interpreter for a LAN monitor can turn to this book's handy reference of protocol numbers, type codes, well-known port numbers, and other sometimes hard-to-find information. The book is also ideal for one planning to take a class in data communications, computer networks (LANs and WANs), or internetworking. For anyone planning to take an advanced class with a prerequisite for understanding telecommunications technologies, Chapter 2 (Physical and Link Layer Interfaces) is a must.

Motivation To Understand TCP/IP

The TCP/IP protocol suite is the most widely used in the world for internetworking LANs and WANs, and there is no end in sight to its prevalence.

Virtually every college student with a computer science major will be familiar with the TCP/IP protocol suite by graduation time.

Learning data communications protocols is similar to learning foreign languages in that each new protocol or language learned makes it easier to learn another. Readers who believe that TCP/IP will be here forever will, of course, think it important to learn this protocol suite. On the other hand, readers who believe that TCP/IP is transient (which it and all protocols are) know that understanding TCP/IP will make it easier for them to learn the next new protocol.

Book Organization

This book is divided into major chapters. Chapters 2–6 correspond loosely to the layers of the TCP/IP protocol suite. Chapter 1 identifies the Internet and the structure and protocols of each TCP/IP layer. It also contains an easy-to-follow example of a typical data flow through the Internet. Chapter 2 describes the link and physical layer interfaces. Most of the revisions to this book are concentrated in Chapter 2, which covers many interfaces to the Internet. In addition, Chapter 3 has been rewritten and now contains a thorough treatment of IP forwarding. Chapter 4, which was not changed significantly, describes the transport layer protocols. Chapter 5 describes the utility protocols used in the Internet; it was revised to include new material, including the dynamic routing protocols. The material from Chapter 7 in the original book was enhanced with a description of HTTP and moved to Chapter 6. Hence, all protocols described herein that provide a direct user service (e.g., FTP, Telnet, Mail, and HTTP) are now contained in Chapter 6. Chapter 7 discusses changes I expect to occur within the TCP/IP protocol suite in the next five years.

Appendixes A–D supply additional information related to the TCP/IP protocol suite:

1. Appendix A provides some historical aspects of networking and discusses the Internet evolution, with comparisons to ITU-T and ISO Standards.

2. Appendix B provides a primer for numbering systems.

3. Appendix C provides the official Internet Protocol Standards.

4. Appendix D lists type codes, protocol numbers, well-known port numbers, and other hard-to-find reserved numbers within the Internet community.

The book also offers readers a glossary of hard-to-find terms. In addition, there are endnotes following each chapter. Readers should generally ignore these endnotes unless they want to obtain the specification or reference material for that topic. (Some endnotes reflect additional information and opinions.) RFCs and other specifications are not typically cited in the textual descriptions.

Acknowledgments

While I had only limited contact with members of the Internet community, any requests I made for assistance were greeted with an enthusiastic response. The normal secrecy associated with vendor-specific, proprietary protocols is absent from communications with "Internetters."

I appreciate the help of Ned Freed, who I contacted because he was the listed co-author of RFC 1341, the MIME Specification. He not only explained why the MIME header was not put in the text (as I believed it could have been), he explained various new changes being made by Working Groups in the IETF. I also contacted SNMP Research Incorporated where Jeffery Case was busy implementing multilingual SNMP-2 agents.

Although I have read several data communications and TCP/IP related books, the source for details provided in this book has primarily been the RFCs and other standards such as IEEE, ANSI, and ITU-T. Some of the examples are taken directly from the RFCs.

Dr. Vinton Cerf was especially helpful with the first revision of this book. He exhibited great patience with my failure to understand the subtleties of his protocols.

I would also like to extend gratitude to Jim Conard for directing me to Artech House and for providing me with technical guidance over the years (about 30 now).

After the first publication of this book, I was employed by Advanced Computer Communications (ACC) as the course developer for internetworking products. My perspective of the Internet has certainly changed since working for ACC. I previously viewed the Internet more from the outside (as a host). My view is now more from the inside (as a router). Since there are gray areas of router operation that are not defined by RFCs, many details vary from vendor to vendor. My knowledge of these gray areas is ACC-biased, for which I am appreciative. Several ACC employees have contributed to my general, and biased, understanding of the Internet. Art Berggreen has been a last resort for information, only because he seems to know everything and the queue to see him is long. Others who I have queried for help include Alan DeMars, Barbara

Berggreen, Dan Valeska, Derek Rosen, Evan Caves, George Wayne, John Mirk, Ken DuPar, Neil Beard, Paul Harding-Jones, Randy Sprague, Steve Johnson, and Todd Dewell.

In conclusion, the Internet is truly an open system—and that quality of openness extends to the people who develop, maintain, and enhance it. I enjoy and appreciate being a part of it.

1

Internet Overview

1.1 Internetworking Evolution

Only 20 years ago, state-of-the-art data communications involved transmitting commands and data, at 2,400 bits per second or less, over analog telephone lines to a remote host for processing by an application program. This was the era of the host-centric network, and all hardware providers used it, as illustrated in Figure 1.1.

To simply send data to another terminal in the same city (or next door), it was first sent to a remote host and then relayed to the other terminal in a process called terminal-to-terminal communication. The problem was that the two terminals could be in Los Angeles while the remote host would be in New York. The situation was made worse by the fact that the communications protocols were proprietary. That is, a "brand X" host would only communicate with a "brand X" terminal. It was a great time for hardware mainframe manufacturers and the telephone companies.

The climate was right for an innovative person like Bob Medcalfe. During the early 1970s, he invented Ethernet, a *local area network* (LAN) technology that eliminated the need for a host in terminal-to-terminal data communications between local subscribers. In addition, Ethernet raised the speed of data communications by a factor of 5,000 (2 Kbps to 10 Mbps.) Ethernet was jointly developed by Digital Equipment Corporation (DEC), Intel Corporation, and Xerox Corporation. After making minor modifications to this LAN technology, the IEEE made it into a standard called IEEE 802.2/IEEE 802.3. As modified, the technology is referred to as Ethernet. To differentiate the new standard from the original, the original (as invented by

1

Figure 1.1 Host-centric networking.

Bob Medcalfe) is referred to as Ethernet Version 2, Ethernet II, or DIX Ethernet (for its developers).

The climate for innovation remained fertile (as it will always be). IBM developed token ring as its LAN technology; this was standardized by the IEEE as IEEE 802.5. These dissimilar LANs (Ethernet and token ring, illustrated in Figure 1.2) would not interwork, nor would they work with any nonproprietary *wide area network* (WAN) connection.

In the late 1970s, Vinton Cerf invented *transport control protocol/internet protocol* (TCP/IP), a nonproprietary network and transport protocol that enabled dissimilar LAN and WAN protocols to interwork. TCP/IP was developed and used in the *Advanced Research Projects Agency (ARPA) Network* (ARPANET). TCP/IP made it possible for the dissimilar network topologies of ARPANET to interwork. In hindsight, it does not seem such a complicated invention. Basically, it provides a transport and network protocol that will work with any physical medium for communications (including copper, coaxial, fiber, satellite, and microwave).

The device used to connect dissimilar networks was originally called a gateway, because it provided a gateway function from one communications medium to another. It is now called a router. The old ARPANET was split in two parts: The first, intended for military use, was called MILNET. The second part was the Internet, which we all use today that has grown well beyond what was originally thought possible.

Although the dissimilar networks could interwork, distributed processing and file sharing (file servers) could not be done because multiple users would access the same file on a server. (See Figure 1.3.) Since there was no file-locking mechanism, when two users opened the same file, one user would corrupt it for

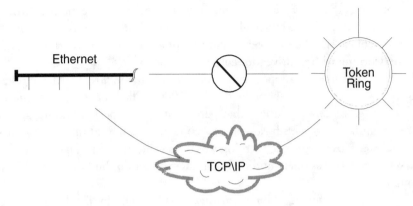

Figure 1.2 Dissimilar networks.

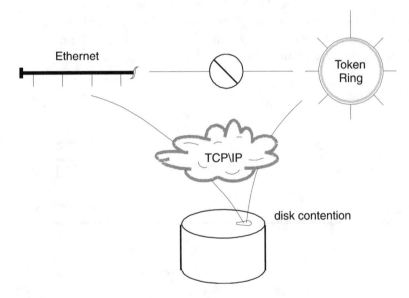

Figure 1.3 File contention for multiple users.

the other. Corvus was the original developer of a *network operating system* (NOS) that provided a file-locking mechanism to resolve file contention. This was first improved by Banyan Vines and later by Novell. By any index of success, the Novell NOS has been a winner.

The *International Standards Organization* (ISO) attempted to standardize an internetworking protocol to replace TCP/IP but failed. ISO's most significant contribution is the *Open System Interconnection* (OSI) reference model. This, or at least the concept, is used throughout the communications industry to identify the protocol layers. The concept put into practice (the Internet) has made internetworking with multivendor equipment possible.

IBM continued to support its *system network architecture* (SNA) and token ring but with an enhancement in the late 1980s called *advanced program-to-program communications* (APPC) that permitted peer-to-peer communications. Other manufacturers with host-centric architectures followed the leader, but it was too late. Host-centric networking was out, as illustrated in Figure 1.4. The market floodgate was open, and manufacturers could not build routers quickly enough.

Advanced Computer Communications (ACC) was in the lot of router manufacturers. Its chief architect, Art Berggreen, was a graduate student at the University of California at Santa Barbara (UCSB) during the invention phase of TCP/IP in the mid to late 1970s. Art and Ron Stoughton were the original developers of the *interface message processor* (IMP), the interface between an IBM 360 host and a gateway, now called a router. Appendix A provides more information on the evolution of networking.

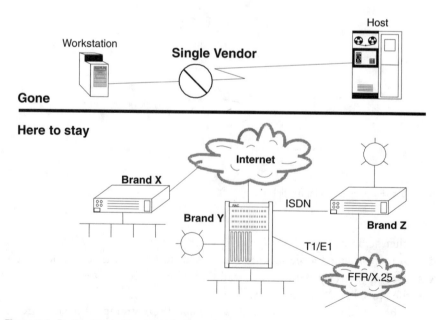

Figure 1.4 Replacement of host-centric networks.

1.2 Internet

Internetworking (or internetting) is the connection of multiple networks. The Internet is the connection of multiple diverse networks with different hardware technologies. The protocol used by the *Defense Advanced Research Project Agency* (DARPA) in the 1970s to construct the original Internet backbone (called ARPANET) is called TCP/IP. The name TCP/IP is taken from the names of the Internet layer protocol (IP) and one of the transport layer protocols (*transmission control protocol* [TCP]). All the protocols used by the Internet, which are collectively described as the TCP/IP protocol suite, are standardized under the auspices of the *Internet Society* (ISOC), a nonprofit, international body.

The Internet connects many government agencies, colleges, and corporations on a worldwide basis. Government agencies connected by the Internet include the following:

1. *National Science Foundation* (NSF);
2. *Department of Energy* (DOE);
3. *Department of Defense* (DOD);
4. *Health and Human Services* (HHS);
5. *National Aeronautics and Space Administration* (NASA).

The most used upper layer programs of the Internet are the following:

1. *World Wide Web* (WWW) *browsers:* Used to search data stores on the Internet;
2. *Electronic mail* (e-mail): Used to exchange electronic memos with individuals or groups with acknowledged delivery;
3. File transfer: Used to send (or exchange) large files;
4. Remote login: Allows a user connected to one computer to log into an application located at a different computer;
5. Remote procedure call: Used to execute remote procedures.

There are many private connections of multiple diverse networks that use the TCP/IP protocol suite. The most popular connection is of multiple Ethernet LANs and serial links to *Internet service providers* (ISPs). Implementations of the TCP/IP protocol suite in private networks are referred to as *internets* (lowercase), or more recently, as *intranets*. Specifications are the same as those for the Internet (RFCs). The descriptions herein pertain to both the Internet and intranets. Section 1.2.1 summarizes the agencies controlling the Internet. The structure and process it describes were put into place during the early part of 1993. Current changes are reflected in RFCs identified in Table 1.1.

1.2.1   Internet Architecture

The Internet architecture is of a layered design, which makes testing and future development of Internet protocols easy. The architecture and major protocols of the Internet are controlled by the Internet Architecture Board (IAB) [1]. Figure 1.5 illustrates the Internet architecture, with a comparison to other popular stacks, and the OSI reference model.

   The Internet provides three sets of services. At the lowest level is the IP, a connectionless delivery service. The next level is the transport layer service. There are multiple transport layer services that use the IP service. The highest level is the upper layer service. Layering of the services permits research and development on one layer without affecting the others—in theory, anyway.

   The physical/link layer envelops the IP layer header and data. If the physical layer is an Ethernet LAN, the IP layer places its message (datagram) in the Ethernet (physical/link) frame data field. The transport layer places its message (segment) in the IP data field. The application layer places its data in the transport layer data field. (Not all application layer protocols have headers.)

   With multiple transport layer and upper layer protocols, what dictates how a particular combination of protocols is chosen? As will be detailed in Section 3.2, a field in the IP header designates which transport layer protocol will act on the IP datagram. Likewise, a field in the transport layer header designates which upper layer program will act on the transport layer segment. An upper layer program, based on its need for reliability and throughput, selects the ap-

Table 1.1
Specifications for ISOC Structure and Process

RFC	Title
2031	ETF-ISOC relationship
2028	The organizations involved in the IETF standards process
2027	IAB and IESG selection, confirmation, and recall process
2026	The Internet Standards Process
1601	Charter of the Internet Architecture Board (IAB)

TCP/IP Internet	Netware	Apple	ISO
Upper	NDS	AFP	Application
Upper	NCP		Presentation
Upper	SAP	ASP	Session
Transport	SPX	ATP	Transport
Internet	IPX	DDP	Network
Link	Data Link	Apple-LAP	Data Link
* Physical	Physical	Physical	Physical

* Physical Layer - Ethernet, 802.3 (10BASE2, 10BASE5, 10BASET),
Token Ring, 802.5, RS232, RS449, RS422, RS423, V.35, X.21

Figure 1.5 Conceptual layering of the Internet protocols.

propriate transport layer protocol to use. Since there is only one IP layer, going down the stack from a transport layer protocol does not involve a selection. The physical/link layer selected by the IP layer is dictated by an interface table associated with the IP address in the IP header.

In theory, any layer may be replaced with a new technology without affecting the other layers. The clear separation of layers was not maintained during the implementation of the Internet, and each layered protocol has dependencies. That is, the functionality dependence was not maintained. Furthermore, some of the functions are redundant, there is no unique protocol identifier field at the transport layer (port numbers are used instead), and the overhead for each layer is relatively high compared to some proprietary protocols. These are problems that the purest might bring to light. It is difficult to take them too seriously when the Internet can handle 800 Mbps throughput rates routinely. In fact, throughput rates at the gigabits per second level have been measured.

The major TCP/IP Internet protocols and their associated layers are illustrated in Figure 1.6. The upper layer protocols are divided into two groups—those that provide a utility function to the Internet and those that provide a service directly to the user. Those that provide a direct user service are the following:

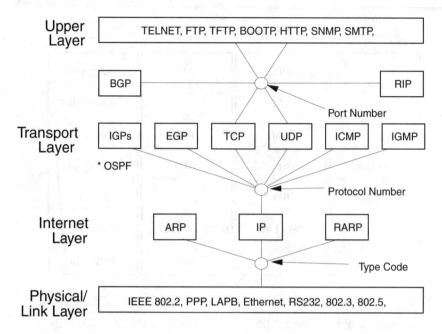

Figure 1.6 Internet architecture and protocols.

1. *Hypertext transfer protocol* (HTTP): Allows for WWW information retrieval, accounting for over one half of all Internet traffic;
2. *Simple message transfer protocol* (SMTP): Provides e-mail capability;
3. TELNET: Provides remote logon capability;
4. *File transfer protocol* (FTP): Provides reliable file transfer capability;
5. X Window System: Provides a graphical interface to applications;
6. *Trivial file transfer protocol* (TFTP): Provides a simple file transfer capability;
7. *Network file system* (NFS): Provides remote virtual storage capability.

Upper layer programs that provide a utility function are the following:

1. *Simple network management protocol* (SNMP): Provides network management information;
2. *Boot protocol* (BOOTP)/*dynamic host configuration protocol* (DHCP): Provides remote loading and configuration capability for diskless workstations;

3. *Domain name service* (DNS): Provides a directory assistance function for using names instead of Internet addresses;

4. *Address resolution protocol* (ARP): Provides a link layer address given an IP address;

5. *Reverse address resolution protocol* (RARP): Provides an IP address given a link layer address;

6. *Inverse address resolution protocol* (IARP): Provides an IP address from a hardware (e.g., DLCI for PVC) address.

The architecture and various protocols are identified in Figure 1.6. The designation of a protocol to one layer or another is more complex than is illustrated.

For example, the *Internet control message protocol* (ICMP) and *Internet group management protocol* (IGMP) are an integral part of the Internet layer. However, each receives data and control in the same manner as a transport layer function, namely, by an assigned protocol number contained in the IP header. Hence, they are illustrated here based on data flow and control. For the same reason, some other protocols may be identified in the Internet literature differently than illustrated in Figure 1.6. For example, the *routing information protocol* (RIP) has an assigned port number making it an upper layer protocol. Yet, another routing protocol, *open shortest path first* (OSPF) has an assigned protocol number, making it a transport layer protocol, as illustrated. Similarly, the *border gateway protocol* (BGP) uses a port number from the TCP header for data flow and control. This is discussed in more detail in Chapter 5.

In theory, all upper layer protocols could use either *user datagram protocol* (UDP) or TCP. Both provide a transport layer function. The reliability requirements of the upper layer program dictate which transport layer protocol is used. For example, some upper layer programs, such as the DNS, may use either UDP or TCP, while others use one or the other exclusively.

UDP provides an unreliable, connectionless transport service, while TCP provides a reliable, in-sequence, connection-oriented service. Because UDP is unreliable, many of the upper layer programs (e.g., FTP and TELNET) only use TCP. The upper layer protocols that do not require a reliable service (e.g., TFTP and SNMP), use only UDP.

IP is a connectionless network protocol providing an end-to-end, datagram delivery service that is unreliable. That is, datagrams may arrive at the destination host damaged, duplicated, or out of order, or they may not arrive at all. This is a major justification for having the transport layer service.

The most popular of the many physical and link layer protocols are identified in Chapter 2. Chapter 2 provides specific information for interfacing the

Internet with the most popular physical networks. It also identifies the appropriate RFC (specification) for these and other physical/link interfaces with the Internet.

1.3 Client-Server Model

The fundamental pattern of activity on the Internet consists of one program requesting another program to provide a service. The two programs may be on the same network or on different networks. The requesting program is called the client and the program providing the service is called the server.

The process of the client is (typically) simple in comparison to the server, which is illustrated in Figure 1.7. The client simply interfaces the user (by way of a device driver) and makes a request to the server. The request is made on a "well-known" port that only the desired server monitors.

The general flow of activities for the server is divided between master and slave functions. The master function receives the request and places a work request in queue for the slave function. It follows that the slave function does all the work, including building the response for the client. The server performs the following master functions:

1. Opens a fixed (well-known) port for the service;
2. Waits for a new client request;
3. Places any received client request in queue for the slave and returns to wait for a new client request.

The server does the following slave functions.

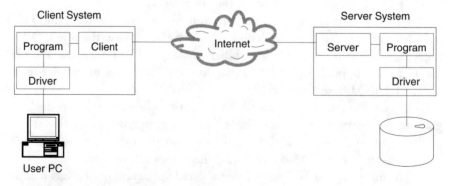

Figure 1.7 Client/server model.

1. Performs service requested;
2. Builds and sends message to client.

Simple client/server examples are the UDP echo server, time server, TFTP, and RARP server, which are described in Chapters 5 and 6. There are complex examples of the client/server model in Chapter 6 (Upper Layer User Service Protocols). A brief description of the TFTP server follows in Section 1.4.

1.4 Internetworking Example

Internetworking is all about using routers to make different, diverse networks interoperate. The illustration in Figure 1.8 has a user connected to an Ethernet (on the left) and communicating with a server on a token ring network. The connection may be with a public network, such as X.25, frame relay, ISDN, ATM, or SMDS, or a private serial line between router A and router B using the PPP protocol.

In Figure 1.8, the goal is simply to transfer a file from the Ethernet-connected laptop to the token ring-connected PC. The protocol selected is the TFTP, and the process is the same for any of the interfaces connecting router A and router B. This requires the TCP/IP protocol stack to be present in both laptop (client) and PC (server), although it could also be done with IPX, Apple-Talk, or other networking protocols.

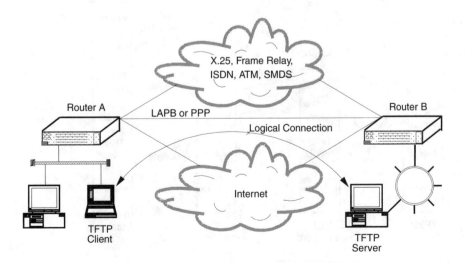

Figure 1.8 File transfer between dissimilar networks.

In the fundamental pattern of activity on intranets (and on the Internet), client programs obtain services from a server program. The client program will send the file to the server program. These programs have only logical interfaces to one another, since the only physical connection to the stations is Ethernet and token ring, which are incompatible. The link layer of the laptop presents the data to IP in the same format as the link layer of the PC presents its data to IP. That format is a datagram and the end of the incompatibility.

Figure 1.9 illustrates the flow of data going from the laptop to the PC—first down the stack at the laptop, then up the stack at the PC—and the opposite for data going from the PC to the laptop.

An upper layer program in the laptop makes a service call to the TFTP client program. The call contains information that enables the TFTP client to construct a descriptor of the file name to be sent, file type, and the destination IP address (TFTP server). The descriptor and user file (message) are given to the UDP.

The UDP attaches its header to the TFTP message that contains the source port number (identification of the TFTP client) and a destination port number (identification of the TFTP server) [2]. Each major upper layer program has a fixed, unique port number called the *well-known* port number. The well-known port number for TFTP data is 69 [3].

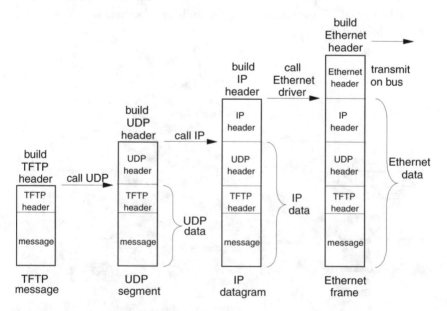

Figure 1.9 Data flow using TFTP/UDP/IP.

IP is central to the entire TCP/IP protocol suite, whether the communication is local or network-wide (excepting some utilities that interface the link/physical layer directly). IP attaches its header, which contains the IP address of the TFTP server and the protocol number for the next transport layer program to process the datagram. A unique protocol number is assigned to each transport layer program, and the number assigned to UDP is 6. IP uses a memory resident routing table to determine which physical interface should be used. In our example, this is a default route to router A that uses the Ethernet interface.

The 32-bit IP address for router A is converted to an Ethernet address (48 bits) by the ARP. An Ethernet header is attached to the IP datagram to form an Ethernet frame. The frame is broadcast on the bus as a unicast packet targeted at router A. While all devices on the Ethernet bus will see the frame, only router A will accept the frame because it recognizes its own 48-bit address. All others (e.g., the PC) will discard the frame.

Router A accepts the frame and after stripping the Ethernet header, passes it to IP, based on a type code in the Ethernet header. Each Internet layer program such as IP or IPX is assigned a unique type code. The type code for IP is equal to 800 hexadecimal. Just as before, IP evaluates the IP address in the IP header to select the next hop closer to the destination IP address. Any of the interfaces illustrated in Figure 1.8 could be used, if configured. After an interface is selected, the above process is repeated until the frame is delivered to the token ring-based TFTP server.

When it is received at the TFTP server, the token ring header is removed and the datagram is passed to IP (based on the type code in the token ring frame). IP removes the IP header and evaluates the protocol number to determine which transport layer program should receive the segment. In our example this is UDP and the protocol number is equal to 6.

UDP receives the segment, removes the UDP header, and examines the destination port number in the removed header to determine which upper layer protocol is to receive the message. The destination port number 69 identifies it as a message for TFTP.

The above describes the process and flow of a single (logical and physical) communication between the TFTP client and the TFTP server. TFTP was selected for this example because it is a simple protocol; it will be explained in more detail in Chapter 6. The first communication from the TFTP client to the TFTP server is a write (WRQ) command. If the TFTP server accepts the command, it sends an acknowledgment (ACK). The TFTP client then sends a data block of 512 octets, which is responded to by the TFTP server with an ACK. If there are more than 512 octets in the message being sent, the process is repeated until the entire message is transmitted and a data block less than 512 octets is sent. This terminates the TFTP session.

Considering that multiple upper layer programs may be busy building a queue of messages for UDP, with only a single path to router B, router A is performing a statistical multiplexing function. Multiple connections by other TFTP clients (e.g., the PC in Figure 1.8) to the same TFTP server are made possible by the source and destination numbers. Both the source and destination ports occupy separate 16-bit fields in the UDP header. Although the destination port number is fixed for TFTP, the source port number is dynamically selected. Theoretically, up to $2^{16}-1$ other TFTP clients at the same IP address could have simultaneous activity with the same TFTP server.

1.5 Illustration Convention

There are international standards and industry standards for the definition of data. (Reference specifically ISO 8824/8825–ASN.1, RFC #1014–XDR by Sun Microsystems, ISO 8613 ODA, and generally ANSI, CCITT, and IEEE.) The nice thing about standards is that you can pick the one you like. The presentation format varies from vendor to vendor, within the same vendor, from standard to standard, and within the same standard. For example, bit numbering in the data link layer description is flipped 180 degrees from the network layer description in the same ITU-T X.25 Recommendation [4].

The illustration convention in this book is intended to be consistent, although the material referenced in different standards (such as ISOC, ITU-T, ANSI, and IEEE) varies. It is not identical to the Internet standards, which are different than other standards (e.g., IEEE, ISO, and ITU-T). It is well-defined and compatible with all others.

The octet and bit order of transmission is identical to TCP/IP specifications. However, the bit sequence labeling is precisely 180 degrees out of synchrony with TCP/IP specifications, which makes it partially resemble some other references, but not all. That is, the least significant bit of an octet is labeled with the smallest number, and it is the first bit of the octet to be transmitted [5]. The purpose of this convention is to identify the octet value of each bit and correlate the hexadecimal numbers displayed on an analyzer to the illustrations and descriptions in this book. Mapping between the various display conventions will help prepare the reader for a graduate course in modern geometry! A comparison to other display conventions is illustrated in Figure 1.10 for comparison.

All numbers are presented in base 10 unless noted otherwise. Base 16 numbers are noted as hex or with a subscript. For example, 3C hex or $3C_{16}$.

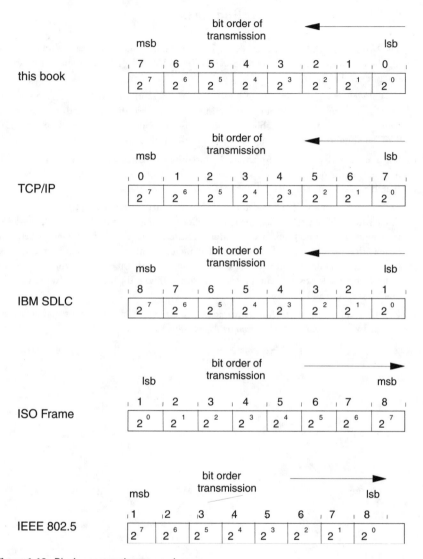

Figure 1.10 Display convention comparisons.

Binary numbers are presented in the same manner as decimal numbers with the least significant bit of an octet on the right (the first bit transmitted). The least significant bit is labeled "+0," and the most significant bit is labeled "+7." Since the term *byte* can be used to describe a different number of bits, the term *octet* is used to precisely describe eight contiguous bits.

Endnotes

[1] The IAB is described in Appendix A, TCP/IP Evolution, and the official document defining the areas of responsibilities of the IAB is RFC 1160. The Internet standards process is described in RFC 1310.

[2] If TCP were being used instead of UDP, it would memorize the parameters in a memory resident table to allow continuing communications between the client and server by specifying only a name of the established association—a connection number. By using UDP, TFTP maintains the association and performs other functions that would be done by TCP.

[3] For a quick reference, see Appendix D for other assigned *well-known* ports, protocol numbers, and Ethernet type codes. To see the latest of these numbers, see the Assigned Numbers RFC, currently RFC 1700.

[4] TCP/IP specifications (RFCs) define only the Internet, transport, and upper layers. Although the octet order of transmission is defined throughout the RFCs, the bit-order of transmission (a physical/link layer function) is scarcely mentioned. For the bit picker, confusion arises because most of the external documents defining the bit-order of transmission label the least significant bit with the smallest number (either 0 or 1), and the RFC specifications label the least significant bit (first transmitted) with the highest (7). Reference RFC 1042 and RFC 1340.

[5] For more information on this controversial subject, see "On Holy Wars and a Plea for Peace," by D. Cohen in *IEEE Computer Magazine,* October 1981.

2

Physical and Link Layer Interfaces

Before describing individual interfaces, it is important to make a distinction between the different layers providing the service. Actual connectivity, electrical or optical, is provided by the physical layer services; this is illustrated in Figure 2.1 with the outer, thin lined cloud. These services are defined by other standards such as RS232, V.35, RS449, RS422, IEEE 802.3, and ISDN BRI and PRI. The TCP/IP protocol suite operates at the Internet layer, and all functions required to interface between the physical layer and the Internet layer are defined as the link layer. Link layer protocols, which are illustrated in Figure 2.1 with the thick, inner-lined cloud, include WAN serial links (e.g., the *point-to-point protocol* [PPP]) and LANs (e.g., Ethernet and token ring). Link layer protocols provide logical services such as framing, congestion control, and link management.

Every link layer protocol *must* use a physical layer 1 medium to communicate. This is akin to saying that every vehicle (car, boat, or train) must have a medium (road, river, or tracks) for operation. Each physical layer may be biased to a particular link layer. For example, the link layer protocol *link access procedure balanced* (LAPB) is normally associated with the network layer of ITU-T X.25. That is because they were developed together [1]. The PPP was designed specifically for multiprotocol operation over serial links. These protocols are normally associated with WANs and geographically unrestricted. The protocols Ethernet and token ring are associated with LANs and geographically restricted.

The next distinction to make is between the TCP/IP protocol suite and the lower layers. The TCP/IP protocol suite is independent of the physical media, because all link layer protocols provide the same interface to IP. The output to and the input from the link layer protocols is the same, an IP datagram. The

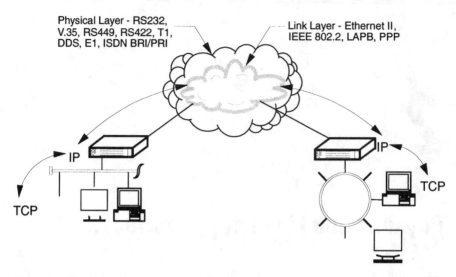

Figure 2.1 Physical and logical interfaces.

format of the IP datagram is the same for a LAN interface as it is for a WAN interface. The TCP/IP protocol stack rides on top of the two clouds illustrated in Figure 2.1 and primarily treats them as a single cloud.

The Internet specifications do not specify which link layer and physical layer interfaces to use. The interface specifications are noted as elective [2]. The specifications do, however, note how any given interface will operate and how it will support multiprotocol operation. Not only does the Internet support multiple physical interfaces, it supports multiple protocol stacks. This is truly an open system, and the support of a particular physical interface and upper layer stack is based purely on demand [3].

The philosophy of the ISOC regarding multiprotocol operation is as follows:

> The coexistence of different protocol stacks is okay, and no effort will be made to find an integrated scheme. Further, if an integrated forwarding scheme should evolve, it will be because of its technical superiority and demand.

Table 2.1 provides a list of the elective interfaces that have precise specifications (as noted) for use with the Internet.

The purpose of Chapter 2 is to provide an introduction to the most common interfaces used by the TCP/IP protocol suite, which is the focus of the book. The level of detail in the description of each LAN or WAN technology

Table 2.1
Physical and Data Link Layer (DLL) Interfaces

Protocol	Name	State	RFC#	Std
IP-ATM	Classical IP and ARP over ATM	Prop	1577	
IP-FR	Multiprotocol Over Frame Relay	Draft	1490	
ATM-ENCAP	Multiprotocol Encapsulation Over ATM	Prop	1483	
IP-TR-MC	IP Multicast Over Token-Ring LANs	Prop	1469	
IP-FDDI	Transmission of IP and ARP Over FDDI Net	Std	1390	36
IP-X.25	X.25 and ISDN in the Packet Mode	Draft	1356	
IP-ARPA	Internet Protocol on ARPANET	Std	BBN1822	39
IP-WB	Internet Protocol on Wideband Network	Std	907	40
IP-E	Internet Protocol on Ethernet Networks	Std	894	41
IP-EE	Internet Protocol on Exp. Ethernet Nets	Std	895	42
IP-IEEE	Internet Protocol on IEEE 802	Std	1042	43
IP-DC	Internet Protocol on DC Networks	Std	891	44
IP-HC	Internet Protocol on Hyperchannel	Std	1044	45
IP-ARC	Transmitting IP Traffic Over ARCNET Nets	Std	1201	46
IP-SLIP	Transmission of IP Over Serial Lines	Std	1055	47
IP-NETBIOS	Transmission of IP Over NETBIOS	Std	1088	48
IP-IPX	Transmission of 802.2 Over IPX Networks	Std	1132	49
IP-HIPPI	IP Over HIPPI	Draft	2067	

that follows should be adequate to explain how each can be used to accomplish multiprotocol operation. For more detail, see the applicable RFCs noted in Table 2.1 and the additional reference material identified in each chapter.

Sections 2.1 and 2.2 describe LAN technologies and how they are used in conjunction with the TCP/IP protocol suite. Sections 2.3–2.11 are related to WAN technologies and how they are used in conjunction with the TCP/IP protocol suite.

2.1 Ethernet

Ethernet has, by far, been the most popular technology used for LANs. So what is it? Ethernet is the name given to a bus technology by inventor Bob Medcalfe in 1973 [4] by concatenating the words ether, which fills all space, and net from network. In an Ethernet, the bus may be either a coaxial cable or a twisted pair wire. The technology is best characterized by its contention access method, which is termed *carrier sense multiple access with contention detection* (CSMA/CD). (This is one of those times when an acronym is appreciated!)

Ethernet was an industry standard by 1979, and Version 2.0 was made a hot product by Xerox, Intel, and DEC in 1982. It contained a jamming sequence to advise other stations of a collision, a maximum frame size detection (jabber indicator), and a low-level heartbeat function for peer communications.

The IEEE standardized Ethernet Version 2.0 with only minor changes in 1985 and called it IEEE 802.2 (DLL) and IEEE 802.3 (physical layer). The standardized version is simply called Ethernet. To differentiate between the two, the original Ethernet, which is supported by most router manufacturers, is called Ethernet II or DIX Ethernet for its developers. Also, the physical component names of Ethernet Version 2.0 were changed, as illustrated in Figure 2.2. While the components' functions remain identical, their names are interchangeable, and it is common to see them mixed. For example, the names NIC and controller are commonly interchanged.

There are three layers to the IEEE Ethernet architecture, the physical layer signaling (PLS), the media access control (MAC), and the logical link control (LLC). The function of the physical layer is to lay, or encode, data from the

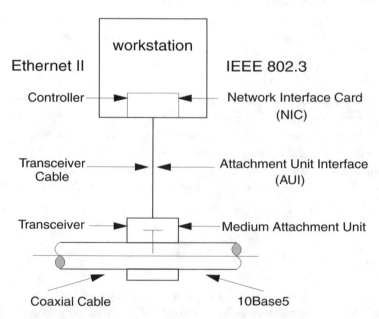

Figure 2.2 Ethernet naming variations.

LLC on the physical medium. The function of the MAC is to serve as an interface between the LLC and the PLS. It passes data read from the physical medium to the LLC and from the LLC to the PLS.

There are two varieties of LLC: LLC type 1 (LLC-1), which is supported by all physical layer standards, is a datagram-like service in that it does not provide frame assurance. LLC type 2 (LLC-2), which may also be used by consenting systems, provides a virtual circuit type connection with frame acknowledgment. Figure 2.3 illustrates the architecture of the IEEE 802.x family and compares it to the architecture of the OSI reference model and TCP/IP with Ethernet II. Ethernet II, the most popular implementation of Ethernet, accommodates multiprotocol operation (explained in Section 2.1.4) without the use of IEEE 802.2.

2.1.1 Ethernet Operation

Control in Ethernet is distributed since there is no authority that grants access to the bus. When a device has something to transmit, it listens on the bus to see whether it is being used. If not, it transmits. Otherwise, it tries again after a waiting period based on an exponential backoff.

Ethernet operation is analogous to a group of people debating in a dark room. The sound waves compare to electrical pulses. In addition, each person can hear the others talking (carrier sense) and can tell if someone else is talking when he or she is ready to speak. This results in the debater pausing for a random period and trying again (collision detection). Courteous debaters never filibuster, nor do they speak when someone else is talking. Otherwise, they are asked to leave the debate (kicked off the net). Each debater has a unique name (Ethernet address) and prefaces each statement (frame) with his or her own name and the name of the person being debated. A debater can speak to an individual in the room (unicast address), to all persons in the room (broadcast address), or to a select group of people in the room (multicast address) [5].

Since Ethernet is an open-ended bus with terminators to prevent reflections, a question often asked is, "How does a transmitter see its own data being sent?" The transmitting station's receive register is filled with the data being placed on the bus. That is, its receiver is reading during transmission. An analogy for this is a person with a glass window between him or her and other potential speakers (tunnel to the bus). If the lighting is such that the window forms a partial mirror, the person sees only his or herself while speaking. If the image of another speaker appears through the window (tunnel to the bus), a collision has occurred. A technical description of collision detection and handling is covered in Section 2.1.2.

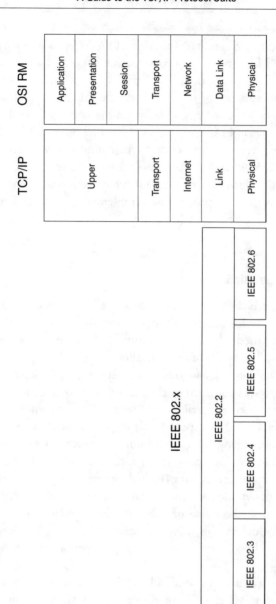

Figure 2.3 IEEE 802 family architecture.

2.1.2   Collision Detection and Handling

CSMA/CD does not have a carrier in the sense of channels being modulated on different frequencies. There are multiple components to the physical layer of Ethernet and *carrier sense* is a signal (on or off) created by the PLS component for the *transmit media access management* (TMAM) component.

Before commencing transmission on the bus, the TMAM attempts to avoid contention by monitoring the carrier sense signal from the PLS component. The PLS generates the electrical signals on the medium and simultaneously monitors for a voltage threshold representative of more than a single transmitting station.

There is a small window of time before the signal can propagate to the furthermost receiver at which a collision can occur, even when the TMAM is monitoring. When a collision is detected by the PLS, it sends the "collision-detect" signal to the TMAM, which sends a jamming signal to advise all stations of the collision. After a period of time, all stations resume contention mode by monitoring.

To avoid the possibility of a collision occurring without being detected by the sender, a minimum frame size is used to insure that there is adequate time for propagation. This minimum frame size, which is referred to as the slot time, must be greater than the round-trip propagation time. At the 10-Mbps rate, a frame size of 64 octets (512 bit times) meets the slot time requirement.

2.1.3   Ethernet Addressing

There are two addresses in an Ethernet frame. The first is the address of the source device, and the second is the intended receiver of the frame. Each address consists of 48 bits, or 12 hexadecimal digits, which are typically denoted by six, two-digit hexadecimal numbers (e.g., 00-80-2C-00-19-20). See Figure 2.4. The first six digits (high order) identify the manufacture and the last six digits (low order) are used for a serial number.

Multicast addresses are identified by the low order bit (+0) of the first octet transmitted. This is easily spotted by the second hexadecimal digit of the MAC address being an odd number. Warning, this may be illustrated as the high order bit in IEEE specifications. This is IEEE binary arithmetic with the low-bit illustrated in the high-order bit position. The IANA has an assigned Ethernet address block, which is used for multicast address assignments. In Internet form the address is 01-00-5E-0X-XX-XX. Note that the high-order bit position or the 4th octet transmitted is equal to zero to denote multicast. The IANA has reserved this bit position for other uses when equal

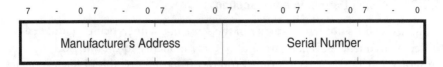

Figure 2.4 MAC address format.

to 1. (See Appendix D, Important Assigned Numbers, for more information on Ethernet addresses.)

Since each transmission on the ether is a broadcast, each device receives all transmitted frames. If the destination address in a received frame is not for that device, the frame is discarded.

2.1.4 Ethernet II Frame Format

The format of an Ethernet II frame is simple; it is illustrated in Figure 2.5. The transmission commences with a preamble consisting of 64 bits of an alternating bit pattern of 1's and 0's. This provides timing synchronization and is typically not shown as part of the frame on a LAN analyzer. The remainder of this description appears just as a LAN analyzer would see the frame:

- The Ethernet II frame commences with the source and destination addresses, each occupying six octets (12 semi-octet, hexadecimal digits). Each address is a 48-bit, field ranging from zero to 2^{48}-1. For example, the least significant bit of the destination address field is the +0 bit of the +5 octet, and the most significant bit is the +7 bit of the +0 octet. An address is expressed as a hexadecimal number with a hyphen separating each pair of digits. The Ethernet addresses are also expressed in other formats, such as dotted decimal and dotted hexadecimal.

- The type code field (two octets beginning with the +12 octet) is equal to 0800 hex for normal IP datagrams. There are other valid Internet type codes. For example, if the type code is equal to 0806 hex, the data field contains an ARP message, or if the type code is equal to 8035 hex, the data field contains a RARP message. Type codes for other protocol stacks are ignored by an IP station, just as a station with a non-IP protocol stack ignores frames with a type code equal to 0800 hex, 0806 hex, and 8035 hex. This is fundamental to the operation of multi-protocol routers.

  The maximum transmission unit (MTU) with Ethernet II is 1,500 octets—that is, the size of the IP datagram (payload). A key difference between Ethernet II and IEEE 802.3 is the use of the +12 and

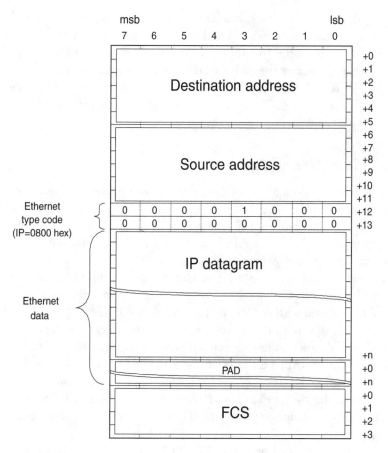

Figure 2.5 Ethernet II frame format.

+13 octets, which are used for a 16-bit length field instead of a type code field. IEEE 802.3 defines a maximum frame size of 1,518 octets, which only provides 1,492 octets for the MTU. Reference, 1,518 minus the MAC header and trailer (18), minus the LLC and SNAP header (8) is equal to 1,492.

The data field of the Ethernet frame contains the IP datagram and has a minimum field size of 46 octets. Hence, the data field of the Ethernet frame may contain a number of octets, each equal to zero, following the IP datagram to meet the minimum length requirement. The zero octets are the frame pad and are not to be confused with padding in the IP datagram.

For collision detection purposes, the minimum packet size is the same for IEEE 802.3 as for Ethernet II, namely 64 octets. The number of octets in the frame header (14), plus the *frame check sequence* (FCS) (4) and the minimum IP field size (46) totals 64 octets, which is 512 bit times. At the 10-Mbps rate, 512 bit times is 51.2 μs, which is longer than two times the end-to-end delay in the maximum length network (49.16 μs). This calculation assumes the propagation velocity of the coaxial cable is .77c (c = 300,000 km/s) and .65c for twisted pair transceiver cables.

2.1.5 IEEE 802.3 Frame Format

IEEE 802.3 alone does not have a mechanism to provide multiprotocol operation. (As noted in Section 2.1.4, the type code field is used for the length field.) Multiprotocol operation is handled by IEEE 802.3 with additional headers. (See Figure 2.6) The first is called the IEEE 802.2 LLC. It contains an 8-bit *destination service access point* (DSAP) address, an 8-bit *source service access point* (SSAP) address and either an 8-bit or 16-bit control field. (The 16-bit control field is used with LLC-Type 2, which provides sequence numbers. The 8-bit control field is used with LLC-Type 1, which does not have sequence numbers.) When the DSAP and SSAP are each equal to 170 (AA hex) and the control field is equal to 3 (denotes an unnumbered information frame) [6], a *subnetwork attachment point* (SNAP) header [7] follows the LLC header. When the three-octet organizational code of the SNAP header is equal to zero, the following two octets are used for the type code field, the same as in Ethernet II. For example, if the Ethernet type code is equal to 0800 hex, an IP datagram, possible pad and FCS follow.

Another difference between IEEE 802.3 and Ethernet II is the address length. IEEE 802.3 supports both 16-bit and 48-bit physical address lengths, while Ethernet II supports only a 48-bit physical address.

IP datagrams to be transmitted by an IEEE 802.3 are encapsulated in a SNAP header, which is encapsulated in an IEEE 802.2 LLC header before being encapsulated in the IEEE 802.3 MAC envelope (network-specific). The receiving entity of such an IEEE 802.3 frame removes the IEEE 802.3 MAC header (and possibly trailer), the IEEE 802.2 LLC header and the SNAP header such that IP sees each frame beginning with the type code just as it does with Ethernet II frames. That is, a type code followed by the IP datagram, pad, and FCS.

Since the FCS calculation, placing the bits on the medium, retrieving the bits from the medium, and verification of the correct FCS is done within the hardware, there is little concern for this departure from the standard bit order of transmission. However, if you are trying to verify the value of a FCS from the display on a LAN analyzer, the special bit ordering must be observed to obtain

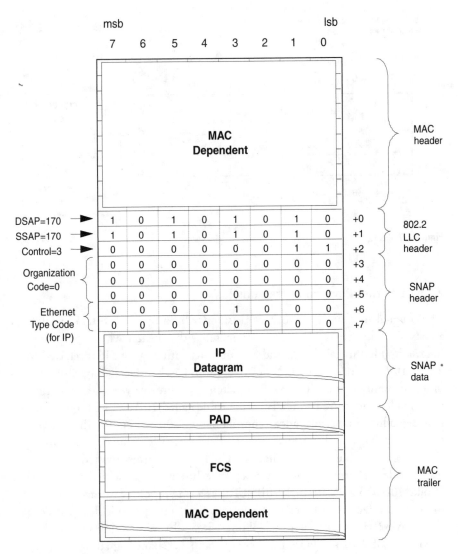

Figure 2.6 IEEE 802.3 frame with SNAP header.

the correct FCS. See Figure 2.7 and IEEE 802.3 for a description of the calculation of a 32-bit FCS.

Normally, bit +0 of the +3 octet of the FCS would contain the least significant bit. As illustrated, this bit position represents the 2^7 position and the least significant bit is found in the +7 bit of the +3 octet (least significant octet). Similarly, all other bits of the FCS are *not* ordered as in other fields.

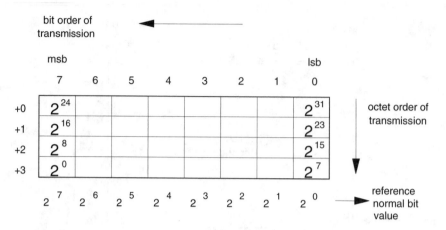

Figure 2.7 FCS.

2.1.6 Ethernet Physical Characteristics

The original medium for Ethernet II, standardized by IEEE 802.3 in 1980, is a coaxial cable that is 1/2 inch in diameter with a center wire, a polyethylene filler, a braided metal shield, and an outer insulating jacket. It is referred to as TYPE 10Base5, denoting 10-Mbps operation, baseband signaling, and a maximum segment span of 500 (5x100) meters. Within five years, IEEE defined a new smaller and more flexible Ethernet cable called 10Base2 with only a minimal degrading in performance (approximately 200 meters for the maximum segment span). For the next five-year offering, IEEE defined a standard wire (22–26 gage) cable system called 10BaseT (T denoting twisted wire). This system offers nearly the same performance but uses unshielded twisted wire. Unlike 10Base5 and 10Base2, 10BaseT uses a ring topology formed with point-to-point connections to a central hub (repeater) using unshielded twisted pair (UTP) wires (four) similar to telephone wire. All three cabling systems have a maximum of 1,024 devices per network and baseband signaling with Manchester encoding. Sections 2.1.6.1–2.1.6.3 describe the 10Base, 10Base2, and 10BaseT cable types, respectively.

2.1.6.1 10Base5

Each end of a coaxial cable segment has a terminator that provides an impedance equal to the characteristic impedance of the cable, which minimizes reflections. The braided metal shield attachment to one terminator of a coaxial cable segment makes contact with an effective Earth reference (ground). As illustrated in Figure 2.8, only one ground per network is required.

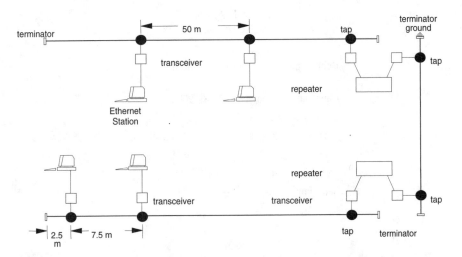

Figure 2.8 Ethernet 10Base5 cabling.

The maximum length (distance between terminators) of a coaxial cable segment is 500 meters (1,640 feet). A maximum of two repeaters may be used to connect coaxial cable segments. This extends the maximum distance between Ethernet taps to 1,500 meters.

Up to 1,000 meters of cabling may be used to connect coaxial cable segments. Thus, the maximum distance between Ethernet stations is up to 2,500 meters. Both Ethernet stations and repeaters are connected to a coaxial cable segment with a transceiver, which may each have a cable length of 50 meters. Hence, with two repeaters (two transceivers each) and two Ethernet stations (single transceiver each), the absolute maximum distance between any two Ethernet stations is 2,800 meters [8].

The maximum number of physical taps per coaxial cable segment is 100, and each tap is physically spaced in multiples of 2.5 meters. The coaxial cable is marked every 2.5 meters to aid in the placement of taps. This spacing minimizes the reflections adding in phase.

Devices on the bus architecture share a common channel, and the transmissions are broadcast. That is, each device on the bus receives all transmissions, and those not addressed to the receiver are filtered out. Each device on a 10Base5 bus is connected with a transceiver, which applies signals to the cable and detects collisions. Transceivers are typically connected to the Ethernet segment with twisted pair wire, which may be shielded to reduce noise. Twisted pair wire can easily support 10 Mbps over limited distances, but the propagation velocity is reduced by at least 10 percent. The characteristics of 10Base5 are summarized in Table 2.2.

Table 2.2
10Base5 Characteristics

Cable type	RJ-8
Maximum segment length	500m
Devices per segment	30
Device spacing	2.5m minimum
Connector type	DB15
Propagation velocity	.77c
Cable impedance	50 Ohm

2.1.6.2 10Base2

When a smaller and more flexible 10Base2 bus is used, the connection is made with a BNC, and the signaling is done in the device being attached [9]. That is, the device being attached performs the transceiver function and a separate transceiver is not used. A 10BASE2 bus (sometimes called Cheapnet, Thinnet, or Thin-wire Ethernet) may be connected to a 10Base5 bus but may not be used to connect two 10Base5 segments. The characteristics of 10Base2 are summarized in Table 2.3.

2.1.6.3 10BaseT

The entire LAN may be constructed with twisted pair wire. A popular twisted pair wire LAN is AT&T's Starlan (with central control). The IEEE standard

Table 2.3
10Base2 Characteristics

Cable type	RJ-58
Maximum segment length	185m
Devices per segment	30
Device spacing	5m minimum
Connector type	BNC
Propagation velocity	.65C
Cable impedance	50 Ohm

describing a twisted pair wire LAN is called 10BaseT. Table 2.4 summarizes the operational characteristics of 10BaseT.

There are many advantages to using a UTP LAN over either coaxial cable. It uses standard telephone wire, is cheaper, and makes it easier to isolate problems. If a run from a station to the repeater hub fails, it does not affect the other stations as with coaxial. The transceiver function is in the NIC, as with 10Base2. The hub may be either passive or active. Active hubs can be very sophisticated, offering features such as an SNMP agent and MIB [10].

2.1.7 Ethernet Bridging

The most efficient way to use an Ethernet interface is to bridge the traffic. Bridging is essentially forwarding the data received to the destination MAC address in the frame. The first beauty of bridging is automatic configuration (setting up addresses for forwarding). That is, the bridging protocol defined by IEEE 802.1 (normally supplied with routers) automatically constructs a forwarding database (FDB) by analyzing messages received on each interface. The second beauty (which can also be a downside) is that bridging forwards all network types, including nonroutable protocols such as NetBios. Bridging Ethernet traffic is referred to as transparent bridging because the data is transparent to the bridging protocol.

Bridges in a simple, nonlooped network go from the down state to learning mode when powered on (unless bridging is manually turned off). In the learning mode, it examines the source MAC address in each received message to see if it is recorded in the FDB with a destination port. If not, it stores the association for future messages received with a destination MAC address equal to this source MAC address. The FDB is then checked to see if there is an entry

Table 2.4
10BaseT Characteristics

Cable type	22-26 gage
Maximum segment length	100m
Devices per segment	1
Device spacing	—
Connector type	RJ45
Propagation velocity	.6C
Cable impedance	75–50 Ohm

for the received destination MAC address. If so, it forwards the message on the port number obtained from the FDB. If not, it floods the message to all ports, except the port number where the message originated. The next bridge receiving the message will perform the same action (i.e., flooding). This is called *forwarding mode*. Bridges are fast learners, but the learning mode may be skipped if forwarding addresses are manually entered in the FDB. This should only be necessary for output only devices to minimize flooding.

That is the scenario for simple, nonlooped bridged networks. What about bridged networks with multiple paths (meshed)? Possible looping and duplicate delivery may be avoided by running a companion protocol to IEEE 802.1d called *spanning tree protocol* (STP) or extended STP in IEEE 802.1g (an enhancement to this is available in draft form). STP consumes only a small percentage of the bandwidth and automatically configures the active topology into a single spanning tree between any two stations, which eliminates data loops. It provides fault tolerance by automatically reconfiguring the active topology for failed stations or new stations added. STP is transparent to end stations, and the only manual configuration requirement is to ensure that each bridge has its own MAC address and that bridging is turned on.

A route bridge is dynamically determined by each bridge sending a *configuration bridged protocol data unit* (BPDU) to all other bridges on all the LANs to which it is attached. The configuration BPDU announces the senders claim to be the root bridge and other information about its interfaces. The configuration BPDU, which is different from a normal BPDU sent on the bridged network, is identified by the destination MAC address, which is a multicast address. (Normal BPDUs contain a unicast address and are flooded.) The IEEE reserved a block of MAC multicast addresses for STP. The STP multicast MAC address published in the Assigned Numbers RFC (RFC 1700 presently) is 01-80-C2-00-00-00. Configuration BPDUs are never flooded (forwarded by a receiving bridge), nor are they relayed to a higher layer protocol, as are regular BPDUs. However, it does use the information from a received configuration BPDU to construct its own configuration BPDU to be sent (at regular intervals). A receiving bridge either accepts the root bridge, or it may challenge with a response configuration BPDU if its MAC address is lower than the MAC address in a received configuration BPDU.

Since all bridges see other bridges' configuration BPDUs, the bridge with the highest priority (lowest MAC address) quickly surfaces, and a leaf topology is constructed from the root bridge. (A *root port*, the port connecting to the root bridge, is also defined.) A bridge offering a duplicate path to the root bridge evaluates its cost compared to any other bridge's cost offering the same path. The bridge with the highest port cost (some of the other interface information) for the duplicate path will set that port in the FDB to blocking mode. The blocking mode affects regular BPDUs only.

Topology changes are handled by regular interval configuration BPDUs. A timed-out configuration BPDU from the root bridge results in the election of a new root bridge, and a new topology.

The downside to transparent bridging is that there is only one network, no matter how big. Hence, it is normally used only for smaller networks. Filtering may be used to keep some traffic from flooding. For example, traffic received from the finance department bridge will be filtered from being forwarded to the engineering department bridge.

Transparent bridging was designed by IEEE to be applicable to IEEE 802.3, IEEE 802.4, and IEEE 802.5, although it is only implemented on IEEE 802.3. See Section 2.2.4 for a discussion of bridging with IEEE 802.5 (token ring).

2.2 Token Ring

IBM developed the token ring LAN technology in 1985 to compete with the DEC, Intel, and Xerox product Ethernet II. The products share several common characteristics. The main differentiating feature of token ring is the access and control. No station transmits unless it is invited to do so. Permission is even required to become part of the token ring to receive a frame. Ethernet is first-come-first-served raw contention, while token ring is all about control. Providing this level of control makes the protocol much more complicated than Ethernet. Although Ethernet outperforms token ring in light to medium load, under a heavy load, token ring continues to perform and support its maximum throughput, while Ethernet drops off rapidly due to collisions.

Token ring was made a part of the IEEE 802 family with minimal changes as IEEE 802.5. It may use the DLL IEEE 802.2 as does Ethernet. The topology of token ring, which is illustrated in Figure 2.9, is similar to a stack of Ethernet hubs (10BaseT).

2.2.1 Token Ring Topology

At the heart of token ring is the multistation access unit (MAU). Clearly the MAUs form a ring topology, but the workstations connected to a single MAU have a star topology. The original IBM version is the 8228, which has 10 ports. Eight of the ports control a workstation, and two are used to connect other MAUs with wire between the ring in (RI) port on one and the ring out (RO)

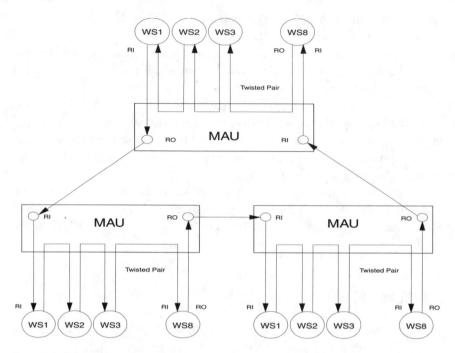

Figure 2.9 Token ring topology.

port on another. With a maximum of nine MAUs, the maximum size token ring was 72 workstations. Newer IBM MAUs (and third party supplier equipment) support 16 workstations each. A workstation is connected to a MAU with a lobe wire that may be shielded or unshielded. The absolute maximum number of workstations in a token ring using a shielded twisted wire varies by vendor. It ranges from 160 to 260. At the higher end of the range, each port must function as a repeater (active hub versus passive hub). The original speed of token ring was 4 Mbps, but it may now be run at either 4 Mbps or 16 Mbps.

2.2.2 Access Method and Control

In the initial state, the loop from the MAU to a workstation is closed. A workstation joining the star topology must supply a 5-volt signal, which causes the MAU to open the switch and engage the workstation. A phantom voltage keeps the switch open, and when the workstation is powered off the switch is automatically closed. This assures that inactive workstations are disengaged from the star configuration, which reduces the total length of wire in the maximum configuration.

A four-phase initialization procedure occurs when a workstation is first connected to the MAU. During this procedure, the workstation obtains configuration parameters and learns its role on the token ring. The workstation could be assigned anyone of five roles that essentially assist the LAN network manager. (LAN Network Manager is a PC-based application that communicates with IBM's host-based network management product, called NetView.) A description of each of these roles follows.

1. Active monitor: Resolves the lost token condition and looping frames;
2. *Ring parameter server* (RPS): Provides configuration information to new workstations and registers the new workstation with the LAN network manager;
3. *Ring error monitor* (REM): Assembles error reports from workstation statistics;
4. LAN bridge server: Active in bridge reconfiguration and recording statistics for traffic handled by a bridge;
5. *Configuration report server* (CRS): Responds to commands from the LAN network manager for reports (such as bridge traffic and errors) and to remove a workstation from the ring.

Access by a workstation to a MAU is by invitation only—that is, by receiving a 3-octet token, which is periodically sent from one workstation to the next on the ring. If a workstation receives the token and does not have traffic to send, it simply sends the token to the next station. If it has traffic to send, and the traffic has a priority as high as required by the token, it sends the traffic to the next station, followed by the token. Each workstation receives the frame sent and if it is addressed to that workstation, the workstation will read the frame and set an ACK indicator for the originating workstation before sending to the next workstation. The originating workstation eventually receives its own frame and breaks the loop. The token is illustrated in Figure 2.10.

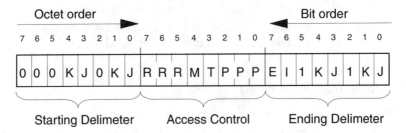

Figure 2.10 Token.

There are three octets in the token fields. The first transmitted is the starting delimeter, the second is the access control, and the last is the ending delimeter (ED). The bit order illustrated is consistent with other standards in this book, but 180 degrees out with the convention of the IEEE 802.5 specification [11]. A description of each octet follows.

- Starting delimeter: Bits +0, +1 and +3, +4 are transmitted as pairs of differential Manchester violations to signal the beginning, which maintains dc balance. These are labeled J and K and described as nondata. The remaining bits are each equal to binary zero.

- Access control: The priority field is contained in the +0, +1, and +2 bits and is flipped, as though seen through a mirror. That is, bit +0 (the first bit of the octet transmitted) has the binary value 2^2, bit +1 (the second bit of the octet transmitted) has the binary value 2^1, and bit +2 (the third bit of the octet transmitted) has the binary value 2^0. Since zero has the lowest priority, the field value, in descending order or the highest priority is 7, 3, 5, 1, 6, 2, 4, and 0. Fortunately, there is no direct interaction between IEEE 802.3 and IEEE 802.5. This mismatch is made operational (translated) by routers. Please be aware that the same condition prevails in the address field. Should the message to be sent have a lower priority than the priority in the token, the workstation must refrain and forward the token.

  The token bit (+3) is always equal to zero in a token and equal to one in a frame.

  The monitor bit (+4) must always be zero. If it is found equal to one by the active monitor, the frame is aborted.

  The reservation field (+5, +6, and +7) is used by a workstation to request a token of lower priority (again with the same bit significance as the priority field).

- ED: The ED, like the starting delimeter, contains a double pair of differential Manchester violations, noted as the J and K bits. The I bit (+6) is equal to zero for intermediate frames and equal to one on the last frame of a sequence. The E bit (+7) is always initially set equal to zero. If an error is detected en route, the bit is set equal to one.

2.2.3 Token Ring Frame Format

The token ring format commences with the same first two octets of a token (SD and AC) and is followed by an octet used for frame control (FC). See Figure 2.11 for illustration. When the first two transmitted bits of the FC octet are

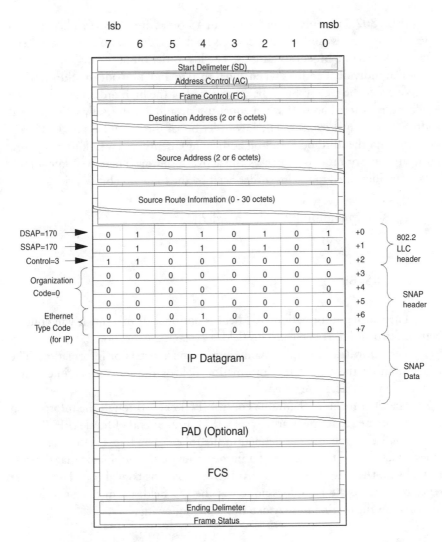

Figure 2.11 Token ring frame format.

each equal to zero, the frame format is MAC only. When the first bit transmitted is equal to zero and the second bit transmitted is equal to one, the frame also contains an IEEE 802.2 LLC header. In addition, depending on the values in the LLC header, the frame could also contain a SNAP header. This permits multiprotocol operation in one of the same methods used with Ethernet. With the correct DSAP and SSAP, the SNAP header is not required for multiprotocol operation.

IEEE 802.5 permits either a 16-bit or 48-bit address format. There is both a destination address and a source address field. The destination may be either a unicast address or a broadcast address. The low order bit of the high order octet defines an individual address when equal to zero and a broadcast address when equal to one. The second bit to be transmitted of the high order octet defines a locally administered address when equal to one and a universally administered address when equal to zero. The low order bit of the second octet (16-bit addressing) and the third octet (48-bit addressing) identifies a well-known address for a supervisory function, such as the active monitor. These well-known addresses are identified in the low order octet by the following bit positions.

1. Active monitor: Bit +7 (high order);

2. RPS: Bit +6;

3. REM: Bit +4;

4. CRS: Bit +3.

The routing information (RI) field is optional. If present, it will contain between two and thirty octets. This field is explained in Section 2.2.4.

The information field contains zero or more octets of information. The factor limiting the size of the information field is the maximum time that a workstation may keep the token.

The FCS includes the frame control (FC) octet through the information field. It does not cover the start of frame (SF), ED, or end of frame (EF) fields.

The ED is the same as defined in the token. The frame status octet provides a method for the receiver of a frame to advise the originator that the correct address (A) was reached and that the message was copied (C). The bits are repeated because the FCS does not cover the ED field. From these two bits of information the originator may determine the following.

1. Addressed station is nonexistent or not active;

2. Addressed station was reached but the message was not copied;

3. Frame was copied.

Figure 2.11 is illustrated with the IEEE 802.2 and optional SNAP headers. Supervisory messages (e.g., from the RPS to the LAN network manager) use only the MAC header. The LLC/SNAP header may be used in multiprotocol environments.

The SNAP header in Figure 2.11 has the individual field values adjusted for IEEE 802.5 transmission, that is, with the high order bit transmitted first.

For example, the unnumbered information (UI) identifier is normally seen as 3, or 00000011 binary. Here it is illustrated as 11000000 binary, or C0 hex.

2.2.4 Source Route Bridging

The RI field in the header (illustrated in Figure 2.10) is never present, unless bridging is used instead of routing. When bridging is used, the routing is predetermined by the source workstation before sending the message. The scheme is simple in concept and never uses stored tables as does transparent bridging. Moreover, unlike transparent bridging, it never learns. Before sending a message, the workstation broadcasts the desired destination MAC address in a *single-route explorer* (SRE) frame to all workstations on the ring. If a workstation recognizes its address, it responds to the originating workstation with the path by which it can be reached. If the originating workstation does not obtain a positive response, it expands the search from its own ring to all other rings, assuming there is one or more bridges attached. This is done with an *all-route explorer* (ARE) frame sent to all bridges on the ring. Each bridge receiving the ARE frame will record its path (bridge number and source ring number) in the RI field. When a workstation recognizes its address, it switches the direction bit in the route control (RC) field (first two octets of the RI field), which causes the ARE frame to be returned to the originating workstation by the same path. This is called single-route return and is typically used with TCP/IP implementations. An all-routes return option may also be used.

The first 16 bits of the RI field are used for the RC field, illustrated in Figure 2.12. Each entry posted by a bridge in the RI field occupies 16 bits, 12 bits for the ring number and 4 bits for the bridge number.

The remaining RI fields are identified as follows.

1. Length: Total length of the RI field;
2. BBB: Broadcast indicators (xx0: unicast, x01: all-routes, x11: single-route);
3. rrrr: Reserved;

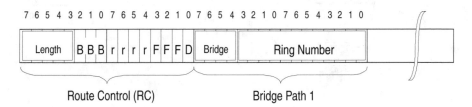

Figure 2.12 RI field.

4. FFF: Largest frame size (000: up to 516 octets, 011: up to 17,800 octets);

5. D: Direction bit (0: from top down, 1: from bottom up).

2.2.5 Bridge Path Discovery

Sending traffic between two workstations connected to the same ring involves only the broadcast of an SRE frame. See, for example Figure 2.13, which illustrates workstation A sending to workstation B on Ring 1. Only Workstation B will respond to the explorer frame indicating that it is on the same ring.

Now consider the case in which workstation A wants to send a message to workstation Z. As there are many paths between the two workstations, there will be no response to the SRE frame and workstation A will send an ARE frame, which will be propagated throughout the network. Workstation Z will receive one such ARE frame with the path in the RI field indicating 1/1, 4/2, and 3/0. This indicates that the ARE frame traversed ring 1 and bridge 1, ring 2 and bridge 4, and ring 3. Workstation Z simply sets the *direction bit* (D) in

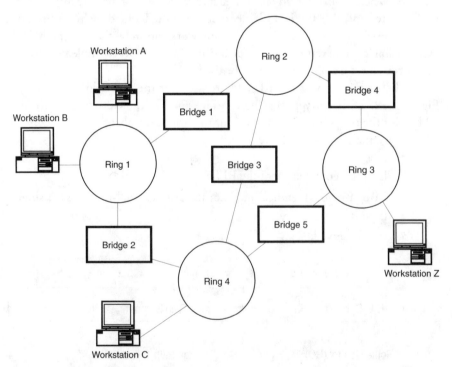

Figure 2.13 Multiple bridge path discovery.

routing control (RC) of the RI field and returns the ARE frame to workstation A by the same path as it was received. Workstation Z would also receive the ARE frame from several other paths and return each in the same manner. So, which path does workstation A select as the proper one? This may be determined by selecting the path with the least hops or by simply taking the first ARE returned, assuming it is the shortest path.

It is apparent that possible loops may be attained with the configuration in Figure 2.13. Token ring has a similar protocol to STP in Ethernet that may block some paths to prevent looping. Transparent bridging from IEEE 802.3 is also designed to operate with token ring. However, the converse is not true. Source route bridging will only work with token ring.

2.3 Serial Links (General)

All WAN interfaces used by the TCP/IP protocol suite have a serial line for the physical medium. The typical link layer interface between TCP/IP and a serial line is provided by the PPP, described later in Section 2.5. The serial line is used to connect routers within the Internet and to connect users to the Internet. The TCP/IP protocol stack may be used to either connect routers or to connect users to routers. The serial line may be dedicated or dial-up, asynchronous or synchronous, and analog or digital. Typically, home users will connect to an ISP with a dial-up, asynchronous analog line. The ISP provides a similar function to that provided by the *Service Bureau Corporation* (SBC) in the 1970s. If the PC has a TCP/IP protocol stack, the user may enjoy full Internet service. If not (when the PC uses PPP only), the user is limited to the upper layer programs provided by the ISP (e.g., e-mail). LAN users with a router are typically connected to an ISP with a dedicated, synchronous digital line. Although this is a faster service, it is more expensive, whether it uses DDS or ISDN.

One of the early standards for the physical layer of serial lines is the *Electronics Industries Association* (EIA) RS-232. This standard specifies the physical attachment and electrical interface between the *data terminal equipment* (DTE) and *data circuit-terminating equipment* (DCE) [12]. The standard limitations are 50 feet on cable length and speed of 20 Kbps. The designed replacements for RS232 are RS449, RS422A, and RS530—each with a balanced signaling technique using two wires with opposite polarities for both the transmit and receive functions. The operation of *automatic call units* (ACUs) is defined in RS366. The ITU-T defined interface specifications X.20 and X.21 in 1976. These specifications are balanced, handle dialing, and use IA5 (ASCII) character commands for control (compared with raising voltage on dedicated wires for control with the EIA standards).

2.3.1 Line Coding

A line code is the representation of a set of binary digits for transmission on a communications line. The type of transmission medium may restrict (or dictate) the type of line code used. For example, a fiber optic medium does not have a plus and minus voltage. Most interfaces specify the line coding to be used, although other line coding could be used. If optional, consideration should be given to both the medium and the intended use. For example, the simple *non-return-to-zero* (NRZ) line coding works well for short distances and is not dc-balanced. The six major types of line coding—NRZ, return-to-zero, *alternate mark inversion* (AMI), ternary code or *alternate space inversion* (ASI), Manchester, and Differential Manchester—are illustrated in Figure 2.14.

The simplest line code to explain is NRZ. With NRZ, the voltage is always zero except when a one bit is being transmitted. For ease of description, assume that the line speed is such that the slot time for each bit is 2 ms. Then for the sample binary value in Figure 2.14 (11001010 binary) [13], the voltage is raised for 4 ms, dropped for 4 ms, raised for 2 ms, dropped for 2 ms, raised for 2 ms, and dropped.

The return-to-zero line code is the same as NRZ except that voltage is raised for only one-half of the slot time when the value being transmitted is equal to one. The first half (1 ms) has the voltage raised, and the voltage is dropped during the second half of the slot time. When ones are not being transmitted, the voltage remains equal to zero.

AMI is a bipolar, 50% pulse train used by T1. It changes voltage and polarity for ones and returns to zero. It remains at zero voltage for zeros.

ASI is used with ISDN BRI. It is described in more detail in Section 2.8.1.

Manchester coding provides the ability to derive the clock speed from the transmitted signals because every time slot contains a change in polarity. Upon examination of Figure 2.14, it may appear difficult (busy). When the following rules are applied, however, Manchester coding is no more difficult than NRZ.

- Rule 1: A voltage change must occur at the midpoint of every time slot.
- Rule 2: The change is positive for transmitting the value one and negative for transmitting the value zero.

The arrows in Figure 2.14 illustrate the changing of polarity at the midpoint of each time slot. Differential coding also has the quality of being dc-balanced. That is, the area under the curve is the same as the area over the curve.

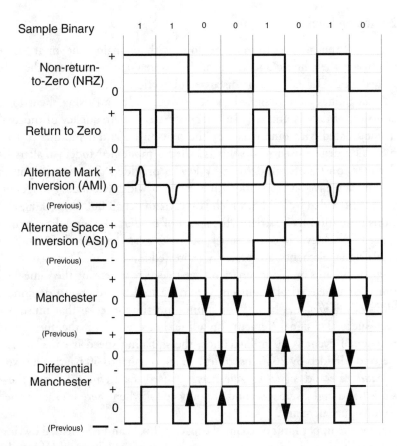

Figure 2.14 Line codes.

Differential Manchester, like Manchester, is relatively simple when the following rules are applied.

- Rule 1: There is a polarity change for every time slot.
- Rule 2: The required polarity change occurs at the beginning of time slots transmitting the value zero. The direction of change is determined from the previous time slot.

The serial line may be digital as provided with DDS, T1/E1, ISDN, SMDS, and ATM, or it may be analog as provided by the *plain old telephone service* (POTS). Since the analog lines were here first, and probably the type most used at home, they are described first.

2.3.2 Analog Lines

Most users of the Internet from home do so with an analog line, or at least the path between their home and the local ISP is analog. From there, the ride is typically digital. Digital lines are discussed in Section 2.3.3.

Analog lines were designed to carry voice only and using them for the transmission of data is not easy. In essence, the carrier frequency of the analog line is modulated with binary data by the transmitter and demodulated by the receiver. The device used to do this is called a modem. Modem makers have faced many constraints, all defined by POTS providers. The most significant constraint they have had to work with is the bandwidth (frequency range) of an analog line provided by POTS, which is approximately 3,000 hertz (Hz). In the past, this limitation has restricted the speed of modems because the maximum signaling rate achievable over a serial line of ω hertz bandwidth is 2ω distinguishable signal elements, algebraically expressed as, $C = 2\omega$. This assumes that each signal contains only one binary bit capable of representing the value 0 or 1. A term used by electrical engineers, often misused by others, to describe the signaling rate is *baud*. A baud is the minimum time interval that must elapse between successive signal elements on a line. For example, if the duration of a discrete signal event is 20 milliseconds, the signaling speed is equal to 1,000 milliseconds divided by 20 milliseconds, which is equal to 50 baud. The expressions *baud rate* and *signaling rate* may be used synonymously, but the expressions baud and bits per second are only equal when there is exactly one bit per baud (signal element).

Since the upper limit of channel capacity (bps) can be increased by simply using more than one bit per baud, modem makers quickly explored this to keep pace with the demand for increased speed of analog lines. The general algebraic expression, including the number of bits per baud, for the capacity of a serial link is, $C = 2\omega \log_2 L$ [14], where $\log_2 L$ is the number of binary bits. Hence, if L is equal to 2 the number of bits is equal to one; if L is equal to 4, the number of bits is equal to 2; and if L is equal to 8, the number of bits is equal to 3 (and so on).

What is so difficult about designing fast modems? Simply pick a large value for L. For example, use L equal to 65,536, which yields a channel capacity of 76,800 bps (2 x 2,400 x 16 = 76,800). The limiting factor is noise. Although a transmitter was able to send signals containing 16 bits, the receiver could not distinguish them because of noise. Noise was taken into account in the equation $C = \omega \log2 (1 + S/N)$, where S/N is the signal to noise ratio [15]. Many techniques were used to increase the number of bits per baud, including modulating the carrier frequency phase angle, amplitude, and frequency.

The latest breakthrough in modem speed was defined in the ITU-T V.34 bis standard. It uses 9 bits per baud and attains a speed of 33.6 Kbps in each direction. Previously, the total bandwidth had to be shared for full duplex operation. A feature called *echo canceling* permits the transmitter and receiver to occupy the same bandwidth. A compression protocol (such as V.42biz) provides even greater throughput, although the actual signaling rate is unchanged. This effectively compresses the data before transmission and, using the same compression algorithm, decompresses it at the receiving end.

2.3.2.1 Analog Signaling

Signaling is the communication between the user and the *local exchange* (LE), between the local LE and the remote LE, and between the remote LE and the destination user. These communications control the call set up and take down at the call conclusion. In-band signaling (2,600 Hz) was originally used, but it has been replaced with out-of-band signaling. Out-of-band signaling uses the frequency 3,825 Hz (as defined by ITU-T) and 3,700 Hz in the United States. The LEs used *channel interoffice signaling* (CCIS) in North America; in Europe, the ITU-T defined *signaling system 6* (SS6) was used. Both of these analog systems have been replaced by ITU-T *signaling system 7* (SS7), which handles analog and digital calls.

SS7, a layered protocol offering both connection and connectionless service, defines two types of signaling, *line signaling* and *register signaling*. Line signaling, which performs line supervision, is used for the call setup and takedown process. It is used for features such as call holding, transfer, and conferencing and metering for billing purposes. It is also used to convey dial pulse addressing information for rotary telephones, which are still in use internationally. Register signaling, which is used to convey addressing information during call setup, handles *calling line identification* (CLI); pay phone, billing, and circuit information such as the use of pulse or tone dialing; and multifrequency dialing.

SS7's flexible design accommodates different country requirements. The flexibility is provided with variants to both line and register signaling. That is, an implementation of SS7 may pick and choose options and features. Such an implementation constitutes a *regional signaling system*. The following are examples of regional signaling systems.

1. Regional signaling system R1: A system of telephony line and register signaling used predominantly in North America. It uses two of five multifrequencies 700 Hz, 900 Hz, 1,100 Hz, 1,300 Hz, and 1,500 Hz. 1,700 Hz and 1,900 Hz are used for control purposes.

2. SOCOTEL: A system of telephony line and register signaling used in France and Spain. It uses two of five multifrequencies and is similar to R1.

3. Regional signaling system R2: A system of telephony line and register signaling used in most geographic regions outside of North America [16]. R2 handles both analog and digital. It uses two of six multifrequencies, both forward and backward.

4. Regional signaling system C1: A system of telephony line and register signaling used in China. It contains mostly derivatives of R2.

5. Regional signaling system P7: A system of telephony line and register signaling used in Sweden. It contains mostly derivatives of R1, such as DTMF.

While the bandwidth source for signaling is a configuration option, the signaling system and variants used must be built into the hardware as required by the country providing carrier service. Providers of international carrier class equipment must configure (hardware or firmware) to meet the requirements of the regional signaling system. Hence, a *remote access server* (RAS) sold in Sweden would not function properly in China, without a firmware change.

2.3.3 Digital Lines

The fastest way to use a telephone line is digitally. The limitations presented by analog lines, such as analog modems, noise, and the Shannon speed limit, are gone. Since humans cannot speak digitally, however, voice has to be converted to digital signals—ideally at the origin.

A digital line service does not have a carrier frequency. Accordingly, binary bits of information are simply transmitted (and received) as on or off line voltage. Just as with the analog line, digital transmissions may be either synchronous or asynchronous. Both are bit synchronous, but asynchronous operation has start and stop bits for each group of five or eight bits (usually) to maintain synchrony as there is no synchronized clocking between sender and receiver. Asynchronous operation also allows a variable amount of inactivity between bit groups (characters) to accommodate a less than perfect human typist. Although the sender and receiver clocks may not be identical, by only sending a few bits before synchronizing on the stop and start bits, bit synchrony is maintained. In the synchronous mode of operation, both the sender and receiver constantly adjust their clock speed based on a common (synchronized) clock, usually supplied by the modem. In this manner, an entire block (or frame) of data is sent continuously without start and stop bits.

In 1974, AT&T offered an all-digital line service called *Dataphone Digital Service* (DDS). DDS is a subchannel of a much earlier technology called T carrier system (T1). DDS service is first offered at 2.4 Kbps, 4.8 Kbps, 9.6 Kbps, and 56 Kbps. The voice service offered on the same T1 network is called *Direct Distance Dialing* (DDD). With these services, up to 24 mixed voice and data channels may be transmitted on the same T1 link. For example, twelve voice lines and twelve 64-Kbps lines are on the same T1 link.

Although T1 has changed since its first introduction by AT&T in the late 1950s, its data capability has always been present since it is an all-digital service. What has changed is the addition of digital local loops from the central office to users. The initial offering was for DS-0 only, which is a designation given to a single 64-Kbps channel. This enabled the transmission of data in its original digital format (baseband) on the same T1 link carrying voice in digital format.

In the 1980s, the speed of the digital local loop from the telephone company was increased to the full DS-1 rate (1.544 Mbps), which enabled the local user to drive an entire T1 link. With this new speed, however, came the responsibility to interface the T1 circuit in accordance with AT&T Publication 54016. That is, the user must develop (or buy) the hardware and software to interface the T1 link as defined in the specification. The equipment used by AT&T (and available to users) to accomplish interfacing T1 is a DSU/CSU, as defined earlier.

Later on, a similar *time division multiplexer* (TDM) technology with improved characteristics and performance called E1 was developed. E1 is based on the *European Postal and Telecommunications* (CEPT) Level 1 standard. E1 is used in most areas of the world outside of North America. Since the CEPT Level 1 standard has been adopted by ISDN, E1 is explained in Section 2.8, ISDN.

2.3.4 T Carrier System (T1)

T1, which was developed in the 1950s by Bellcore, originally provided an interoffice trunk for telephone company voice circuits. That is, it collected 24 analog voice calls and transmitted them on a single, four-wire circuit using TDM technology.

The equipment used for T1 is illustrated in Figure 2.15. The device used for collection of 24 low-speed circuits (64 Kbps, voice or data) is called a *channel bank* (CB), so named by Bellcore. The function of the CB is to convert up to 24 analog signals to a digital format, NRZ. It then multiplexes channels onto a single data stream and sends to a *data service unit* (DSU) via a 4-wire interface such as V.35 or RS422. The DSU performs signal conversion from NRZ line code to the bipolar bit pulse train AMI. The AMI pulse train is sent to a *channel service unit* (CSU), which provides circuit protection and the actual framing for T1. In addition to framing, the CSU enforces the ones density requirement.

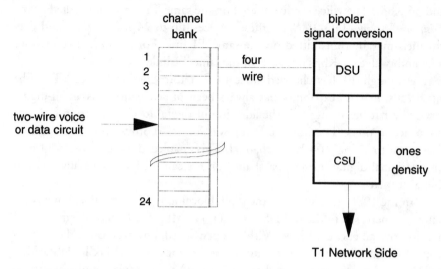

Figure 2.15 Channel bank and CSU/DSU.

This amounts to assuring that sufficient one bits are included to maintain synchronization. This process is explained in more detail in Section 2.3.4.4.

2.3.4.1 T1 Frame Format

The T1 frame format contains 24 channels. A single framing bit is illustrated in Figure 2.16. The sampling rate for T1 is 8,000 frames per second, which yields 1.544 Mbps. That is, 8,000 x ((24 x 8) + 1). Each slot may or may not contain a pulse with a magnitude of three volts. If a slot contains a pulse, its polarity is opposite from the polarity of the last pulse. The binary value of the pulse, when present, is one: when absent, it is zero. The framing bit is used to provide synchronization and amounts to an 8-Kbps channel (8,000 frames per second).

  Extended super frame (ESF) is a group of 24 frames. The purpose of grouping the frames is to allow the framing bit to be used for other functions, since only a 2-Kbps channel is needed for frame synchronization. To provide syn-

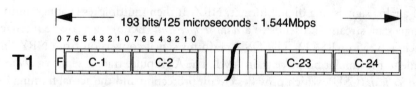

Figure 2.16 T1 frame format.

chronization, only channels 4, 8, 12, 16, 20, and 24 are used, which results in a 2-Kbps channel (6 x 8,000/24). See Figure 2.17.

The remainder of bits (as defined by AT&T 54016) used for framing are used for a *data link* (DL) channel and a 6-bit *cyclic redundancy check* (CRC). The DL channel uses every other frame commencing with the first frame, which provides a 4-Kbps channel (12 x 8,000/24). The DL channel originally used a protocol called BX.25, which is essentially LAPB. The DL channel is used to convey network management information, particularly fault detection and performance monitoring.

The 6-bit CRC uses the remainder of the previous framing bits to provide a 2-Kbps channel (6 x 8,000/24). (ISDN PRI defines the same use of the framing bit, except the DL channel and the 6-bit CRC, which are left for further study.)

Frame #	Fe	DL	BC	* Source of Robbed Bits
F1	-	m	-	Channel 1 through Channel 24
F2	-	-	C1	Channel 1 through Channel 24
F3	-	m	-	Channel 1 through Channel 24
F4	0	-	-	Channel 1 through Channel 24
F5	-	m	-	Channel 1 through Channel 24
F6	-	-	C2	* Channel 1 through Channel 24
F7	-	m	-	Channel 1 through Channel 24
F8	0	-	-	Channel 1 through Channel 24
F9	-	m	-	Channel 1 through Channel 24
F10	-	-	C3	Channel 1 through Channel 24
F11	-	m	-	Channel 1 through Channel 24
F12	1	-	-	* Channel 1 through Channel 24
F13	-	m	-	Channel 1 through Channel 24
F14	-	-	C4	Channel 1 through Channel 24
F15	-	m	-	Channel 1 through Channel 24
F16	0	-	-	Channel 1 through Channel 24
F17	-	m	-	Channel 1 through Channel 24
F18	-	-	C5	* Channel 1 through Channel 24
F19	-	m	-	Channel 1 through Channel 24
F20	1	-	-	Channel 1 through Channel 24
F21	-	m	-	Channel 1 through Channel 24
F22	-	-	C6	Channel 1 through Channel 24
F23	-	m	-	Channel 1 through Channel 24
F24	1	-	-	* Channel 1 through Channel 24

The Framing Bit - The high order bit of each frame.
The Robbed Bit - Low order bit of the 6th, 12th, 18th and 24th frames.

Figure 2.17 T1 framing subchannels.

2.3.4.2 Framing Subchannels

Prior to ESF, the only error checking was between adjoining network elements. This is because the errors flagged (e.g., bipolar violation) were not relevant except between two nodes, since they were corrected by the repeaters (or regenerators). In addition, when multiple T1s are collected and transmitted on a fiber link, the bipolar significance is lost. The CRC permits error checking between the source and destination and provides a comprehensive view of the T1 transmission quality. With the ESF format, diagnostic error checking is accomplished online without disruption of user service. Error checking with the D4 format required offline procedures. The subchannels are described as follows.

- DL control: The largest of the framing bit subdivisions is the DL subchannel. It is composed of the framing bits from every odd-numbered DS-1 frame. Since there are 8,000 DS-1 frames per second, the capacity of the DL subchannel is 4 Kbps. The function of the DL subchannel is the same as the D-channel for *integrated services digital network* (ISDN) *primary rate interface* (PRI), except AT&T Publication 54016 defines the protocol as BX.25 instead of LAPD. BX.25 is, for all practical purposes, the same as CCITT X.25 link layer; the two protocols are basically compatible. Their differences are subtle. For example, in the handling of a disconnect response without the P-bit set to one, BX.25 returns a UA response rather than a DM response (per X.25) to avoid having the response misinterpreted as a solicit for setup. Since it is only layer 2 of X.25, it will be called LAPB here.

    The function of the DL channel is to provide fault detection, fault reporting, and general network management. The DL channel, as previously noted, has a capacity of 4 Kbps. This is consistent with all three major specifications for the ESF DL channel—AT&T Publication 54016, ANSI T1.403, and CCITT I.441. The functions performed by the DL channel are the primary differences between the three specifications. AT&T 54016 defined a protocol called the *telemetry asynchronous block serial* (TABS) to provide management information from end to end. The protocol defines the content of the messages sent in the information section of the BX.25 frames. There are two types of TABS messages, a request message and a response message. The following are the different types of command messages.

    1. Activate DTE loopback;
    2. Deactivate DTE loopback;
    3. Send one-hour performance data;

4. Send performance monitoring counters;

5. Send errored ESF data;

6. Reset errored ESF data;

7. Send 24-hour errored second performance data;

8. Send 24-hour failed second performance data.

Response messages first repeat the command message with a status indication, followed by a maintenance message. The maintenance messages provide the requested reports or confirm those requests that do not require a report—activate loopback, for example. To understand the commands and reports, it is necessary to have a basic understanding of the acronyms and terms used. Definitions follow:

1. CRC6 error event: A CRC6 error event occurs when the calculations made by the DTE, based on the incoming DS-1 signal do not agree with the CRC-6 field transmitted by the network.

2. *Out of frame* (OOF) state: An OOF state begins when any two of four consecutive frame synchronizing bits received from the NI are incorrect. An OOF state ends when reframe occurs.

3. ESF error event: Each ESF is examined for an error event, and an ESF error event is determined by the logical ORing of a CRC6 error event and the OOF state. ESF error events are also processed into *errored second* (ES) and severely ES. Definitions follow.

4. ES: An ES is a second with one or more ESF error events.

5. Failed signal state: A failed signal that is declared when 10 consecutive severely ESs occur.

6. *Failed second* (FS): A FS is counted for every second a failed signal state exists.

Another variation of the DL control channel is defined in ANSI T1.403 called *zero byte time slot interchange* (ZBTSI), which is an optional method to B8ZS of providing clear channel [17].

- Framing: The framing function is performed with one-half of the remaining overhead bits. Namely, the framing bit from every fourth DS-1 frame commences with the fourth frame of an ESF frame. Since there are six DS-1 frames out of every 24 DS-1 frames used for framing, the bandwidth for framing is 2 Kbps. The value of the framing bit sequence is a binary constant, "0 0 1 0 1 1." The major specifications for ESF (AT&T, ANSI, and CCITT) describe the framing bit identically.

- CRC6: The remaining DS-1 framing bits (every fourth bit commencing with the second frame) form a new error-checking mechanism. There are six C-bits per ESF frame, yielding a subchannel of 2-Kbps bandwidth (6/(24 x 8) = 2) which are used for error checking. Error checking is accomplished by the transmitter calculating a six-bit, CRC sum (called CRC6) for each ESF frame and transmitting it in the immediately following ESF frame. The CRC6 sum is calculated for an ESF frame and compared to the value contained in the next received ESF frame. That is, the CRC6 sum in an ESF frame is for the previous frame. If the received and calculated check sums agree, the probability is nearly 99 percent (63/64) that there is no error in the transmission. If they do not agree, an error event is recorded. The major specifications for ESF (AT&T, ANSI, and CCITT) describe CRC6 identically. A detailed description of CRC6 is contained in AT&T Compatibility Bulletin #142. Although the calculation and checking is identical between specifications, the reporting of error information represents a major difference between AT&T Technical Publication 54016 and ANSI T1.403.

2.3.4.3 T1 Signaling Channel

Signaling is used to identify the called and calling stations, to set up and release connections, and to provide billing information and other supervisory functions. There are four options offered to accomplish signaling.

The first option uses the entire 24th channel (8 bits) from each T1 frame for signaling but provides the end user with a clear channel. That is, all eight bits of each of 23 user channels are available for data. This is called *common channel signaling* (CCS) and or the transparent option (option T). CCS can be implemented with either a *bit-oriented signaling* (BOS) where each bit has a designated meaning (e.g., on/off hook) or with a *message-oriented signaling* (MOS) where bits are grouped to form messages.

The other three options, which use in-band signaling, or *robbed bit signaling* (RBS), are referred to as Option 2, Option 4, and Option 16. The name *robbed* indicates the payload is reduced. RBS uses the low order bit of frames 6, 12, 18, and 24 with ESF. The source bits are identified in Figure 2.16 with an asterisk. (Note that these options are not available with E1.) When used, the capacity is reduced by 1.333 Kbps (4/24 x 8,000). Note that when *super frame* (SF) format is used, only frames 6 and 12 are robbed. The robbing options are described as follows.

1. Option 2: This option robs the low order, least significant bit (lsb), of each user channel (DS0) in frames 6, 12, 18, and 24 of each ESF frame. The bits are used to reflect the onhook and offhook status. An offhook signal is interpreted as either an immediate start (engage trunk without a dial signal) or a wink start (engage trunk after a short specified time). All four bits of each ESF frame are called A signaling bits for Option 2.

2. Option 4: This option uses the same low-order bits as Option 2, except that the bits from frames 6 and 18 are A signaling and the bits from frames 12 and 24 are B signaling. Note that there are two A signaling bits and two B signaling bits per ESF frame.

3. Option 16: This option uses the same low-order bits as Option 2 and Option 4, except the bit from each of the four frames is used for a separate signaling channel. That is, the bit from frame 6 is used for A signaling, the bit from frame 12 is used for B signaling, the bit from frame 18 is used for C signaling, and the bit from frame 24 is used for D signaling. These are often referred to as the ABCD bits.

Note that E1 also offers CCS (common to T1, but with a different source) and *channel associated signaling* (CAS), which is unique to E1 and uses ABCD bits similar to T1. *Non-facility associated signaling* (NFAS) is applicable to either T1 or E1 using CCS. The regional signaling systems described in Section 2.3.2.1 are divided as would be expected (most supporting E1). That is, R1 uses T1, R2 uses E1, C1 uses E1, and P7 uses E1 with some R1 characteristics such as CAS.

2.3.4.4 Clear Channel

To get a feel for the timing of an individual bit, divide the number of nanoseconds per second (1×10^9) by the total number of bits per second, which yields 648 nanoseconds per bit. Clock synchronization can only be maintained with a minimum density of one bits in the data stream. The AT&T Publication 54016 specifies that there should not be more than 15 zeros in a row or fewer than three bits equal to one in every 24-bit sequence. When too many subslots are sent as zero (no pulse), the transmitter and receiver get out of synchrony. The initial resolution to this restriction was to permanently allocate one bit (the high order bit) of every octet (eight bits) for the ones density requirement. This reduces the individual channel speed from 64 Kbps to 56 Kbps (-8000 frames per second).

The most popular solution to this problem, specified by both AT&T and ANSI, is a technique called *bipolar with 8 zeros substitution* (B8ZS) [18]. This technique provides clear channel (nonjammed with a bit forced to one) by replacing any zero octet with an invalid zero (one containing a polar violation). This is done by replacing the zero bit value in bit positions 4 and 7 with a pair of pulses that violate the bipolar scheme. See Figure 2.18.

A receiver of such a polar violation recognizes it and simply corrects the violation. The polarity of the first pulse added to create this violation is determined by the polarity of the previously transmitted pulse. The inserted viola-

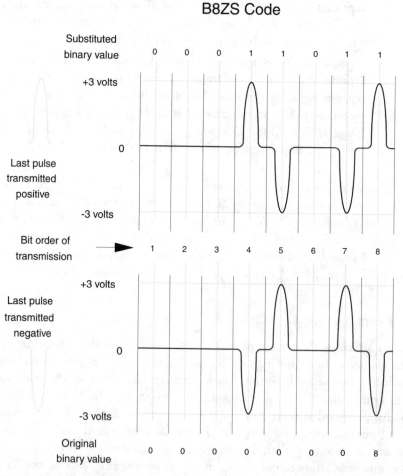

Figure 2.18 Bipolar violations in B8ZS\par.

tion is corrected by the insertion of the second violation. (Two wrongs make it dc-balanced!)

Figure 2.18 represents a 24-channel T carrier system. A similar technique is used for the CEPT lever 1, or E1 carrier, called *high-density bipolar order 3* (HDB3). This scheme works on semi-octets instead of octets. That is, it searches for any contiguous string of four nonpulses and replaces them with bipolar violations that are dc-balanced. HDB3 replaces the first and fourth bit positions with pulses—as in B8ZS. Both techniques produce a clear channel.

2.3.4.5 Voice Over T1

The T carrier system is a pure digital service that necessitates converting analog signals (voice) to digital signals for transmission, a function opposite to that performed by modems. These devices are simply called *analog/digital* (A/D) and *digital/analog* (D/A) devices. The original technique used to convert analog signals to digital data is called *pulse code modulation* (PCM). PCM is inefficient compared to some of the recently developed techniques. However, it was convenient when it originated, as it required only seven bits of each data octet per analog signal sample. The spare bit (high order) from each data octet could be set equal to one to satisfy a ones density requirement for equipment operating without B8ZS.

2.3.4.6 T Carrier Medium and Inherent Problems

The medium for the T1 carrier is a four-wire circuit. Because of the baseband signal, the range is limited to a short distance (less than a mile). This necessitates repeaters, usually every one-half to one mile. Furthermore, due to the high speed of T1, older equipment is unable to maintain synchrony without frequent one bits in the data. There are multiple methods of overcoming this problem. The first method developed utilized every eighth bit as an overhead bit that is always set to one (jamming a pulse in each octet). This worked fine for voice communication since only seven bits were required for the technique used to digitize the analog voice signal. However, as newer methods were developed, they could not be implemented because all the old repeaters were set up to enforce every eight bit being equal to one. Other incompatibility problems exist with the framing format. Again, this is partly because newer framing formats have been developed that are incompatible with the older equipment. Hence, a piece of equipment advertised as T1-compatible may not work with a similarly advertised piece of equipment purchased from a different vendor. Note that not only does the basic framing format vary, the technique of digitizing analog voice signals varies. The most serious incompatibility problems result from multiple T1 specifications.

2.3.4.7 Migration of T1 to Integrated Services Digital Network (ISDN)

The only specification used by vendors for T1 until 1989 was AT&T's Publication 54016. The ANSI T1.403 specification released in 1989 had superior capabilities to the AT&T, LAPB protocol used for the DL channel. The committee (T1E1) developing the ANSI T1.403 standard is represented by 19 LECs, 11 IECs, 42 manufacturers, and 17 general interest groups. AT&T's Technical Reference 54016 of September 1989 supported the ANSI T1.403 specification and made recommendations for a transitional migration to the addressing specified in CCITT I.441 and the CSU management facility as described by ANSI T1.403. Although AT&T is a voting member of the T1E1 committee, it plans to continue with the services described in AT&T Technical Publication 54016 (ESF). AT&T's goal is to skip the interim stages (improvements) of ANSI T1.403 and go directly to ISDN, which is based on T1 technology and yet a further improvement to ANSI T1.403. AT&T's Technical Reference Publication 41449, dated March, 1986, describes the AT&T ISDN PRI. This specification references the 1984 CCITT Recommendations I.211, I.411, and I.431 (Layer 1); I.440 and I.441 (Layer 2); and I.450 and I.451 (Layer 3) [19]. The basis of ISDN today is the international standards noted above, modified only as necessary because of areas where the standards were vague or undefined. The 1988 CCITT set of recommendations contained similar clarifications (Blue Books).

So where does this leave T1? Since ISDN is based on the fundamentals of T1, characteristics of ESF will be around for many years. Within a few years there should not be a problem determining which T1 standard to implement. Instead, the challenge will lie in determining which ISDN standard to implement, and as with ESF, more than one standard will exist! This is the positive picture for ISDN.

2.4 Link Access Procedure Balanced (LAPB)

LAPB mode is a class of HDLC procedures that has a balanced operation, compared to the older modes such as the normal response mode, which has a master and slave operation, and asynchronous response mode, which has both a master and slave function in each node. LAPB defines the options of ISO 7776 (HDLC) used for the DLL of ITU-T X.25. This is layer 2 on the OSI reference model, as defined by ISO 7498. The architecture of LAPB is illustrated in Figure 2.19. LAPB has HDLC option 2 (two-way reject) and option 8 (delete response I-frames), as defined in ISO 7776, set by default. LAPB can also use options 1, 10, and 14.

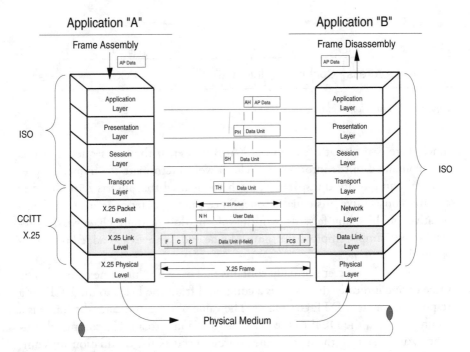

Figure 2.19 LAPB architecture.

Although the mode is balanced, the two entities using LAPB have different roles. The role of the user side is referred to as DTE and the network side is referred to as DCE. The address of the DTE is always equal to 3 and the address of the DCE is always equal to 1. Either the DTE or DCE may initiate link setup and frame transmission, but the response to various error conditions may be different. For example, if both sides solicit the other to setup up the link at the same time, the DTE will always comply by sending a link setup command.

LAPB in conjunction with the physical layer provides a vehicle for delivery of packets to the adjacent node. It is not concerned with what happens to the packet after it is delivered to the connected node—only that it arrives there safely and is acknowledged. Although LAPB was not designed for stand-alone multiprotocol operation, it has been in widespread use as the DLL of X.25 and was easily adapted by router manufacturers.

HDLC is the basis for other popular protocols used, such as LAPD for the D-Channel of ISDN and LAPF used for frame relay. It was developed from IBM's *synchronous data link control* (SDLC) protocol.

The general format of the LAPB frame, illustrated in Figure 2.20, commences and ends with a flag octet. The flag octet has the value of 126 (7E hex), which begins and ends with a binary zero and has all other bits equal to one.

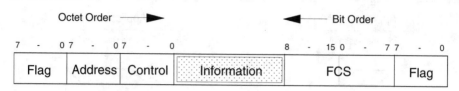

Figure 2.20 LAPB frame format.

Since there is no length field in the LAPB header, if the flag sequence of bits were contained in the information field, it would terminate the frame. To prevent this, and assure transparency, a scheme is used that inserts zero bits after any five consecutive bits equal to one. Since the flag octet contains 6 bits consecutive one bits, any false flags in the information field are changed. The receiving side simply employs the same scheme and removes any added bits. This is called *bit stuffing*.

The address octet normally contains one of two values, one or three. The value of one indicates the frame is a command from the DTE to the DCE, or a response frame to a DTE command. The value of three indicates the frame is a command from the DCE to the DTE, or a response to a DCE command. Another way of stating this is that the address specifies the destination for commands or the source for responses.

The control octet is used for frame acknowledgment, flow control, and defining the format of the remainder of the frame. This is a powerful octet and the topic of Figure 2.21.

Since LAPB can only address two stations (self and physically attached neighbor, or DCE and DTE), it can only be used as a trunk protocol—that is,

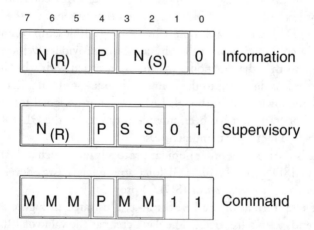

Figure 2.21 LAPB control octet format.

only between two directly connected stations. Directing the information in the frame to a network layer protocol at the destination requires a network header. Some router manufacturers use a proprietary protocol (short stack) above LAPB (DLL) to direct the recipient of the type of information carried in the frame, or the next higher layer to process the data. This typically consists of placing the Ethernet type code in the first two octets of the information field. For example, if the value is equal to 800 hex, it denotes an IP datagram and if equal to 8137 hex, it denotes an IPX datagram. Since this is a proprietary enhancement not described by a standard, it may not interoperate between routers from different vendors for multiprotocol operation.

The type of frame is determined from the first two least significant bits, as illustrated in Figure 2.21. If the lsb (+0) is equal to zero, it is an information carrying frame. If bit +0 is equal to one and bit +1 is equal to zero, it is a supervisory frame. If both bits +0 and +1 are each equal to one, it is a command frame [20].

N(R) is the expected send sequence number of the next I-frame to be received. The number cycles sequentially from the value zero to minus one. Bit +5 of the octet is the lsb of the N(R) field. N(S) is the sequence number of this frame. Other options such as modulo 128 may be supported, depending on the router vendor. Bit +1 is the least significant of the N(S) field. The window size determines the number that may be sent before awaiting a response. In an information frame, the P-bit bit is always set equal to one and solicits a response.

A transmitter (DTE or DCE) can never send more frames than the window size permits. The physical window size is determined by the size of the N(S) and N(R) fields. By default the size is three bits, which is referred to a module 8, because there are eight numbers possible in a three-bit field. The logical size, determined from a setup parameter called K, is defaulted to the value 2. This means that a sender may only have two outstanding, unacknowledged information frames. The ACK is accomplished by the receiver returning either an information frame with the N(R) field set to the value of the last received N(S)+1, which acknowledges the receipt of N(S). The ACK process makes LAPB a reliable, but slow protocol.

If the +0 bit of the control octet is equal to one and the second bit is equal to zero, it is a supervisory frame. Bits +2 and +3 form the type field, which is used to describe the supervisory frame. The field value zero is for a *receive ready* (RR) frame, the field value of one is for a *receive not ready* (RNR) frame and the field value of two is for a *reject* (REJ). The P-bit in a supervisory frame solicits an immediate status response.

If both the first and second bits are each equal to one, it is an unnumbered command frame. The bits noted as M define the type of unnumbered command frame. There are five unnumbered command frames, described as follows.

1. *Set asynchronous balanced mode* (SABM): Used to place the receiver (DTE or DCE) in link asynchronous balanced mode;

2. *Disconnect* (DISC): Terminate the mode previously set and suspend operation;

3. *Disconnect mode* (DM): Response to any command other than SABM while in disconnect mode;

4. *Unnumbered ACK* (UA): Acceptance of a mode change;

5. *Frame reject* (FRMR): Response used to report an error not recoverable by retransmission of the identical frame.

After setting up the link with an SABM command and soliciting a UA response from the other side, the process of sending information frames commences with receiving data from the network layer to be transmitted. This process, which is illustrated in Figure 2.22, is described in the following list (whose numbers correspond to those used in Figure 2.22).

1. Receive data from the packet layer.

2. The address octet (with the value of 1 or 3) and the information control octet are added to the beginning of the data provided by the packet layer (information).

3. The 16-bit FCS is calculated and placed after the information. The entire bit stream is bit stuffed by adding one bits for every sequence of five zero bits.

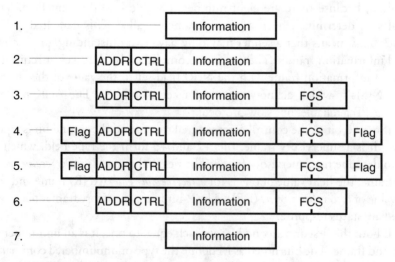

Figure 2.22 LAPB process.

4. The flag octets are added and the entire sequence is transmitted by passing it to the physical layer.
5. The receiving physical layer passes the frame to the DLL.
6. The DLL removes the flag octets, de-stuffs any added one bits, and verifies that an error was not made during the physical layer process by checking the FCS.
7. The identical information as received from the transmitting packet layer is passed to the receiving packet layer.

Steps 3 through 6 are normally performed by the hardware.

There are several variables and parameters that may require adjusting for proper operation of LAPB. The two most visible ones are the T1 timer and the N2 counter. T1 specifies the maximum time allowed for a response to a transmitted frame. The N2 counter specifies the maximum number of times that the T1 timer can expire before taking the link down.

2.5 Point to Point Protocol (PPP)

The PPP was developed by the *Internet Engineering Task Force* (IETF) of the ISOC and the specification may be found in RFC1661. It was developed to replace an older protocol called the *serial line Internet protocol* (SLIP). SLIP essentially bounded the serial bit stream by delimeters and nothing more.

The PPP provides a standard method for encapsulating and transporting multiprotocol datagrams over point-to-point links. It provides a *link control protocol* (LCP) for establishing, configuring, and testing the data link connection, and many *network control protocols* (NCPs) for peer binding in multiprotocol environments. PPP can aggregate multiple physical links with dynamic bandwidth allocation, illustrated in Figure 2.23.

LCP involves setting up the encapsulation format, negotiating frame size, verifying various configuration options, and terminating the link. Configuration options include compression options, user validation (password), link monitoring, and IP address registration or assignment. See the PPP logical flow in Figure 2.24.

PPP uses an HDLC-like structure and provides a mechanism to bind many of the popular protocols with PPP in order to ride PPP frames and to be identified by the receiver. Where the network layer protocol was specified in a LAPB frame with a proprietary protocol, PPP has a family of standard NCPs that bind PPP to handle different network layer protocols. Each NCP is provided with a negotiation mechanism to configure and establish its

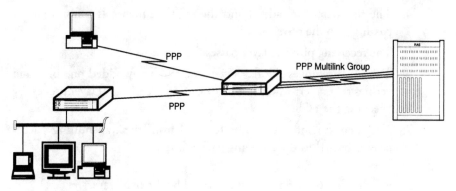

Figure 2.23 PPP serial links.

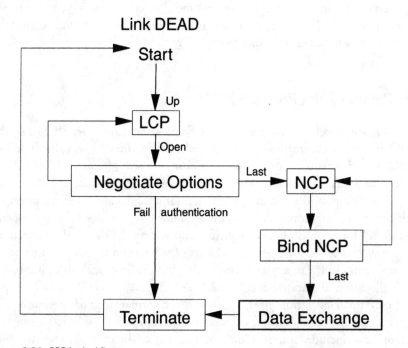

Figure 2.24 PPP logical flow.

unique handling. The network header occupies what would normally be the beginning of the information field of a LAPB frame. That is, the payload portion. Note that the term *packet* in PPP literature is used to describe the basic unit of encapsulation, which may be equated to *frame* except when link layer fragmentation is being performed. This could be confusing to those coming from an X.25 environment, who relate the term packet with *packet switching*.

The information in this header is network-related and since the description here is limited to single packets per frame, without fragmentation, the expressions have the same meaning.

During the initial PPP negotiation phase, the header in the information contains LCP frames that contain command type codes for communication with the peer LCP. Once PPP reaches the network phase illustrated in Figure 2.24, each network layer protocol (such as IP, IPX, and AppleTalk) is separately configured by the appropriate NCP. NCPs are never executed until completion of the LCP phase, including user validation. Examples of NCPs include the *Internet protocol control protocol* (IPCP) and the *Internet exchange control protocol* (IPXCP) [21].

After the link is initialized (and the binding of NCPs is complete), the protocol number in the header contains a value representing the specific protocol data, as illustrated in the PPP frame format in Figure 2.25. This identifies it such that the peer (receiver) can pass the data to the appropriate protocol handler.

Although the general appearance is that of HDLC (flag, address, and control), the content is different. This is why it is described as *HDLC-like*. The address octet always contains the value 255 to indicate a broadcast to all stations. The control octet always contains the value 3, which indicates that this is an UI frame and the *poll/final bit* (P/F) bit (not illustrated) is zero. The FCS field is illustrated as 16 bits, but it can also be 32 bits if negotiated.

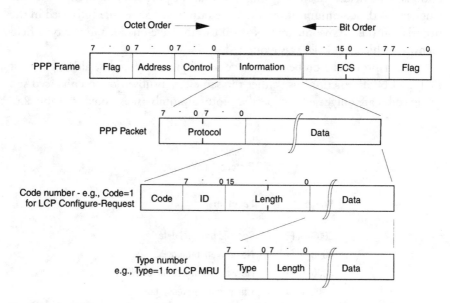

Figure 2.25 PPP frame format.

Transparency is accomplished by bit stuffing, as with HDLC for synchronous operation. Octet stuffing is used for asynchronous links to handle control characters. An escape octet is defined for this purpose.

2.5.1 Protocol Numbers, Codes, and Types

The first two-octet field of the information field is the protocol number, which defines the function and format of the remainder of the frame. The range of protocol numbers with functions is illustrated in Table 2.5. The low order bit of the field must be equal to one for the resemblance to HDLC. That is, if equal to zero, it would continue to extend the field another octet in length until a bit equal to one was found.

If the protocol number is C021 hex, the LCP is identified and the next octet contains one of the codes identified in Table 2.6. Other protocol numbers prior to the network phase include C023 hex (*password authentication protocol* [PAP]), C025 hex (link quality report), C02B hex (*bandwidth allocation control protocol* [BACP]), and C223 hex (*challenge handshake authorization protocol* [CHAP]) [22].

Each LCP packet has a code, identifier, length field, and data field to negotiate with its peer. The data field contains the identification of the configuration option (type field) being negotiated. The LCP codes provide the mechanism, or verbs, to negotiate the various options defined in the configuration options (type field). An analogy might be that the protocol number specifies the language of the communication, the code number specifies the verb used in the negotiation (i.e., "Give me one. No you cannot have one."), and the type field identifies which candy jar was requested.

Each protocol number may have a unique set of codes, although most LCP and NCP functions use either the same set, a subset, or an enhanced set. Some codes are only used for specific protocol numbers, as noted in Table 2.6.

Table 2.5
PPP Protocol Numbers

Range	Function
0000–3FFF	Specific protocol data
4000–7FFF	No network protocol
8000–BFFF	Network control protocol (NCP)
C000–FFFF	Link control protocol (LCP)

Table 2.6
PPP Codes

Code	Description
0	Vendor-specific
1	Configure-Request
2	Configure-Ack
3	Configure-Nak
4	Configure-Reject
5	Terminate-Request
6	Terminate-Ack
7	Code-Reject
8	Protocol-Reject/ LCP
9	Echo-Request/LCP
10	Echo-Reply/LCP
11	Discard-Request/LCP
12	Identification/LCP
13	Time-Remaining/LCP
14	Reset-Request/LCP
15	Reset-Reply/CCP

Code 0 may be used during LCP or NCP negotiation to set or obtain company proprietary information [23]. The mechanism allows proprietary features without encumbering the IANA with proprietary number requests. Vendor-specific packets may be sent at any time, even before LCP has reached the opened state.

The actual LCP configuration options, subject of the negotiation, are identified in Table 2.7 From the analogy above, this is where the particular candy jar is identified. Although each configuration option (type number) is negotiable, most have a default if not negotiated. For example, the default *maximum receive unit* (MRU) is 1,500 octets.

2.5.2 PPP Process

The process of handling PPP frames is similar to LAPB with regard to the flags, transparency, and FCS. The LCP and NCP negotiation and the transfer of datagram can best be visualized by observing the changing 16-bit protocol field.

Table 2.7
LCP Type Options

Type	Descriptior
1	Maximum receive unit
2	Async control character map
3	Authentication protocol
4	Quality protocol
5	Magic number
6	Reserved
7	Protocol field compression
8	Address and control field compression
9	FCS alternatives
10	Self-describing pad
11	Number mode
12	Multilink procedure
13	Callback
14	Connect time
15	Compound frames
16	Nominal data encapsulation
17	Multilink MRRU
18	Multilink short sequence number
19	Multilink endpoint discriminator
20	Proprietary
21	DCF identifier

From Table 2.5 note that when the numbers are in the range C000 to FFFF hex, the packet belongs to the LCP, and the next eight bits contain an LCP code. When the range of the protocol field is 8000 to BFFF hex, the datagram belongs to an NCP and just as with the LCP datagram, the next eight bits contain an NCP code. When the protocol field is in the range 0000 to 3FFF, it identifies a packet belonging to a specific network layer protocol, such as IP or IPX.

As an example, pretend you are a data analyzer looking at frames on a PPP link where you observe that the protocol field of a PPP frame is equal to C021 hex. This indicates that LCP is negotiating link parameters. Then, a few frames later, the protocol field changes to 802B hex. This would indicate that the Novell NCP was negotiating with the host and peer to use the PPP link. Finally, observe that the following frames contain a protocol field equal to 002B hex to identify datagrams belonging to the Novell network layer, IPX. The receiving host will pass these datagrams to the IPX/SPX stack for processing.

If the PPP link were being used for both IP and IPX, the frames with the protocol field equal to C021 hex would be followed by frames with the protocol field equal to 8021 hex (IP) and 802B hex (IPX) to bind both IP and IPX network layer protocols with their peers. Then, the following frames could be mixed, depending on what happened to be in queue for the link, with frames containing a protocol field equal to 0021 hex (IP) and frames containing a protocol field equal to 002B hex (IPX) [24].

2.5.3 NCPs

While the LCP was designed for reasonable feature extensibility, the NCP strategy is to add complete new *data link layer* (DLL) protocols that initialize the link for operation in a multiprotocol environment. For example, the new IPv6 was recently added. There are many DLLs registered with the IANA, and each has an individual NCP to bind that protocol. NCPs have a protocol number in the range 8000 hex to BFFF hex. See Table 2.5 where an overview of the IP NCP is provided, and general capabilities of a sample of other NCPs are identified. Also see the summary of RFCs pertaining to PPP in Table 2.9.

2.5.4 Internet Protocol Control Protocol (IPCP)

IPCP is the NCP for IP and has no special privileges. It, like all other NCPs, cannot be executed before the completion of the LCP phase. The primary function of IPCP is to enable the link to handle IP datagrams. IPCP packets are identified with the protocol number 8021 hex.

IPCP type number 3 is used for negotiation of the IP address to be used. The length field is equal to six, which provides four octets for the IP address. If the IP address field is non zero, it represents the IP address that the peer wants to use. Otherwise, if it is equal to zero, an IP address is being requested in the configure-request code (see Figure 2.23 and Table 2.8). The peer can either return an IP address with a configure-ACK code, or send a configure-nak code. The default is to not assign an IP address. Hence, the PPP frame to request an IP address would appear as 0X7EFF03802101000A030600000000. As transmitted, this is interpreted as follows.

> 7E - Flag, FF - Address, 03 - Control (UI)
>
> 0821 - Protocol number for IPCP, 01 - Code number for Configure Request

Table 2.8
Other NCP Type Numbers

Bridging NCP		IPX NCP	
Type	**Description**	**Type**	**Description**
1	Bridge identification	1	Network number
2	Line identification	2	Node number
3	MAG support	3	Compression protocol
4	Tinygram compression	4	Routing protocol
5	LAN identification	5	Router name
6	MAG address	6	Configuration complete
7	Spanning tree protocol		

000A - Length of Code and Type fields, 01 - Identifier (arbitrary)

03 - Type number, 06 - Length, 00000000 - 32-bit binding for an IP address

Although IPCP is best known for IP address registration, it also provides for selection of the compression algorithm to be used. Note the difference between the LCP compression and IPCP compression. The LCP compression, option 8, simply omits the frame address and control octets, while the IPCP compression (code 2) compresses the TCP/IP headers to as little as three bytes. When header compression is used, the protocol number in the PPP header is set to either 002D hex or 002F hex for IP datagrams, instead of the normal 0021 hex.

2.5.5 Other NCPs

The design of PPP has allowed for over 16,000 NCPs, and the possibilities are essentially unlimited. Table 2.8 identifies the type numbers used for the bridging NCP and the IPX NCP. To get a feel for all the NCP and the related RFC, see Table 2.9. (There is no summary available on the RFC library, but RFC 1700, Assigned Numbers RFC, identifies the allocated NCP numbers.)

The type numbers in Table 2.8 are predictable. The bridging protocol needs agreement on such things as the MAC address and whether or not STP is used—factors unique to bridging. With IPX the network number and node number are needed, which are unique to IPX (and XNS) addressing.

Table 2.9
PPP LCP and NCPs—Related RFCs

RFC	Title
2153	PPP Vendor Extensions
2125	The PPP Bandwidth Allocation Protocol (BAP) and The PPP Bandwidth Allocation Control Protocol (BACP)
2097	The PPP NetBIOS Frames Control Protocol (NBFCP)
2043	The PPP SNA Control Protocol (SNACP)
2023	IP Version 6 Over PPP
1994	PPP Challenge Handshake Authentication Protocol (CHAP)
1993	PPP Gandalf FZA Compression Protocol
1990	The PPP Multilink Protocol (MP)
1989	PPP Link Quality Monitoring
1979	PPP Deflate Protocol
1978	PPP Predictor Compression Protocol
1977	PPP BSD Compression Protocol
1976	PPP for Data Compression in Data Circuit-Terminating Equipment (DCE)
1975	PPP Magnalink Variable Resource Compression
1974	PPP Stac LZS Compression Protocol
1973	PPP in Frame Relay
1969	The PPP DES Encryption Protocol (DESE)
1968	The PPP Encryption Control Protocol (ECP)
1967	PPP LZS-DCP Compression Protocol (LZS-DCP)
1963	PPP Serial Data Transport Protocol (SDTP)
1962	The PPP Compression Control Protocol (CCP)
1915	Variance for the PPP Connection Control Protocol and the PPP Encryption Control Protocol
1877	PPP Internet Protocol Control Protocol Extensions for Name Server Addresses
1841	PPP Network Control Protocol for LAN Extension
1764	The PPP XNS IDP Control Protocol (XNSCP)
1763	The PPP Banyan Vines Control Protocol (BVCP)
1762	The PPP DECnet Phase IV Control Protocol (DNCP)
1663	PPP Reliable Transmission
1662	PPP in HDLC-Like Framing
1661	The Point-to-Point Protocol (PPP)
1638	PPP Bridging Control Protocol (BCP)

Table 2.9 (continued)
PPP LCP and NCPs—Related RFCs

RFC	Title
1619	PPP Over SONET/SDH
1618	PPP Over ISDN
1598	PPP in X.25
1570	PPP LCP Extensions
1552	The PPP Internetwork Packet Exchange Control Protocol (IPXCP)
1378	The PPP AppleTalk Control Protocol (ATCP)
1377	The PPP OSI Network Layer Control Protocol (OSINLCP)
1376	The PPP DECnet Phase IV Control Protocol (DNCP)
1332	The PPP Internet Protocol Control Protocol (IPCP)

2.5.6   PPP Feature Highlights

This section provides a high-level description of several popular PPP features. For a detailed description of these features, refer to the RFC identified for the feature.

2.5.6.1   Authentication Protocols

After link establishment, PPP provides the capability to authenticate peers on a dial-up (or dedicated) link. The three options available are PAP, CHAP, or RADIUS (explained in Section 5.16). PAP is a simple two-way verification of a clear text, eight-octet password. CHAP uses a three-way handshake based on a "secret" key that is not transmitted across the link, and the 16-octet password is transmitted in MD5 encrypted format. The type of authentication is requested in a LCP packet (protocol number=C021 hex) with a configure request (option number=1) and type number equal to 3. Then, the last four octets specify either PAP (C023 hex) or CHAP (C223 hex).

Assuming a configure ACK is received from the peer for the configure request sent, new option numbers and type numbers are used for authentication. The sequence for CHAP, determined from the new option codes, is 1) challenge, 2) response, 3) success, or 4) failure [25]. Challenges are made randomly throughout the connection (with different identifiers and challenge values) to limit the possible attack exposure time.

2.5.6.2 Link Quality Monitoring

Link quality monitoring is provided throughout the connection if the quality protocol is negotiated. This protocol maintains counters for the packets and octets transmitted and received on the link. It periodically sends this information to the peer in the form of a *link quality report* (LQR). The LQR tells when and how often the link failed or dropped data. The LQR may be used to select other routes when excessive data is being dropped.

Link quality monitoring is selected with a LCP configure request (protocol number=C021 hex) containing option number=1 and Type number equal to 4, followed by the protocol number C025 hex and the reporting period. The reporting period is four octets in length and represents the interval time in hundredths of a second the LQR is sent [26].

2.5.6.3 PPP Multilink

PPP *multilink* (ML) on the OSI reference model falls on about two and one half. That is somewhere between the network layer and the DLL. The purpose of multilink is to create an ML bundle, a single logical connection above the DLL that controls multiple physical layer connections [27]. The ML bundle has many benefits, including the following:

1. It allows for the aggregation of multiple physical circuits to provide increased bandwidth, increased reliability, and reduced cost.

2. It permits the addition and deletion of physical circuits without interruption of service and without changing the routing characteristics.

3. It makes it possible to achieve graceful degradation of service when a physical circuit fails.

4. It provides a single logical interface for all circuits in the ML group to upper layer protocols.

5. It provides segmentation and reduced latency and potentially increases the MRU.

It is not as though the aggregation of multiple physical links in one logical unit is a new invention. The design is similar to that used with LAPB ML in ITU-T CCITT X.25; it is also defined in ISO 7776. Before that, moreover, a similar operation was used during the mid 1960s to connect multiple low-speed trunks (by *Aeronautical Radio, Inc.* [ARINC]).

When ML is configured, the physical links are transparent to network management, data forwarders, and dynamic routing programs. That is, they only see the ML logical port. The program bound for a particular data forwarder (e.g., IP, IPX, FFR, X.25, ISDN, or AppleTalk) adds a header to a datagram to be sent by PPP (e.g., protocol number equal to 0021 hex for IP). Since only the ML bundle is known, it transfers the packet to the ML program for processing where the ML header and framing are added before passing to an individual ML physical link member for transmission. The particular physical link member is selected in a manner that accomplishes load balancing. Hence, the first fragment of each stream contains two headers. The receiving peer uses the ML header for reassembly and the second header to determine the network layer handler, just as though multilinking operations were not configured. Of course, if the datagram is not fragmented, every ML packet contains two headers—the first for control of the ML bundle and the second for multiprotocol operation. See Figure 2.26.

PPP ML is initiated for each physical link at the end of the LCP phase and prior to any NCP binding. The negotiating LCP protocol number is C021 hex, with a configure request (code equal to 1) containing a type number equal to 12 and other characteristics, such as the following.

1. *Maximum receive reconstruct-unit* (MRRU): Type 17;

2. Short (12-bit) or long (24-bit) sequence numbers (24 bits): Type 18;

3. Discriminator: Type 19.

The discriminator number is unique to an ML bundle. The option is used to ensure that new links are added to the correct ML bundle. Authentication

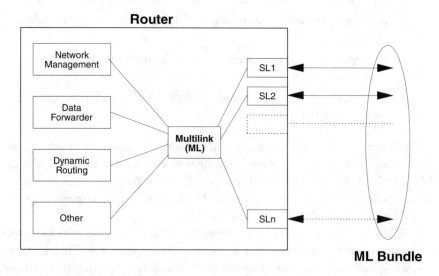

Figure 2.26 Multilink organization.

may also be used to protect against hostile joining of an ML bundle. The default size of the sequence number field is 24 bits, but it can be negotiated to the short size field of only 12 bits. The MMRU option allows the ML MRRU size to be larger or smaller than the LCP MRU (type 1, minimum of 1500 octets).

Receipt of a configure ACK completes the peer binding for ML operation. A peer responding with a configure ACK has agreed to ML processing of frames received with the protocol number equal to 3D hex.

Immediately following the protocol number is a four-octet ML header used to resequence frames (or frame fragments). Each frame, or frame fragment, has a sequence number to assure that it is reassembled in the same order as disassembled by the sending peer. The size of the sequence number field is either 12 bits or 24 bits, as negotiated during the LCP phase (default size is equal to 24 bits).

Another important part of the ML header is the *beginning bit* (B-bit) and the *ending bit* (E-bit). The first frame fragment of a stream has the B-bit set, and the last frame fragment has the E-bit set. Intermediate frame fragments have both the B-bit and E-bit each set equal to zero. A frame that is not fragmented has both the B-bit and the E-bit set, indicating that it is the only frame of the stream.

Figure 2.27 illustrates the ML frame format for a nonfragmented IP datagram with a 12-bit sequence number. Note that the B-bit and E-bit are each equal to one indicating that the IP datagram is not fragmented.

2.5.6.4 Bandwidth Allocation Protocol (BAP) and Control Protocol

The *bandwidth allocation protocol* (BAP) is an extension to the ML protocol that was developed nearly a year later [28]. It provides refinement to some of the features touted earlier for ML. For example, it enables the graceful adding and

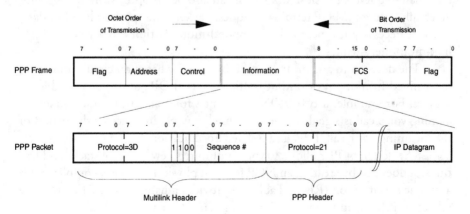

Figure 2.27 Multilink frame format.

deleting of physical circuits to accommodate increased or decreased traffic, without loss of traffic.

The ability to gracefully allocate and remove physical links from an ML bundle is called *bandwidth on demand* (BOD). BOD is initiated in ML with the new protocol BACP. The primary function of BACP is to bind the ML peers to use BAP. It does this by identifying (negotiating a unique discriminator) the SLs associated with each ML bundle. It is negotiated after the completion of the LCP phase. That is, as part of the NCP phase. BACP used a new type code (23), which identifies each end of an SL with a unique discriminator, identifying it as belonging to a particular ML bundle. BACP uses the protocol number C02B hex and the same format of LCP negotiation, but only with the seven code options beginning with configure request. The only ML option for BACP is the negotiation of the favored peer status. The result of this negotiation is the identification of one peer as the favored peer in event of a later race condition when simultaneous requests are made. Once negotiated for one SL of an ML bundle, it is in effect for all SLs of the ML bundle. From here, BOD is accomplished with BAP.

Once setup properly using BACP, the BAP is used to negotiate the graceful adding or dropping of an SL from (or to) an ML bundle. BAP is identified in the PPP packet with a new protocol number, C02D hex. Using the BAP, a peer may also request that its peer initiate the adding or dropping of an SL (shy peer). As illustrated in Figure 2.28, the peer may optionally defer to the favored peer with a *call back request.* Figure 2.28(a) illustrates adding a link to the bundle, and Figure 2.28(b) illustrates gracefully dropping a link.

Using the BAP to add and drop a link is graceful in that each peer completes any work in process (e.g., fragmented datagram) before actually dropping the link. After activity ceases on the link, LCP is used to drop the link with a terminate-request A physical link which can also be dropped without BAC (not gracefully) using the LCP terminate-request packet. This may result in lost data, but it is necessary if a peer fails to respond multiple times to a graceful BAC link-drop-query-request.

The decision to add or drop an SL by a peer is based on circuit utilization determined from input or output traffic reports (LQR) or on the available resources. For example, if two ISDN B channels form an ML bundle and an incoming voice call is signaled, one B channel may be dropped for the period of the incoming voice call, PPP, and Related Specifications.

The design of PPP allows for near unlimited new protocol stacks and features. It does this by creating an NCP for each physical interface, complete with a unique specification (RFC). Table 2.9 provides a handy list of RFCs that are related to PPP, including the NCPs. (Summary RFC not available.)

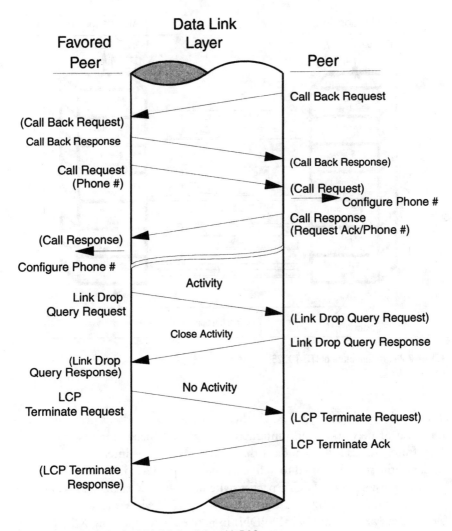

Figure 2.28 (a) Adding and (b) deleting links with BAC.

2.6 ITU-T X.25 (Formally CCITT X.25)

ITU-T X.25 is a network layer interface between the DTE and DCE, as previously identified in Section 2.4. Where LAPB only addressed and delivered to an adjacent node, X.25 addresses can deliver to the end points of the network. The architecture of X.25 is illustrated in Figure 2.29. Once the end points are established, multiple logical channels may be used to multiplex virtual circuits [29].

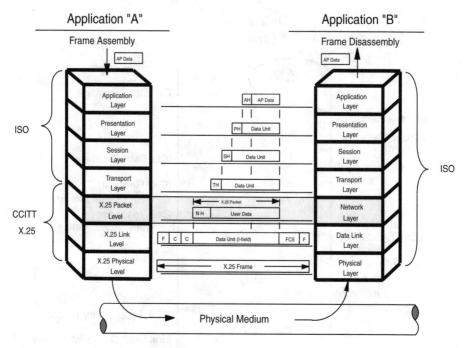

Figure 2.29 Architecture of ITU-T X.25.

In the world of ITU-T X.25, there is a stack of standards describing various areas of networking and internetworking. A challenge for new users is to pick from the stack the standard that covers the area of interest. This section discusses the standards used to define the interfaces to X.25.

At the core of the ITU-T packet switching network is X.25. This describes both the DLL (LAPB) and the network layer. Devices accessing a *packet-switched public data network* (PSPDN) in native mode (speaking X.25 packets) are called packet mode DTEs. Those that cannot are start-stop mode asynchronous terminals (dumb), in the view of ITU-T. They must use an assistance operator called a *packet assembler disassembler* (PAD). The interface between a dumb terminal and the PAD is covered in X.28, and the interface between the PAD and the packet mode DTE is covered in X.29.

The function of an X.29 PAD is to establish and clear virtual calls. It also identifies Recommendation X.3, which defines the options to be used by a dumb terminal. Examples of the options are line speed, parity, line feed insertion, data forwarding (on CR for example), echo, break action, and line folding.

The physical interfaces to a PSPDN are X.20 (asynchronous) and X.21 (synchronous). To provide compatibility until the world adapted to these interfaces, there is a category called *bis,* which essentially grandfathered standards such as RS232 and V.35. These are called X.20 bis and X.21 bis.

The interface used by dial-up packet mode DTEs is described in X.32, and the interface used between X.25 networks is X.75. The interface recommendation between a PSPDN and a *private packet data network* (PvtPDN) is X.327 and X.25. The interface recommendation between a packet mode DTE and the ISDN is X.31.

2.6.1 Virtual Circuits and Logical Channel Numbers

The purpose of X.25 is to perform call establishment, multiplex logical channels, flow control, packet assurance and integrity, and call clearing. Although all these features are important, multiplexing virtual circuits with logical channels is the main feature, and it should come to mind when thinking of X.25. There are two types of virtual circuits, as illustrated in Figure 2.30. One corresponds to a dedicated circuit and is called a *permanent virtual circuit* (PVC). The other corresponds to a dialed circuit and is called a *virtual call* (VC). The expression virtual circuit is used to describe either a PVC or VC.

When a calling (originator) DTE places a call, it is assigned a *logical channel number* (LCN) (e.g., "m"), which is associated with the calling DTE and the network. When the network forwards the call to the called (destination) DTE another LCN that is associated with the called DTE and the network (e.g., "n") is assigned. The network forms yet another number (e.g., "x"), which is unique in the network and defines the path between the pair of LCNs "m" and "n." The number "x" then represents the virtual circuit between the two DTEs. The network will, until the virtual circuit is cleared, associate all input from LCN "m" with LCN "n" and form LCN "n" with LCN "m." At the same time, either DTE may place other calls with different LCNs to form different virtual circuits.

There are 12 bits allocated to carry the LCN in the packet header, as will be seen, which provides for 4,096 LCNs. This is the maximum, and it is not reasonable to use this many. Remember, the more LCNs, the bigger the bill. Furthermore, the throughput on a serial link actually decreases after 10 LCNs are active, depending on the line speed.

LCN zero is reserved for restart, diagnostics, and registration. The first partition is used for PVCs. If there are no PVCs, LCN 1 may be used for the first LCN of the *incoming calls* (LICs), or if there are no PVCs or incoming only calls, it may be used for the first *two-way call* (LTC). If only outgoing calls are

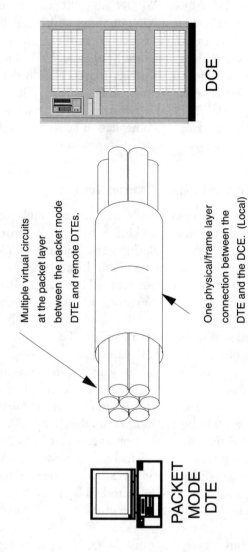

DCE

Multiple virtual circuits
at the packet layer
between the packet mode
DTE and remote DTEs.

One physical/frame layer
connection between the
DTE and the DCE. (Local)

PACKET
MODE
DTE

Figure 2.30 Virtual circuits.

used, LCN 1 may be used for *outgoing only calls* (LOCs). The next partition is used for LICs; the next partition is used for LTCs; and the last partition is used for LOCs.

To minimize call collisions, the DTE assigns LCNs for outgoing calls from the highest numbers first, in descending order. The DCE assigns LCNs from the lowest numbers for incoming calls in ascending order. The range of each category is illustrated in Figure 2.31.

A virtual circuit is closed after some period of inactivity (the length of the period depends on the cost associated with an open virtual circuit) [30].

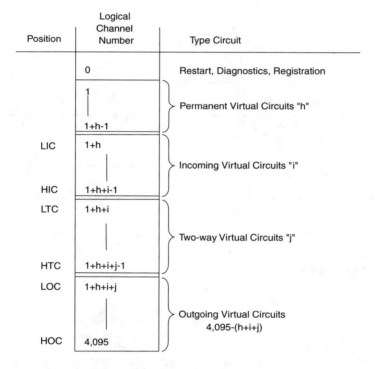

*Channel Positions: LIC=Lowest Incoming, HIC=Highest Incoming, LTC=Lowest Two-way, HTC=Highest Two-way, LOC=Lowest Outgoing, HOC=Highest Outgoing.

*Permanent VCs=h, Incoming VCs=i, Two-way VCs=j. The sum of h+i+j is less than 4,095.

*Incoming Logical Channel Numbers (LCN) assigned by the DCE - Outgoing LCNs assigned by the DTE.

Figure 2.31 Logical channel number assignment.

2.6.2 Frame Format

From a glance at Figure 2.32, one quickly recognizes the frame format. It is LAPB exactly as described in Section 2.4. The information field is called the *packet data unit* (PDU). It is the payload of the information carrying HDLC frame. The PDU contains a packet header and user data. The header is three octets in length, and the data size is fixed but can be negotiated for different sizes, from 16 octets to 4,096 octets. PDUs begin on X.25 data packet boundaries and the M bit (*more data*) is used to fragment PDUs that are larger than one X.25 data packet in length.

The packet header has only three basic fields and is very efficiently organized, as illustrated in Figure 2.33.

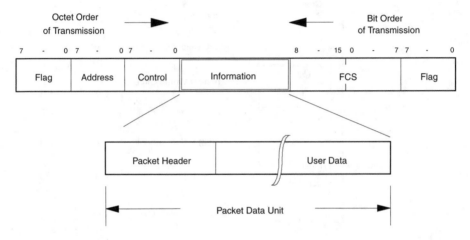

Figure 2.32 X.25 frame format.

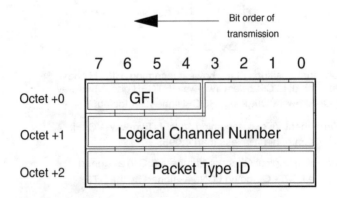

Figure 2.33 X.25 header format.

There are four bits in the *general format identifier* (GFI) that identify all formats. These bits define if there is module 8 or 128 sequence numbers and identify whether there is local or remote significance, the type addressing used (standard is X.121 and by exception, ISDN addressing is specified), and whether it is user communication or with a network function (qualified data). When the communication is with a network function, the *qualified data bit* (Q-bit) is set. The network function is called a PAD and must not be used for IP riding an X.25 network. These devices interface dumb terminals (those not capable of initiating call requests and, in general, not speaking X.25). Compliant terminals are referred to as packet mode DTEs.

The LCN is binary-encoded, and bit 1 of octet 2 is the lsb. This is the field that in conjunction with the address field of both the calling DTE and the called DTE form a virtual circuit.

The *packet type identifier* (PKID) identifies the various packet types, as illustrated in Figure 2.34. The "*" in the hex digit of the first four PKID types is a flag to indicate that these bits are to be ignored as they will vary.

The packet types in Figure 2.34 are ordered by the value of the PKID field and may be used to easily identify the various packet formats. These values are identified in Figure 2.35, which illustrates setting up a call and the transfer of data. The PKID field is identified for the call request, call accept, and data packets. The action starts on the left side of Figure 2.35 with a call request, PKID equal to 0B hex.

All communications, except those with a permanent LCN, are initiated with a call request. Some X.25 networks, including the ISDN under present tariffs in most areas, charge for virtual circuit holding time.

The call request contains the calling DTE address, the called DTE address, facility codes, and, optionally, user data.

There are multiple types of facility codes: international, calling, called, and CCITT-specified facilities. There are also four classes of facility codes (A, B, C, and D). Typical facility negotiation will be for packet size, window size, class, user group, and reverse charging. Yet another type of facility request is actually to the service provider and is called online configuration. This type of facility allows negotiating features like throughput class, default packet size, default window size, and logical channel types and ranges.

Remote significance is set, as illustrated in Figure 2.36. This causes the DCE to pass the call request to the addressed DTE for acceptance. Otherwise, the DCE would send the acceptance and save network time. The bit selecting this action in the GFI is referred to as the *D-bit*.

The PKID field contains both send and receive sequence numbers for data packets, similar to those in LAPB. For data packets, there is also a *more data bit* (M-bit). The M-bit is used when the compete message is not

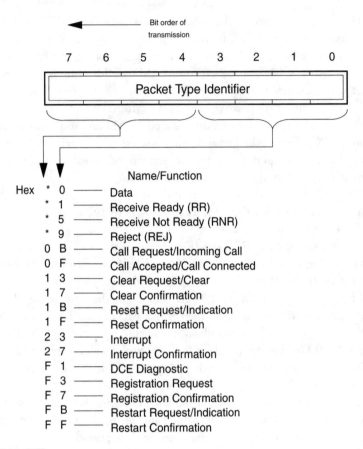

Figure 2.34 PKID.

contained in a single packet. A receiver of the M-bit will know there is more to come.

2.6.3 Call Request Packet Format

The call request allows for user data to be present and is typically used for the fast select option (datagram type delivery and release the VC). The first field checked by the DCE is the two high-order bits of the first octet of user data (called the protocol code [PC]). These two bits can designate different handling. Of interest here is when both bits are each equal to one. This identifies two packet mode DTEs that are using the VC and signifies that the DCE should ignore the remainder of the user data. The field is illustrated in the lower portion of Figure 2.36.

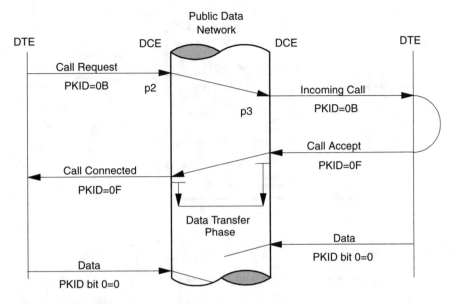

Figure 2.35 Call request and data transfer.

Several values have been reserved for the first octet of user data, and the value CC hex identifies a TCP/IP call request [31]. That is, bits +2, +3, +6, and +7 are each equal to one, and all other bits are equal to zero. Notice that this sets both bits of the PC field on, which identifies the call as DTE to DTE. Once the connection is made, all following data packets will be delivered (transparently) to the addressed DTE.

The value 80 hex (binary 10000000, decimal 128) specifies the protocol code for DTE to DTE and further identifies the use of a SNAP header to encapsulate and identify other network layer protocols. The SNAP header is the same as defined for Ethernet, token ring, and other interfaces. SNAP headers are not included in the subsequent X.25 data packets.

Negotiable facilities such as packet size, maximum packet, size and window size of X.25 may be used. Others such as the fast select option (datagram delivery) and setting the D-bit to indicate end-to-end significance must not be used.

2.7 Frame Relay

Frame relay, the by-product of early ISDN specifications, is based on frame-switching technology. It is the most popular of the fast packet techniques

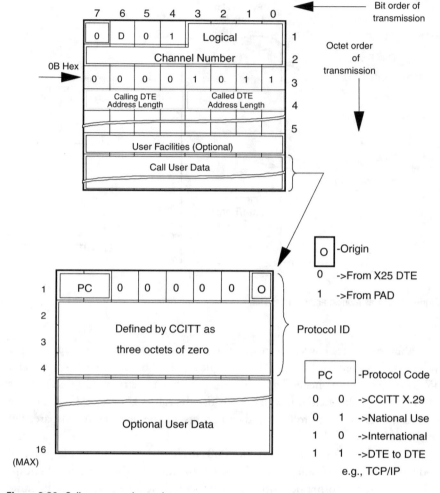

Figure 2.36 Call request and user data.

(originally called *fast frame relay* [FFR]) used for connecting LANs with serial links and a natural migration from an X.25 packet switched network. Although frame relay is offered in many public networks, it is most often used in private networking with leased T1 facilities. Public FFR service is offered over ISDN BRI and PRI, DDS, T1, E1, DS0, DS1, and DS3.

It is easy to see from a glance at Figure 2.36 why FFR is faster than X.25. It provides a virtual circuit at the DLL instead of the network layer. In addition, the overhead for packet assurance, sequence integrity, flow control, and call setup are absent from frame relay. Basically, it behaves like a LAN, but over a WAN.

The primary specifications for FFR are ANSI T1.617/T1.618 and ITU-T Q.922/Q.933. The specifications provide for two types of FFR. Type 1 has only the core functions of Q.922, and the remainder of the DLL is user-defined. This is the common offering. Type 2 defines a full implementation of Q.922 for the DLL of operation between end users. The specifications also provide for both *permanent virtual circuits* (PVCs) and *switched virtual circuits* (SVCs). SVCs nor the full FFR data link (type 2) are being implemented.

FFR can be implemented in a fully meshed network with T1 or E1 lines. This requires $n(n - 1)/2$ lines and is expensive. It is referred to as *just-in-case bandwidth*. A cheaper alternate is a public FFR, which is referred to as *just-in-time bandwidth*.

2.7.1 Frame Relay Format

Just as with X.25 and PPP, the FFR frame format can be recognized at a glance. See Figure 2.37. It has many similarities to LAPB. The flag octets and FCS are identical, and transparency is provided the same way (bit stuffing). The header contains only two octets, although the Q.922 specifications allow for up to four. The user data field consists of an integral number of octets to a default maximum of 262. In typical LAN applications, the maximum of 1,600 is used to preclude unnecessary segmentation and reassembly.

The 2-octet FFR header shown magnified contains a 10-bit *data link connection identifier* (DLCI), which is used to multiplex multiple virtual circuits. As originally defined, the C/R bit denoted a command when equal to one and a response when equal to zero. The command and response mode was never implemented, and the bit is always set equal to zero. The low order bit of each octet denotes the last octet of the header. It is always equal to zero in the high order octet and equal to one in the low order octet (last).

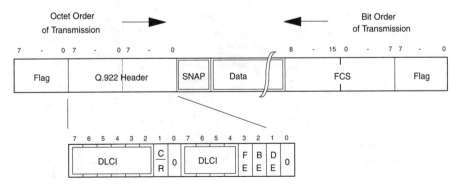

Figure 2.37 Frame relay format.

The three bits FE, BE, and *discard* (DE) are used for congestion control. The DE bit identifies the frame is eligible for discard in a congestion situation. The *forward explicit congestion notification* (FECN) bit advises the destination that congestion was encountered en route. The *backward explicit congestion notification* (BECN) bit notifies the originator that congestion was encountered with a frame it sent—back off! See FECN and BECN illustrated in Figure 2.38. The notification for the BECN is accomplished by setting the bit in any frame going to the originator. The notification may also be made with an out of band signaling by sending a message on a management DLCI, zero, or 1023 depending on which management interface is selected (described in Section 2.7.2).

To accommodate multiprotocol routers, the SNAP header is used so that a receiving router will know what protocol stack is to handle the data. The SNAP header, which is defined by the IEEE, is illustrated in Figure 2.39.

The IEEE defined two classes of protocols for SNAP. The privileged, or preferred, are assigned a unique *network layer protocol identification* (NLPID) and only require a two-octet header (illustrated on the top in Figure 2.39). The control field identifies an UI frame (ISO 7776, HDLC). When the NLPID is equal to CC hex an IP datagram is identified, which immediately follows. (Note that the Q.922 header is repeated for reference.)

All other protocols, such as IPX, must use the *all others* category by specifying the NLPID field equal to 80 hex. A three-octet field called the *organizationally unique identifier* (OUI) is used to identify the organization that administers the *program identifiers* (PIDs) that follow. The value of zero for the OUI specifies an Ethernet type code. Since the OUI has an odd number of octets, a pad octet must be used before the NLPID to make the PID fall on an

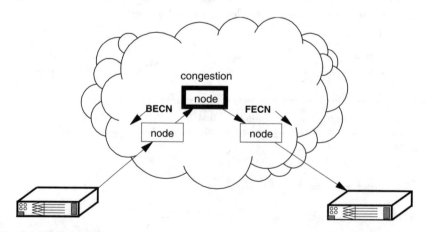

Figure 2.38 Congestion control.

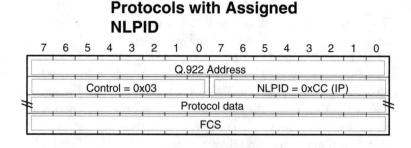

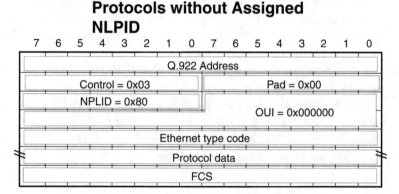

Figure 2.39 SNAP header format.

even octet boundary, as illustrated. Given these conditions, NLPID equal to 80 hex, and the OUI equal to zero, the next two octets contain the Ethernet type code of the protocol sending data. Note that IP can also use this variety of the SNAP header by placing 800 hex in the type code field. For compatibility, many implementations use this form of the SNAP header for IP, although it uses six more octets.

2.7.2 Frame Relay Options

Several options must be selected when subscribing to the FFR service provider. In some cases, however, the service provider will dictate the options, leaving the subscriber no choice. There are at least three varieties (in use) of the *data link control management information* (DLCMI) formats and two varieties of DLCI assignments.

The original developers of the *local management interface* (LMI) were Cisco Systems, Digital Equipment Corp., Nortel, and StrataCom [32]. In

1990, these companies defined extensions to the existing draft ANSI and CCITT specifications and a LMI specification that used DLCI 1023 for communicating management information and signaling. It was the de facto standard. ANSI T1.617 Annex D, which was finalized in 1991, specified DLCI 0 as the management interface. CCITT Q.933 Annex A, published in 1992, identified DLCI 0 as the signaling channel. See Table 2.10 for a comparison of DLCI values. Since the DLCI, like the LCN for X.25, has only local significance, it is tolerable since router manufacturers allow this option to be configured.

Signaling is performed on DLCI zero with both ANSI T1.617 Annex D and ITU-T Q.933 Annex A, while the LMI format uses DLCI 1023. Most router manufacturers support these variations (configuration options) and the ability to turn off the out-of-band signaling when running in a bridged environment that does not support network management. All specifications use DLCI 16 as the first valid user traffic virtual circuit.

The specifications differ on the format of management messages on the signaling DLCI, along the same lines as with the DLCI assignment. Each management message format is bounded by flags (LAPF), and the DLCI immediately following the flag field is equal to either 1,023 or 0. Similar management information, but in a different format, is contained in each type message. Such information includes type, status, and bandwidth for each PVC.

2.7.3 FFR Performance and Pricing

The pricing structure for FFR service was probably modeled after a buffet eatery—and every eatery has a different way of charging for the basics and extra for the big eaters. The pricing for FFR service is usually based on a *committed burst*

Table 2.10
DLCI Assignment

ANSI T1.617-D/CCITT Q.933-A		FRF LMI	
Range	Function	Range	Function
0	Signaling/management	0 - 15	Reserved
1 - 15	Reserved	16 - 1007	Assignable PVCs
16 - 991	Assignable PVCs	1008 - 1018	Reserved
992 - 1007	Reserved	1019 - 1022	Multicast
1008 - 10023	Reserved	1023	Signaling/management

rate (Bc) over a *time interval* (Tc), typically one second. When Tc is one second, Bc is equal to a *committed information rate* (CIR). Then there is an *excess burst rate* (Be), which is an amount over the Bc that may be handled, depending on congestion. See Figure 2.40.

The service provider will guarantee to deliver information submitted within the CIR and to attempt to deliver information beyond the CIR but within the Be. The latter will be marked for discard eligibility in the event of enroute congestion. (See Frame 3 in Figure 2.40) If the service provider has policing disabled, a user could burst to double the CIR (assuming this is less than the access rate) without data being discarded. This is a user living on the edge. With policing enabled, data arriving at a rate exceeding the Bc + Be rate is discarded. (See Frame 4 in Figure 2.40.)

The first two frames in Figure 2.40 have the DE bit equal to zero; the third frame exceeded the CIR and has the DE bit set to one; and the fourth frame may be automatically discarded, depending on the service provider.

The following is a summary of data handling with FFR.

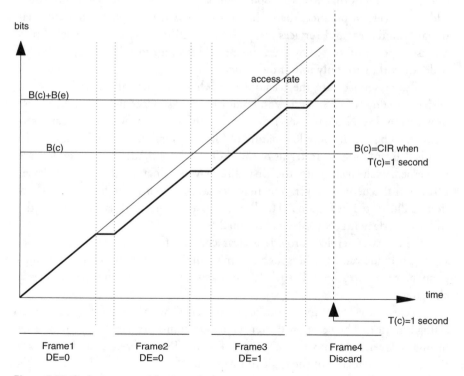

Figure 2.40 Performance model.

1. CIR = Bc/Tc and when Tc = 1 second, CIR = Bc.
2. When the rate is in the range 0<X<Bc, the DE bit is equal to 0.
3. When the rate is in the range Bc<X Bc<+Be, the DE bit is set equal to 1.
4. When the rate is in the range Bc<+<Be<X access rate, data may be discarded.

Some service providers calculate Be as CIR* (1 + Be/Bc). Others may provide a CIR and declare Be is equal to a percent (e.g., 25%) of the CIR.

2.8 Integrated Services Digital Network (ISDN)

ISDN is a worldwide public telecommunications network designed to replace existing public analog telecommunications networks and deliver a wide variety of services. ISDN started with well-defined user interfaces and planned for a set of digital switches that would support a broad range of traffic types and value-added services. In practice, there are multiple networks implemented with national boundaries, and services are provided with existing technologies. From the user's point of view, however, there is a single, uniformly accessible worldwide network, with only minor differences.

Bearer services are the backbone of what has normally been associated with the analog (POTS) service, and many of these services are the same as those provided by ISDN. Bearer services carry the information. Although bearer services may change the very low nature of bit transmission, the higher level content is delivered intact. In terms of the ISO reference model, bearer services are concerned exclusively with the first three layers. That is, the physical layer, DLL, and the network layer. The network layer, which is the focus of this section, is illustrated in Figure 2.41. IP operates at the network layer and uses the PPP to interface ISDN [33]. See Section 2.5.

The bearer services provide a clear channel for user data, and the teleservices provide value features such as e-mail, file transfer, and web surfing. Supplementary services include such features as CLI, call forwarding, and reverse charging.

Teleservices combine the transport function with the information processing function. They employ bearer services to transport data and provide a set of higher layer functions that correspond to the ISO reference model layers 4 through 7, which are not the topic of this section.

In defining standards for users to access ISDN, consideration was given to the likely configurations of user premises equipment. The first step was to group

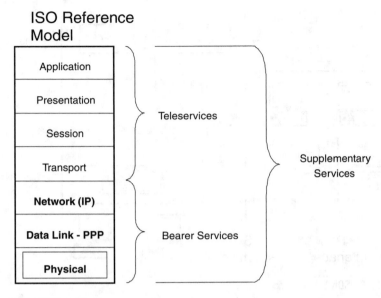

Figure 2.41 ISDN architecture.

functions that would exist on the user's premises in ways that resemble actual physical configuration.

Functional groups are sets of functions that may be needed at an interface. Reference points are used to divide the functional groups. In Figure 2.42, the *network termination* (NT1) function includes the physical and electrical termination of the network at the customer's premises. In the United States, this function is not included with ISDN service, but in Europe it is supplied by the PTT.

The functional elements identified in Figure 2.42 are described as follows:

1. *Terminal equipment 2* (TE2) describes non-ISDN-compatible devices that include analog phones, computers, and terminals. They are compatible with present-day interfaces such as RS232, V.24/V.28, and V.35.

2. *Terminal equipment 1* (TE1) describes ISDN-compatible devices such as PCs with ISDN cards installed.

3. *Terminal adapters* (TAs) convert the physical and electrical interface from a TE2 to an ISDN physical and electrical interface (S interface). They provide the necessary interface and protocol conversion for a subscriber terminal to interface the NT2 function.

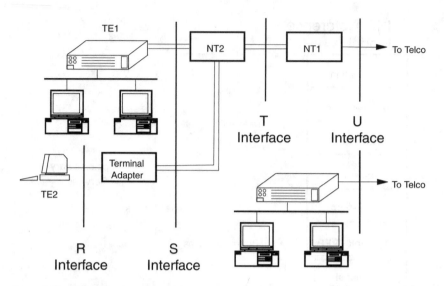

Figure 2.42 ISDN BRI reference model.

4. *Network termination 2* (NT2) provides for switching and concentration of ISDN lines. NT2 corresponds to the OSI reference model layer 2. The function is accessible at the S interface reference point. The TE1 and terminal adapter functions are often contained in the NT2.

5. NT1 terminates the ISDN loop at the customer's premises and converts the 2-wire U interface to the 4-wire S interface. This function is provided by PTTs in Europe.

The reference points identified in Figure 2.42 are described as follows:

1. R interface: This is a non-ISDN-compatible interface on the DTE, such as RS232 or V.35.

2. S interface: A four-wire interface that switches calls between devices at the customer premises. It allows for direct connection from an ISDN-compatible terminal or connection of a non-ISDN terminal via a terminal adapter.

3. T interface: Electronically identical to the S Interface. The T interface does not support local switching. It is simply a standard interface to an NT1 device.

4. S/T interface: A term commonly used when there is no NT2.

5. U interface: The two-wire interface provided by the telephone company in North America.

The basic unit of service offered by ISDN is a bearer channel, referred to as a B-Channel. ISDN defines a channel as a digital conduit for voice or data. B-Channels are 64-Kbps channels used for data or digitized voice. A channel designated as the D-Channel is used to exchange signaling messages for scheduling the use of B-Channels. Other channels offered by ISDN are essentially aggregated B-Channels. The H0 channel aggregates six B-Channels, H10 aggregates 23 B-Channels, H11 aggregates 24 B-Channels, H12 aggregates 30-B Channels, and NX64 is a variable aggregation from 1–24 B-Channels.

There are two types of user access to the bearer services of ISDN. The first is called the *basic rate interface* (BRI), which is the replacement for the normal home POTS. It provides two B-Channels (64 Kbps each) and one D-Channel (16 Kbps). This is referred to as 2B+D. The second, PRI, normally provides 23 B-Channels (64 Kbps each) and one D-Channel (64 Kbps), referred to as 23B+D. NFAS, which uses a different PRI for the D-Channel and extends the PRI to 24 B-Channels, may be used. The D-Channel (16 Kbps or 64 Kbps) is primarily used for call setup, but it can also be used for packet switching when not in use.

There are two classes of BRI service—circuit mode and packet mode. Packet mode provides both switched and permanent virtual circuits (X.25) and a connectionless service is planned. The most general circuit mode service is called clear channel because the full 64 Kbps on each B-Channel is available to the user, for either voice or data. The two B-Channels may be multilinked together to form a 128-Kbps link and still be able to receive voice calls by automatically dropping one B-Channel while the voice call is active. (This is described in detail in Section 2.5.6.3.)

2.8.1 Physical and Electrical Characteristics

The BRI provided at the S/T interface may be either point-to-point or point-to-multipoint. See Figure 2.43. In the point-to-multipoint mode, there may be up to eight devices located anywhere on the passive bus up to 200m from the NT. In an extended mode, with all eight devices closely located at the far end of the bus, the span may be up to 1 km. This is the same limitation placed on the point-to-point mode.

The electrical characteristics at the S/T interface are similar to V.35, which may be characterized as a voltage-limited current interface. In order to accommodate two B-Channels and one D-Channel and framing overhead, the bit rate is defined as 192 Kbps. This could be perceived as a problem since the U

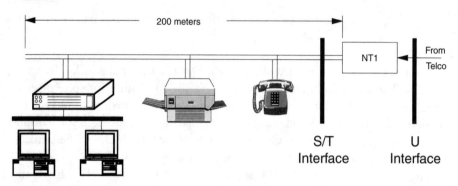

Figure 2.43 BRI physical configuration.

interface operates at 160 Kbps [34]. However, the actual user data rate across the S/T interface is only 144 Kbps (2B + D = 64 + 64 + 16), so no data is lost.

The encoding scheme used at the S and T interfaces is called pseudo-ternary coding. Unlike the T1 carrier scheme, ones are sent as null and zeros are sent as alternating positive or negative pulses (spaces). This produces a bipolar, full pulse train, which is referred to as pseudo-ternary or ASI (compare to AMI of T1). As with AMI, polar violations are used as a method to convey management information. For example, frame synchronization is based on the second violation of the alternate polarity rule. (Polar violations must occur in pairs for dc balancing.)

The S/T interface is a passive bus, much like Ethernet, except the protocol used is *carrier sense with multiple access and contention resolution* (CSMA/CR) [35]. It functions like a looped bus (more like token ring) because the NT echoes any bits sent by a TE on the D-Channel. In idle mode, TEs send all ones and verify the echo bits that they sent. To bid for access, the TE begins transmitting data and monitors the D-Channel echo bits. If the echoed D-Channel bits match the transmitted bits, the TE continues to send. Should the echo bits not match, it immediately stops sending and waits for an idle channel condition. Each TE is assigned a different number of idle bits that must be observed before it can commence transmitting. Once a TE captures the bus via the D-Channel (sees its own, unmodified pattern echoed by the NT), it may send data on the B-Channel.

2.8.2 PRI Frame Formats

The frame format and multiplexing scheme used for ISDN PRI in North America, Japan, South Korea, and Taiwan is based on T1 carrier; it is illustrated in top portion of Figure 2.44. ISDN PRI in Europe, which is based on the CEPT Level 1 carrier and designated as E1, is illustrated at the bottom of Figure 2.44.

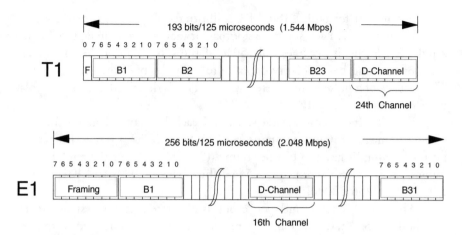

Figure 2.44 PRI frame format for T1/E1.

The T1 format uses the 24th channel for the D-Channel, and the E1 format uses the 16th channel for the D-Channel. (The 0th E1 channel is reserved for framing, which is accomplished with one bit from every frame in the T1 format.) ITU-T Recommendations G.703 and G.704 define the frame formats for T1 and E1, respectively.

Each B-Channel provides 64 Kbps and combining 23 B-Channels with one D-Channel forms a PRI for countries using the T carrier system. Combining 31 B-Channels with one D-Channel forms a PRI for countries using the E carrier system. Basically, the D-Channel is used to make call setups and tell the switch where to send the data from B-Channels. It is no coincidence that this looks like a T1 carrier service. Remember, ISDN is an evolutionary development and started with the proven T1 carrier technology defined by AT&T Specification 54016. ITU-T Recommendation I.431 defines both 1.544-Mpbs and 2.048-Mbps data rates.

If more than one T1 PRI interface is used, all 24 channels may be used, and the signaling is done on a D-Channel from another PRI. This is referred to as NFAS. PRI interfaces also support aggregated groups of 64-Kbps B-Channels or Hn channels (64-Kbps). The H-Channels are predefined groupings of B-Channels for both T1 and E1 PRI. H0 channels group six B-Channels, and with NFAS T1 four H0 Channels are supported (3 without NFAS). E1 can support five H0 channels in addition to the D-Channel. H11 channels use all 24 B-Channels (with NFAS), and H12 channels use time slots 1 through 15 and 17 through 31 (D-Channel always present) for a total of 30 B-Channels on a H12 channel.

A time slot for both T1 and E1 is illustrated Figure 2.45; each contains eight sub time slots. Each sub time slot may or may not contain a pulse. When a pulse is present, it occupies only 50% of the sub time slot, has the binary value of one, and has a polarity opposite to the polarity of the previous pulse. The absence of a pulse is interpreted as zero. This is described as a 50% bipolar pulse train and also called AMI.

Each T1 frame contains 24 slots, and each E1 frame contains 32 slots. The E1 frame has 256 bits per frame (32 x 8). Since each T1 frame also has one bit for framing, it has 193 bits per frame ((24 x 8) + 1)). At the rate of 8,000 frames per second, T1 provides 1.544 Mbps (193 x 8000) and E1 provides 2.048 Mbps (256 x 8000).

The rate of 8,000 frames per second is of interest. It was designed to accommodate voice transmission. The majority of voice is contained in the frequency range of 300 to 3,300 Hz. To assure adequate channel bandwidth for this range (3,000 Hz), the signaling rate must be at least 6,000 Hz. (The maximum channel capacity is equal to two times the signaling rate—Nyquist's analysis.) The number was rounded up to 8,000 per second to assure quality and include signaling.

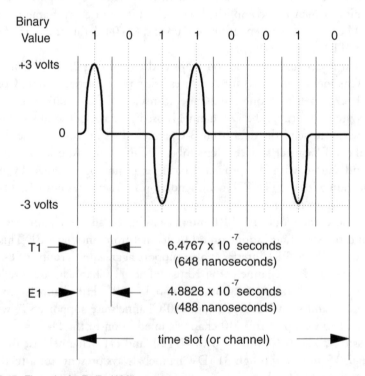

Figure 2.45 Time slot for T1/E1 (AMI).

2.8.3 ISDN Signaling

The bandwidth source (path) of line interface signaling bits is the D-Channel for both T1 and E1. That is, the 16th channel with E1 and the 24th channel with T1. This is called CCS, because a single D-Channel is common to multiple BRI (or PRI) channels. The D-Channel layer 2 protocol used is the *link access procedure on the D-Channel* (LAPD), although some older implementations may still use LAPB. LAPD is referred to as the *digital subscriber signaling system no. 1* (DSS 1) DLL. The D-Channel layer 3 protocol is defined by ITU-T Recommendations Q.930 (general), Q.931 (call control), Q.932 (supplemental services), and Q.933 (frame-mode services). These protocols form the DSS 1 network layer. The network layer header contains a message type field that identifies four groups of signaling messages, which are call establishment, call information phase, call clearing, and miscellaneous. Communications between ISDN LEs is handled by SS7, defined by the ITU-T Q.700 series of recommendations.

2.8.4 Service Anomalies

A TDM is used to collect data from 24 channels (each up to 64 Kbps—DS0) and to convert the analog signal to a digital format, NRZ. The early TDMs developed by AT&T were called CBs. The individual channels are multiplexed onto a single data stream and sent to a digital service unit via a four-wire interface such as V.35 or RS422. The function of the digital service unit is to convert the NRZ data into a bipolar pulse train (AMI). This pulse train is sent to a CSU via a four-wire interface (15-pin connector at CSU) where framing for the T1 carrier system is performed. This includes enforcing the ones density requirement, either by jamming a pulse in each octet, or with B8ZS to provide clear channel. The CSU also provides circuit protection.

So what does this T1 stuff have to do with ISDN PRI or BRI? In North America, AT&T offers an ISDN BRI service called BRI Terminal Extension (BRITE) that is derived from 3 DS0s (2B+D). Some old CSUs still in operation may jam a pulse (high order bit) in every time slot to assure that the ones density requirement is met. When such equipment is traversed, the maximum speed possible for a tail circuit (DS0) is 56 Kbps, even when the sender and receiver use B8ZS. These older CSUs do not recognize B8ZS and treat it as an actual polar violation. Hence, if you configure both ends for clear channel, and it does not work, try dropping back to 56 Kbps by using the option to jam a pulse in the high order bit of each octet.

The technique and process described above is for a T carrier system. The process is the same for an E carrier system, except instead of 24 channels there

are 32 channels and HDB3 is used instead of B8ZS. A subtle difference is that
the CSU/DSU is *customer premise equipment* (CPE) in the United States; in
Europe, these units are supplied by the PTT.

As mentioned earlier, the characteristics of ISDN vary by country and by
the switching system used. Table 2.11 identifies the switch types and the geographic areas in which they are used.

One such variation by country is the requirement to use a *service profile
identifier* (SPID). This is not a telephone number and functions like an account
number. It is required in North America only. The SPID is supplied by the
network provider and ranges from 3 to 20 digits. Some North American ISDN
implementations require a SPID for each B-Channel.

2.9 Asynchronous Transfer Mode (ATM)

Asynchronous transfer mode (ATM) has a lot in common with ISDN, described
in Section 2.8, because the specifications are provided by the same source, the
ITU-T. The specifications are titled Broadband ISDN (B-ISDN), and broadband is defined as a service requiring transmission channels that support rates
greater than the present ISDN PRI rate (1.544 Mbps and 2.048 Mbps) [36].
To avoid confusing B-ISDN with the older ISDN, it is called *narrowband
ISDN* (N-ISDN) [37]. In general discussion, the term ATM is used synonymously with B-ISDN. The word *asynchronous* in the term ATM indicates that

Table 2.11
ISDN Switch Types

PRI	BRI	Location
4ESS	4ESS	North America
5ESS	5ESS	North America
DMS100	DMS100	North America
—	NI1	North America
NET5	NET3	Northern Europe
VN3	VN3	France
TA30	TA2	Germany
NTT	NTT	Japan
—	KDD	Japan
TP013	TP012	Australia

the recurrence of cells from an individual user is not necessarily periodic—not that the transmission is asynchronous [38].

The ATM Forum (ATMF), formed in 1991, provides standards (called implementation agreements) for ATM. The ATMF represents over 600 ATM equipment vendors, service providers and users. Although it is not a formal standards body, the ATMF is the driving force in developing standards for ATM. The differences between ITU-T Recommendations and ATMF Specifications primarily involve naming and slight variations that ATMF injected to accommodate copper or fiber facilities (compared with fiber only with ITU-T). The IETF has a working group called the IETF IPATM WG that is active in developing requirements for ATM. Many of its members are also members of the ATMF [39].

The distinguishing feature of ATM, as compared to N-ISDN circuit switching and packet switching, is its ability to handle both time-sensitive data (e.g., voice and video) and bursty LAN data. It does this with a connection-oriented service based on a very fast cell relay technology that has multiple levels of virtual connections. One virtual connection may have sublevel connections, each with different delivery characteristics. Sublevel A may handle data packets from a router, while sublevel B handles a voice or video transmission. A further refinement is that each sublevel is simplex. That is, it is unidirectional, not bi-directional as with other virtual connections such TCP/IP or X.25. To visualize this, consider a single TCP connection. It consists of two sockets (one for each end point) that define a connection. Now consider two subconnections to the TCP connection, one for the transmit and one for the receive. This is conceptually the connection-oriented service provided by ATM. In this analogy to TCP/IP, the connectionless service is provided by SMDS, described in Section 2.10.

2.9.1 ATM Architecture and Interfaces

ATM provides, in Internet parlance, part of the link layer function of B-ISDN. ATM is responsible for multiplexing/demultiplexing, flow control, and moving cells between devices.

The architecture of B-ISDN is illustrated in Figure 2.46. The ISOC reference model is shown for reference. Neither the OSI reference model nor ISOC reference model align well with the B-ISDN architecture (defined in ITU-T in Recommendation I.432) or the ATMF architecture and is a topic of debate. The *user network interface* (UNI) is the name given by the ATMF to the interface between an ATM network and the CPE. The UNI interface corresponds to the T interface with N-ISDN (see Figure 2.41),

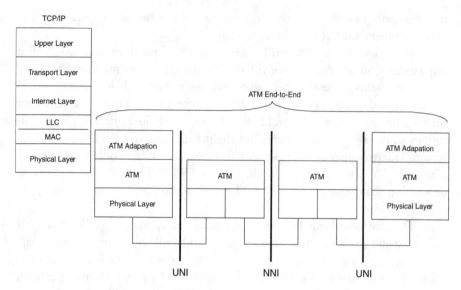

Figure 2.46 ATM architecture and interfaces.

which is labeled T_B for broadband. The ATM interface model is the same as the N-ISDN interface model, except each is labeled with a subscript (B) to denote broadband. The ATMF also defines a *network-to-network interface* (NNI) and a router to ATM-compatible DSU interface called *ATMF data exchange interface* (ADXI). (ITU-T's equivalent is simply called DXI.) Since most router connections to an ATM are through a DXI, it will be discussed in Section 2.9.4.2. The interface names UNI and NNI are preceded with "P-" to denote private ATM connection.

The physical layer has two functions, the *physical medium* (PM) function and the *transmission convergence* (TC) function. The PM is responsible for the actual transmission of bits between ATM equipment, including end user devices, private ATM switches, or the public network. It is dependent on the PM used, which is defined by the ITU-T only in terms of rates supported. The initial rates supported are 155.52 to 622.08 Mbps, which corresponds to STM-1 (OC-3) and STM-4 (OC-12). The ATMF has added support for shielded and unshielded twisted pair wire, coaxial cable, and fiber optic cable (ITU-T originally only supported fiber optic cable). The TC sublayer deals with medium independent functions such as framing and error control (detection and reporting).

The ATM layer is the heart of ATM. It provides multiplexing based on an identifier in the ATM 5-octet header (described later). ATM constructs the header outbound and deciphers it on inbound. The constant bit rate of ATM is ideal for time-sensitive sources but is not ideal for bursty LAN traffic without the ATM layer. ATM maps variable rate sources into constant bit rate cells. This enables ATM to meet the isochronous demands of voice and video and the asynchronous requirements of data with the same transmission system.

The *ATM adaptation layer* (AAL) is the highest layer on the ATM stack. It is primarily responsible for providing an interface to the higher layer programs using ATM. The operative word here is programs. Since programs have different service requirements, the AAL is divided into four service offerings (A, B, C, and D), as illustrated in Table 2.12.

An AAL type code (1, 2, 3, and 4) was originally assigned to each service class A, B, C, and D respectively. Service classes C and D (AAL type codes 3 and 4) were essentially combined, and a new AAL type code was added (number 5). This area is undergoing design changes and a new service class "X" is being added.

Table 2.12 identifies the different service classes. Class A could represent a typical voice connection that has a *constant bit rate* (CBR) and that is delay-sensitive and connection-oriented. Class A applications reserve the entire bandwidth. Class B applications require a *variable bit rate* (VBR), connection-oriented, delay-sensitive service—for example, packet audio and video. Classes C and D do not require end-to-end timing and have VBRs—for example, X.25 packet switching. The ATMF defines the service classes as follows.

1. CBR: A CBR service, which emulates circuit switching, is used for time-sensitive data such as voice or video.

2. VBR: A VBR service allows mixed time-sensitive data with nonsensitive data—for example, teleconferencing and information retrieval containing portions or voice or video.

3. *Available bit rate* (ABR): An ABR service is designed for nonsensitive data, such as a file transfer.

Table 2.12
ATM Service Classes

Service Class	A	B	C	D
Connection mode	Connection	Connection	Connection	Connectionless
Bit rate	Constant	Variable	Variable	Variable
End-to-end timing	Required	Required	Not required	Not required
AAL type code	1	2	"3/4", 5	"3/4", 5

4. *Unspecified bit rate* (UBR): A UBR service is designed for applications with nonisochronous data that are not sensitive to cell loss, such as file transfers and e-mail.

Devices requesting service from the network will specify the desired characteristics in terms of quality of service. The upper part of AAL layer interprets the service characteristics, and the lower part performs *segmentation and reassembly* (SAR). The SAR is responsible for formatting the cell's payload.

2.9.2 ATM Connections

ATM connections are conceptually the same as X.25 connections except that ATM has a connection within a connection. That is, each connection is again multiplexed to form a subconnection. The most basic ATM connection is called a *virtual channel* (VC), which has local significance only and is unidirectional. A VC is contained in a *virtual path* (VP), which also has local significance only but is bidirectional. The connection hierarchy is illustrated in Figure 2.47.

Each VC link has an identifier called a *VC identifier* (VCI) that is carried in the ATM header. It is a unidirectional point-to-point connection between ATM switching points and has only local significance. The sum of VC links between two end points is called a *VC connection* (VCC).

Each VP link contains one or more VCs and has an identifier called a *VP identifier* (VPI) that is carried in the ATM header. It is a bidirectional, point-to-point connection between ATM switching points and has only local significance. The sum of VP links between two end points is called a *VP connection* (VPC).

An ATM switch can switch an entire VP, as illustrated in Figure 2.48 at the bottom. This is called *VP switching*. The source and destination end points contain the same VCs, although the VPI may change since it has only local significance. The more general case involves VC switching. In VC switching the ATM switch associates a single VC with different (one or more) VPs. Hence, the VCC (concatenation of VCs) would be associated with multiple VPs, as illustrated in Figure 2.48 with VP 1, VP 2, VP 8, and VP 9.

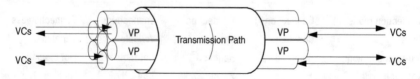

Figure 2.47 ATM virtual channels and virtual paths.

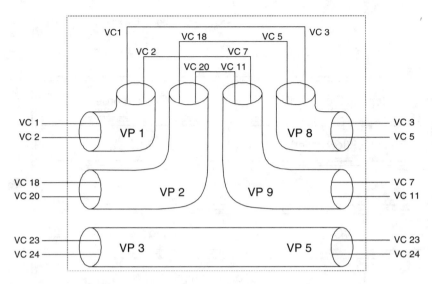

Figure 2.48 Virtual channel switch matrix.

ATM uses ITU-T signaling protocol defined in Q.2931 to establish virtual connections and allocate resources. It is sender-driven, establishes quality of service at setup time for input and output, and relies on hard-state switches to maintain connections.

2.9.3 ATM Cell Formats

The total fixed number of octets in an ATM cell is 53. Why was a prime number chosen? It wasn't. One number was chosen for the payload (48 octets) and another number was selected for the header (5), and they were added. There was controversy on the size of the payload and on the size of the header. Those using E1 favored a larger number for the payload (64), and those using T1 favored a smaller number (32). An even compromise was 48 octets for the payload and five octets for the header. A field description of the ATM cell format across the UNI, which is illustrated in Figure 2.49, follows.

1. The *generic flow control* (GFC) 4-bit field was designed (reserved) to be compatible with the IEEE 802.6 DQDB MAN standard. It is not defined by either ITU-T or ATMF and is always set equal to zero. If defined later, it will have only local significance.

2. The VPI field has 8 bits for cells across the UNI. Bit +3 of octet +0 is the high order bit and bit +4 of the +1 octet is the low order bit.

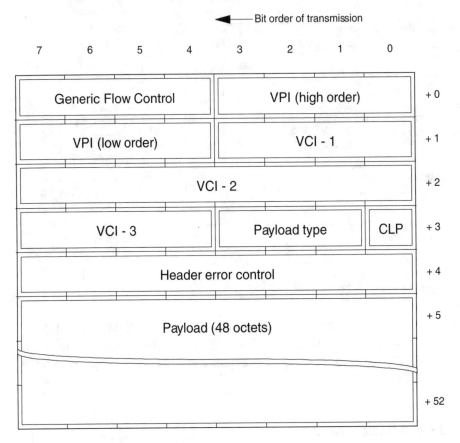

Figure 2.49 ATM UNI cell format.

The VPI field for cells across the NNI has 12 bits, illustrated in Figure 2.50.

3. The VCI field has 16 bits for cells across the UNI and NNI. Bit +3 of octet +1 is the high order bit, and bit +4 of octet +3 is the low order bit. Just as with an X.25 LCN and a frame relay DLCI, there are pre-assigned VPIs and VCIs. The logical function of VPIs and VCIs is to perform multiplexing. They are used to perform cell rate decoupling (unique to ATM). This is accomplished by inserting empty cells (identified by the VPI and VCI each equal to zero) into the cell stream for unassigned cells. This makes the rate that users generate data independent of the fixed rate of the physical medium. Also, pre-assigned VPIs and VCIs are used to convey *operation, administration, and maintenance* (OAM) messages.

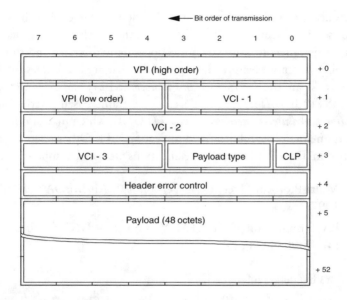

Figure 2.50 ATM NNI cell format.

4. The payload type field has 3 bits to signal a user generated cell (bit +1), forward congestion notification similar to the FECN bit in frame relay (bit +2), and a type indicator (bit +3) for higher layers, such as a last cell flag by AAL5.

5. The *cell loss priority* (CLP) bit may be set by the ATM network to indicate eligibility for discard, which is similar to the DE bit in frame relay.

6. The header error control octet (+ 4) provides CRC protection for the header only.

The ATM cell format across the UNI is identical to the ATM cell format across the NNI except that the 4 bits in the GFC field are not needed and used to lengthen the VPI field from 8 bits to 12 bits, as illustrated in Figure 2.50. Hence, in the NNI bit +7 is the high order bit of the VPI field, and bit +4 of the +1 octet is the low order bit of the VPI field.

2.9.4 Accessing ATM Networks

ATM did not start with a well-defined access method, and several have been tried. This section briefly describes access methods to ATM.

New accesses and applications are being designed to exploit the speed and quality of service features of ATM. IP is a connectionless service that poses a problem for designers to utilize a connection-oriented service and to make use of the quality of service feature. Two protocols that are candidates for utilizing ATM networks are the *Internet streams protocol* (ST2+) and *resource reservation protocol* (RSVP) [40]. The ATMF developed a frame based interface to ATM called *frame-based user-to-network interface* (FUNI), which may also be used for L2TP, a tunneling protocol described in Section 2.11. Other reference material is identified in Table 2.13. ATMF specifications are listed as follows.

1. Multi-Protocol Over ATM Version 1.0, AF-MPOA-0087.000, July 1997;

2. LAN Emulation Over ATM Version 2—LUNI Specification, AF-LANE-0084.000, July 1997.

2.9.4.1 Data Exchange Interface (DXI)

A *data exchange interface* (DXI) developed by the ATMF provided a standard method for routers to access an ATM network via an ATM compliant DSU. DXI is a layer 2, HDLC-like interface that routers are very apt at performing. The DSU (DCE) on one side communicates with the router (DTE) using the HDLC-like protocol and on the other side with the ATM network using either AAL5 or AAL3/4. The physical connection to the router may be V.35, EIA 449, RS 422, and EIA 612/613.

Table 2.13
ATM Reference Material

RFC	Title
2105	Cisco's Tag Switching Architecture
2098	Toshiba's Router Extension for ATM
1932	IP over ATM: A Framework Document
1926	An Experimental Encapsulation of IP Datagrams on Top of ATM
1821	Real-time Service in IP-ATM Networks
1819	Internet Stream Protocol Version 2 (ST2+)
1754	IPATM WG ATM Forum Recommendations V1
1680	IPng Support for ATM Services
1577	Classical IP and ARP over ATM
1483	Multiprotocol Encapsulation over ATM Adaptation Layer 5

2.9.4.2 Multiprotocol Encapsulation

Multiprotocol encapsulation over ATM has been defined by the IETF using an LLC type header and with AAL5 [41]. This provides connection (type 1) or connectionless (type 2) service with multiprotocol operation, which is accomplished with an IEEE 802.1a SNAP header. The value of the fields within the SNAP header are identical to the multiprotocol operation with frame relay. See Section 2.7.1. The IETF also defines how ATM may be used to replace existing LANs and WAN links within classical IP networks [42]. This includes LANs (Ethernet, token ring relay PVC between IP routers, and backbones between existing LANs).

2.9.4.3 LAN Emulation (LANE)

LAN emulation (LANE) is the ongoing direction of the ATMF and enables similar LANs (Ethernet or token ring) to use an ATM network for interconnection without protocol conversion. The actual end systems (workstations, servers, bridges, etc.) may or may not be on the same physical network. The association is logically formed with what is called an *emulated LAN* (ELAN), which is emulated to the point of requiring a router if the end systems are not on the same ELAN (a downside). Broadcast messages are only sent to those end systems that are on the same ELAN. An ELAN may be used to connect ATM-connected end systems to other ATM-connected end systems or to LAN-connected end systems, and vice versa. LANE version 2 supports quality of service connections between ATM-connected end systems and multicast addressing. Each ELAN is composed of a set of *LANE clients* (LECs) and one or more *LANE servers* (LESs). ELANs are used to connect intermediate systems (bridges and routers) and end stations (hosts or PCs). LECs contain the *LANE UNI* (LUNI) and are responsible for data forwarding and address resolution. Each LES is configured to provide a facility for registering and resolving unicast and multicast MAC addresses to ATM addresses. LECs register their addresses with an LES, and should an LEC have an unknown destination address, it uses an ARP-like function to determine the ATM address.

2.9.4.4 Multi-Protocol Over ATM (MPOA)

Multiprotocol over ATM (MPOA) is, more or less, a continuation of LANE and is also client-server oriented. Although the name implies that the major emphasis of MPOA is the operation of multiple protocols over ATM, it is not. (IEEE 802.1d could have been named multiprotocol over Ethernet because it handles multiple protocols, but instead it is called bridging.) MPOA, like bridging, provides an alternate method of Internet layer route calculation and forwarding. It does this without adding to the vertical stack and without end systems participating in a dynamic routing protocol, but it does add parallel functions. MPOA was necessitated because LANE requires a router for intersubnet forwarding. MPOA requires the services of LANE and a new routing protocol called *next hop resolution protocol* (NHRP). NHRP is a query and response protocol between the point of ingress client and its NHRP server. The NHRP server provides a destination ATM address to the client, which uses this information to establish a direct ATM connection to the point of ATM egress (possibly requiring normal Internet layer forwarding from there). This enables IP datagrams to skip the hop-by-hop trip and go directly to the destination. Provision has been made for fall back to Internet layer forwarding for security or policy reasons.

A proposed scheme by Cisco [43] called tag switching requires protocol conversion and a new protocol layer.

2.10 Switched Multimegabit Data Service (SMDS)

SMDS provides a connectionless, cell-relay, packet switching service based on the IEEE 802.6 Distributed Queue Dual Bus (DQDB) standard. This is a competing technology to ATM and *broadband ISDN* (or B-ISDN) developed by ANSI and ITU-T.

Where the FFR service was based on a variable data field, cell relay has a fixed cell size (53 octets) because it is a TDM system. In place of the explicit congestion control mechanism in FFR service, SMDS uses an access class enforcement mechanism both inbound from the user and outbound to the user. Because the cell size is quite small (53 octets), the switching delay is from 20 milliseconds to 80 milliseconds for cross networks with data speeds up to 45 Mbps. Although the cell size is small, the service provides a MTU of 9,180 octets of user data, with no minimum. With a very low error rate (10^{-12}) it is ideal for LAN to LAN bursty traffic.

The architecture of SMDS is illustrated in Figure 2.51. For a part with familiarity note that IEEE 802.2 with SNAP is used for the DLL. Above this is IP and below are two physical layer protocols. The *SMDS interface protocol* (SIP) Level 3 and SIP Levels 1 and 2 work within one *logical IP subnetwork* (LIS), only. Hosts connected to SMDS can communicate with any other host within their LIS. Any communications outside the LIS requires an IP router, which must be configured to handle each LIS. Since the same SMDS service provider has many LISs, it gives the appearance that SMDS was designed by a router salesman.

The SMDS protocol at the physical level (between the user and the service provider), or SIP, has three sublayers (1, 2, and 3). The service above this is the

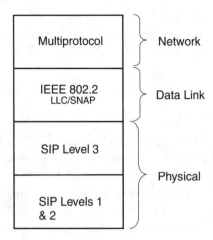

Figure 2.51 SMDS architecture.

MAC service. SMDS can function either as a MAC bridge or as a multiprotocol router by using the SNAP option of the link layer protocol LLC.

The service provider assigns a *subscriber network interface* (SNI) for each interface to the SMDS network. Each host can communicate directly (and only) with other hosts in the same LIS without a router. The hosts may be addressed with a single unicast address or a group address (multicast). Other LAN-like features include source address validation, closed user groups and destination address filtering [44].

SMDS is not considered a long-term service, but it continues to be used by a few in preference to ISDN and FFR because it is fast. In time, users will likely move to ATM.

Similar alternatives to SMDS include DXI and *SIP relay service* (SRS). DXI is only a partial implementation of SMDS that uses an existing link protocol such as HDLC, which is described in Section 2.9.4. (See Bellcore Technical Advisory [TA-TSU-1239].) SIP layer 1 and SIP layer 2 may then be implemented in an external CSU/DSU. SRS uses a Q.922 header to encapsulate the SIP layer frame and send to a SMDS relay center. (See Bellcore TA-TSV-1240.)

2.11 Layer 2 Tunneling Protocol (L2TP)

The *layer 2 tunneling protocol* (L2TP) is a recent innovation designed to permit otherwise ineligible users to traverse the Internet in a transparent fashion. The protocol specifically permits dial-up users (analog or digital) with unregistered IP addresses to connect to their home network using a PPP connection (IP,

IPX, AppleTalk, etc.) via an ISP and the Internet. Neither the remote end system or its home site hosts require special software to use this feature. It is accomplished by tunneling from the ISP to a server on the home network that also speaks L2TP.

2.11.1　L2TP Architecture

Although L2TP is a link layer protocol, it uses UDP to create a connection between the ISP and a L2TP server on the home network. The UDP connection appears as a physical layer interface to the L2TP. Hence, it has its own link layer, Internet layer, transport layer and upper layer programs—all hidden in the UDP packet. Just as with other program stacks, the lower layers of L2TP carry supervisory and payload frames, while the upper layers engage in call acceptance and keep-alive monitoring during the call connected state. The architecture of L2TP is illustrated in Figure 2.52.

2.11.2　L2TP Process and Data Flow

Figure 2.53 illustrates the elements and functions of L2TP and L2TP data flow and control. The following terms are unique to L2TP.

1. *L2TP access concentrator* (LAC): A device such as a *network access server* (NAS) attached to one or more PSTN or ISDN lines capable of PPP operation and of handling the L2TP protocol. The LAC needs only implement the media over which L2TP passes traffic to one or more

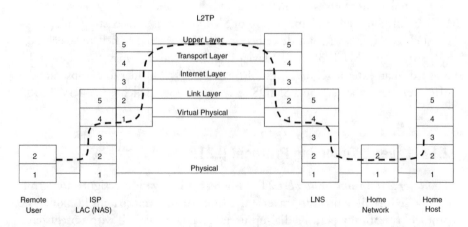

Figure 2.52　L2TP architecture.

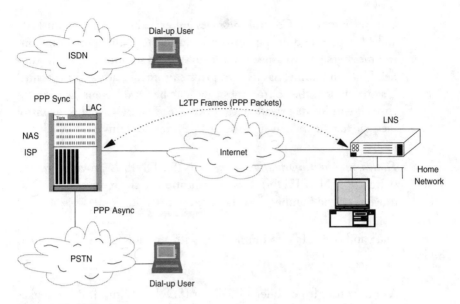

Figure 2.53 L2TP process.

LNSs. It may tunnel any protocol carried within PPP. The NAS provides temporary, on-demand network access to users. The access is
point-to-point using PSTN or ISDN lines.

2. *L2TP network server* (LNS): A LNS operates on any platform capable
   of PPP termination. The LNS handles the server side of the L2TP
   protocol. Since L2TP relies only on the single media over which L2TP
   tunnels arrive, the LNS may have only a single LAN or WAN interface, yet still be able to terminate calls arriving at any LAC's full range
   of PPP interfaces (e.g., asynchronous PSTN or synchronous ISDN).

3. Session: L2TP is connection-oriented. The LNS and LAC maintain
   state for each user that is attached to a LAC. A session is created when
   an end-to-end PPP connection is attempted between a dial user and
   the LNS, or when an outbound call is initiated. The datagrams related
   to a session are sent over the tunnel between the LAC and LNS.

4. Tunnel: A tunnel is defined by a LNS-LAC pair. The tunnel carries
   PPP datagrams between the LAC and the LNS. L2TP uses UDP as a
   logical physical layer interface, which has its own protocol stack including upper layer programs. Many sessions can be multiplexed over
   a single tunnel. A control connection operating in-band over the same
   tunnel controls the establishment, release, and maintenance of sessions and of the tunnel itself.

5. Control messages: Control messages are exchanged between an LAC and LNS pair, and operate in-band within the tunnel protocol. Control messages govern aspects of the tunnel and sessions within the tunnel. Since the tunnel has its own private protocol stack and the tunnel is states-driven, the control messages may be for different layer protocols, all within the private stack, of course. The PDU of a noncontrol message for the link layer of the private stack contains a tunneled PPP frame.

6. *Dialed number information string* (DNIS): DNIS is provided by the switch (PSTN or ISDN). It is an indication to the receiver of a call as to what phone number the caller used to reach it.

The data and control flow commencing with a remote user is described as follows.

1. A remote user from either a PSTN or ISDN dials into ISP and makes a PPP connection, including normal LCP and NCPs negotiation. The ISP may be independent or part of the telephone network. In either case, it is using an NAS that contains the LAC program. The ISP may perform a limited validation to screen out hackers and it may also perform a pre-authorization with the LNS. The remote user may specify the LNS desired or the ISP may determine the LNS from either the DNIS or from a database lookup.

2. The LAC initiates a tunnel with the LNS by opening a UDP connection with the destination port number equal to 1701 and a dynamically assigned source port number (Sa). The tunnel setup will include sufficient information to enable the LNS to authenticate the dial-up user (PAP, CHAP, or RADIUS). Either an LAC (for incoming calls) or an LNS (for outgoing calls) may establish a tunnel. The L2TP link layer exchanges start-connection-request and start-connection-reply messages to establish the tunnel. Following the establishment of the tunnel, the LNS and LAC configure the tunnel by exchanging other L2TP supervisory messages.

3. The LNS either accepts the negotiated LCP options and NCPs, or it may renegotiate. The LNS may also perform PAP/CHAP validation of the dial-up user. A dynamically assigned UDP port number is used for the source port for communication from the LNS to the LAC, and the destination port number is the same as the received source port number (which was determined by the LAC [Sa]). These pairs of UDP port numbers define the tunnel. Multiple dial-up users

may use the same tunnel. Each dial-up user has a unique *call ID* (CID) for the tunnel, which is assigned by the Internet layer of L2TP during the call establishment phase. The L2TP Internet layer exchanges call-request, call-reply, and call-connected messages to establish the session. The CID defines the session for this dial-up user on the tunnel.

4. PPP frames received by the LAC from the dial-up user are stripped of the CRC, link framing, and transparency information. This amounts to destuffing bits for synchronous PPP and removing the control characters for asynchronous PPP. The bare PPP frame is encapsulated in the payload portion of a L2TP frame and sent to the LNS.

5. LNS accepts the frame and strips the L2TP header, which leaves the bare PPP frame. This is processed as though L2TP were not involved. That is, it is passed to the protocol stack identified in the PPP header (e.g., IP, IPX, and AppleTalk). Note that PPP sees the frame as though the LAC was directly connected to the LNS, although the PPP frame could have traversed many nodes (hidden in a UDP packet) before arrival at the LNS. This enables L2TP to handle multiple dial-up users in a single multilink bundle. Hence, LCP and NCP negotiations are conducted via the UDP tunnel and a particular PPP link is identified by its CID within the tunnel.

6. The process is the same for a PPP frame going from the LNS to the LAC. A PPP frame destined for a remote dial-up user is assigned a CID after call setup. It is stripped of the CRC, link framing, and transparency information, then encapsulated in the payload portion of a L2TP frame and sent to the LAC. The LAC strips the L2TP header and processes the PPP frame as though received from a local connection.

2.11.3 L2TP Operation and Message Format

L2TP has a protocol stack that is based at the LAC/LNS transport layer, which is provided by UDP (illustrated in Figure 2.52). It functions as though the UDP tunnel were a physical layer service, and a combination of its link layer and Internet layer provide frame transport, call establishment, and management. The L2TP transport layer and upper layer provide for detection of tunnel connectivity problems.

There are two types of L2TP messages identified by a single bit in the L2TP header, which is called the T-bit and is the high order bit of the first octet transmitted. That is, it is the +7 bit of the flags field in Figure 2.54. A description of each field follows.

1. When the T-bit is equal to one, it is a *management message* (MM) and the remainder of the flags field is not used. Otherwise, equal to zero, it is a *payload message* (PM). PMs are used to transport encapsulated PPP packets. Both the MM and PM have a fixed header containing 12 octets. The MM must use the Nr and Ns fields, which are 16-bit sequence number fields (similar in operation to the Nr and Ns of HDLC). That is, the Ns field is the sequence number of this frame, and the Nr field represents the next expected Ns sequence number to be received. It acknowledges receipt of the sequence number N(r-1). The sequence number fields provide an acknowledgment mechanism and a retransmission capability for MMs. The Nr and Ns fields may be absent in PM types if bit +4 of the Flag field is equal to zero. An MM or PM frame with the length equal to zero is used to communicate the Nr and Ns fields. Since acknowledgments can be piggybacked, the zero-length option is only used when one side is primarily sending and the other side is primarily receiving. Otherwise, the windowing mechanism (used to also provide flow control for MMs and optionally for PMs) would stop transmission when the window was full.

2. The version (Ver) number must be equal to 2, which identifies L2TP. This is bit +1 of the second octet transmitted. The 16-bit CID and 16-bit tunnel number provide 65K tunnels, each capable of multiplexing 65K remote dial-up users.

The MM header may be followed by specific *attribute-value pairs* (AVPs). This field is variable in size, but the entire MM length with attributes must not

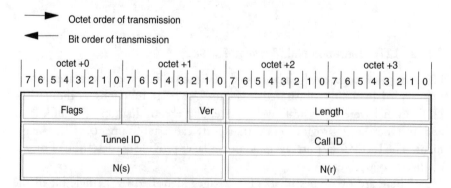

Figure 2.54 L2TP frame format.

exceed 1,500 octets. Each AVP has a 4-octet header followed by a 16-bit attribute number and the value associated with the attribute, as illustrated in Figure 2.55. A description of each field follows.

1. M: The mandatory bit affects the action taken for an invalid AVP. When the M-bit is set and the message type (illustrated in Table 2.14) is a *call management* (CM) type, the session is closed. If the message type is *control connection management* (CCM), the tunnel (and all calls) is closed. If the M-bit is not set, an invalid AVP is ignored.

2. H: The hidden bit determines if the value is passed in cleartext (H-bit=0) or encrypted (H-bit=1). When the H-bit is set, a shared secret and a MD5 hash is used to encrypt the value and the AVP must be preceded by a random vector AVP, identified in Table 2.15 with attribute 36.

3. Overall length: This is a 10-bit field that allows up to 1,024 octets in an individual AVP.

4. Vendor ID: This is the vendor's *private enterprise code* (PEC), which may be used to implement private extensions with assurance they do not collide with other vendor's extensions. PECs are defined in the Assigned Numbers RFC (the latest is currently RFC 1700). For example, the PEC for ACC is 5.

5. Attribute: The attribute field has 16 bits to define an ordinal value of the attribute. There are presently 38 attributes, as identified in Table 2.15. The value of the attribute is a variable length field.

MM type messages will normally have AVPs, and the first one (illustrated in Table 2.15 as Attb=0) must define the message type, which is illustrated in

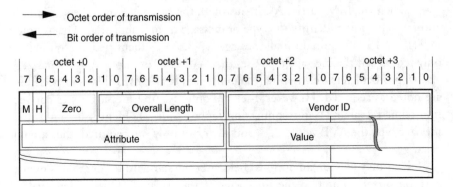

Figure 2.55 AVP format.

Table 2.14
Message Types

Value	Name	Type	Function
1	SCCRQ	CCM	Start-Control-Connection-Request
2	SCCRP	CCM	Start-Control-Connection-Reply
3	SCCCN	CCM	Start-Control-Connection-Connected
4	StopCCM	CCM	Stop-Control-Connection-Notification
5	Reserved		
6	Hello	CCM	
7	OCRQ	CM	Outgoing-Call-Request
8	OCRP	CM	Outgoing-Call-Reply
9	OCCN	CM	Outgoing-Call-Connected
10	ICRQ	CM	Incoming-Call-Request
11	ICRP	CM	Incoming-Call-Reply
12	ICCN	CM	Incoming-Call-Connected
13	CCRQ	CM	Call-Clear-Request
14	CDN	CM	Call-Disconnect-Notify
15	WEN	ER	WAN-Error-Notify
16	SLI	PSC	Set-Link-Info

Note: Control connection management (CCM), call management (CM), session control (PSC-PPP), error reporting (ER).

Table 2.14. That is, the first 16 bits of the first AVP is equal to 0 and is followed by a 16-bit field equal to the message type (1 to 16).

When the MM carries AVPs (normal), the first AVP must be attribute 0, and the value must identify the type of message from Table 2.14. This is noted in Table 2.15 as required for all message types. AVPs identified as required are only required if the message type uses that AVP. For example, AVP values 2, 3, and 4 (protocol version, framing capability and bearer capabilities respectively) are noted as required. However, they are only used by message types connection-request (value 1) and connection-reply (value 2). Hence, for any other message type the AVP values 2, 3, and 4 are not only not required, they are not used.

PMs may contain an optional field after the standard header that is used as a 16-bit offset pointer to the first octet of the payload. This is identified by setting the +1 bit of the flags field equal to 1. When used, there may also be a

Table 2.15
AVPs

Attr	Req	Attribute Name	Usage
0	Y	Message type	All types/16-bit integer indicating message type
1	Y	Result code	16 bit-integer value or ASCII character string
2	Y	Protocol version	8-bit L2TP protocol and revision numbers
3	Y	Framing capabilities	32-bit bitmask of supported framing types
4	Y	Bearer capabilities	32-bit bitmask of supported bearer types
5	N	Tie breaker	8-byte value for connection collisions
6	N	Firmware revision	16-bit integer for vendor's firmware revision
7	N	Host name	ASCII string name (DNS name of issuer)
8	N	Vendor name	ASCII string describing issuing device
9	Y	Assigned tunnel ID	16-bit integer tunnel ID assigned by sender
10	Y	Receive window size	16-bit integer receive window size
11	Y	Challenge	Value issued to peer for challenge
12	N	Q.931 cause code	Cause code or optional ASCII message
13	Y	Challenge response	16-octet CHAP type response
14	Y	Assigned call ID	Sender assigned 16-bit integer ID
15	Y	Call serial number	One or more octet identifier assigned to a call
16	Y	Minimum BPS	Minimum line speed/32-bit integer
17	Y	Maximum BPS	Highest acceptable line speed/32-bit integer
18	Y	Bearer type	Indicates bearer type (analog or digital)
19	Y	Framing type	Indicates framing type (syn or asyn)
20	Y	Packet processing delay	Processing time estimate/16-bit integer
21	Y	Dialed number	ASCII string phone number called or to be called
22	Y	Dialing number	ASCII string phone number of caller
23	Y	Subaddress	Addition dialing information/ASCII string
24	Y	Connect speed	16-bit integer actual line speed of connection
25	Y	Physical Channel ID	Vendor specific physical device/16-bit
26	N	Initial LCP Confreq	Octet string containing initial CONFREQ
27	N	Last Sent LCP	Octet string containing final CONFREQ
28	N	Last Received LCP	Octet string containing final CONFREQ
29	Y	Proxy Authen Type	Authentication type/16-bit integer (PAP/CHAP)
30	N	Proxy Authen Name	Authentication response/ASCII string
31	N	Proxy Authen Challenge	Octet string Challenge presented by LAC to client
32	N	Proxy Authen ID	ID presented to client with challenge/16-bit integer

Table 2.15 (continued)
AVPs

Attr	Req	Attribute Name	Usage
33	Y	Proxy Authen Response	Password or CHAP response/Octet string
34	Y	Call errors	Six 32-bit connection error counters
35	Y	ACCM	Two 32-bit send and receive ACCM values
36	Y	Random Vector	Variable length octet string for optional hiding
37	Y	Private group ID	LAC or LNS identification of a customer group
38	N	Receive connect speed	32-bit identification of receive access speed

pad to assure 32-bit alignment. When the offset option is not used, the payload immediately follows the Nr and Ns fields.

2.11.4 Summary

ACC developed access partitioning to enable remote users to reach its private network through the Internet. The feature uses dialed number routing to funnel traffic from the origination point at an ISP to the home network. Access partitioning optionally supports L2TP. Creating a tunnel with a remote dial-up was explored by Ascend Communications with a protocol called *point-to-point tunneling protocol* (PPTP), see draft-pppext-pptp-02.txt. Cisco Systems developed the *layer two forwarding* (L2F) protocol to create a tunnel similar to L2TP (see draft-valencia-l2f-00.txt). Both have been joined by Copper Mountain Networks, IBM Corporation, Microsoft Corporation, and 3COM Corporation in supporting L2TP. It seems that L2TP, in draft status, has the nod.

L2TP is a multilayered protocol tunneled in (riding on or piggybacked by) the transport layer protocol UDP. That is, its physical layer is layer 4 of the Internet. L2TP link layer uses tunnels between the LAC and LNS. It exchanges a connection-request and connection-reply to open a tunnel. L2TP Internet layer uses a session for each dial-up user. It exchanges a call-request and call-reply to assign a unique CID to each session on a tunnel. L2TP transport layer provides flow control with the sequence numbers Nr and Ns. Finally, L2TP upper layer functions include a keep-alive mechanism to differentiate failures on the tunnel with extended periods of no activity. This is accomplished by injecting Hello CMs after a period of inactivity. An unacknowledged Hello CM results in the L2TP transport layer resetting the session.

2.11.5 L2TP References

L2TP is in *draft* status, and the latest specification may be obtained from the InterNIC with the file name "draft-ietf-pppext-l2tp-08.txt," dated November, 1997.

Other related draft specifications include the following:

1. L2TP headercompression ("L2TPHC"), file name "draft-ietf-pppext-l2tphc-00.txt"

2. L2TP over AAL5 and FUNI, file name "draft-ietf-pppext-l2tp-aal5-funi-00.txt"

Endnotes

[1] LAPB is used in some environments as a link (or trunk) protocol with multiprotocol support but requires a proprietary enhancement. When LAPB is used in this manner, it may not work with different vendors' equipment.

[2] RFC 1250 defines the specific RFC number that is applicable for each type of physical interface. See the Host Requirements (RFC 1122 and RFC1123) and Requirements for IPv4 Routers (RFC 1812) for more information on network-specific, link layer protocols. Also, for the specifications on interfacing point-to-point synchronous and asynchronous using ISO's HDLC, see RFC 1661 (updated by RFC 1570) and RFC 1332.

[3] For more information on the "Future of the Internet architecture," see RFC1287. Also of interest are RFC1185 and RFC1371.

[4] See the Bob Metcalfe guest account in *Network Computing,* May 1991, for an entertaining article on "How Ethernet was Invented—The First Time."

[5] A similar analogy, with more detail, is contained in RFC 1180.

[6] The IEEE reserved the value 6 for the use of IP with LLC. The Internet architects instead chose to implement IP on IEEE 802 using LLC with SNAP. This encapsulates the IP datagram after the Ethernet type code 800 hex. The value 6 is reserved for future IP use.

[7] SNAP headers are defined in IEEE 802.1a.

[8] The IEEE 802.3 specification defines the maximum in terms of total segments, which is five. If two of the segments are link segments, then there may be a maximum of three coaxial segments used on any transmission path. The maximum length of the link segment is only described in terms of the maximum end-to-end propagation delay. Shorter sections may be joined with coaxial connectors to form the maximum-length segment. Lengths of 23.4 meters, 70.2 meters, and 117 meters are multiples of a half wavelength in the cable at 5 Mbps and may be used to minimize reflections.

[9] BNC is the physical connector between two 10Base2 cables. BNC is short for Bayonet Neill-Concelman, after the developers of the connector. It is also commonly called a barrel nut connector.

[10] See *Intelligent Hubs* by Nathan J. Muller, Norwood, MA: Artech House, Inc., 1993.

[11] See IBM Token Ring Reference Manual, part number SC30-3374—02.

[12] Other names used for DCE include data communications equipment and data computer equipment.

[13] This value was selected because it has double ones, double zeros, single zero, and single one.

[14] Analysis by Nyquist, published in 1928.

[15] Claude E. Shannon, published in the Mathematical Theory of Communications, Bell System Technical Journal, in 1945.

[16] Both R1 and R2 are defined in ITU-T Fascicle VI.4, Specifications of Signaling Systems.

[17] Clear channel is an expression used to describe removing the requirement that one bit in every eight be equal to one to satisfy ones density requirement. The most widely accepted method of accomplishing this is with B8ZS, which can be performed on any DS-1 frame. An infrequently used technique called *zero byte time slot interchange* (ZBTSI) requires one half of the DL subchannel. This method is used by US West.

[18] B8ZS is sometimes called bipolar 8 with zero code suppression in other literature.

[19] The layers 2 and 3 Recommendations are also called Q.921 and Q.931, respectively. These Q-Series Recommendations are identical to the respective I-Series Recommendations, but they have been numbered differently by the ITU-T Study Group XI that developed them. The I-Series Recommendations are from Study Group XVIII, which oversees all of the ISDN standardization activities.

[20] The labeling of bits in the ITU-T X.25 Recommendation commence with "1" in the least significant bit position. Zero is used here to be consistent within this book.

[21] IPCP is often associated with the assignment of an IP address. Its primary function is to bind the two peers for handling IP datagrams. Assignment of an IP address is only one of the tasks it performs.

[22] See RFC1570 for LCP extensions.

[23] See RFC 2153, PPP Vendor extensions.

[24] See Standard #1 for the current PPP specification—presently RFC1661. See Standard #2 for the current protocol numbers, options and type codes—presently RFC1700.)

[25] See RFC1994, PPP CHAP, for specification.

[26] See RFC1989, PPP Link Quality Monitoring, for specification.

[27] See RFC1990, PPP Multilink, for specification.

[28] See RFC2125, BAP and BACP, for specification.

[29] PPP may be used to access an X.25 network. See Section 2.x and RFC 1598.

[30] See RFC1356 for detail regarding restrictions on using X.25 and ISDN.

[31] The protocol identification field (containing the Protocol Code field) is administered by ISO/IEC and CCITT.

[32] Originally called the Gang of Four, and later the frame relay forum (FRF).

[33] See RFC1618, PPP over ISDN

[34] The actual signaling rate at the U interface is 80 Kbaud. A four-level line code called "two binary, 1 quaternary" (2B1Q) is used, which results in 160 Kbps. Remember that bps is equal to the baud rate times $\log_2 L$, where L is the number of discrete elements per signal. In this case, L is equal to 4 and the baud rate times $\log_2 L$ is equal to 80 x 2 = 160. (Or, simply, bps = baud rate times the bits per baud.).

[35] This is also called "Perfect Scheduling."

[36] This is not the broadband that uses frequency division multiplexing for LANs, as with WANG for example.

[37] It was also referred to as wideband for awhile. Most literature goes from N-ISDN to B-ISDN.

[38] Asynchronous does not mean asynchronous as in transmission with start and stop bits. Neither does it refer to automatic teller machines (ATMs). It seems that better names could have been used.

[39] Reference RFC 1754, IPATM WG ATM Forum Recommendations VI.

[40] See RFC1821, Real-Time Service in IP-ATM Networks.

[41] See RFC1483, Multiprotocol over AAL5.

[42] See RFC1577, Classical IP and ARP over ATM.

[43] See RFC2105, Cisco's Tag Switching Architecture.

[44] See RFC1209 for specification of IP over SMDS networks.

3

Internet Layer Protocols

The IP receives data directly from a link layer protocol, such as Ethernet, and functions at the architectural level of the Internet layer (functionally equivalent to the network layer of the OSI reference model). Although the protocols ARP, RARP, and IARP receive data directly from a link layer protocol in the same manner as IP, their descriptions are deferred to Chapter 5 where utilities to the Internet are described. By the same rationale, the description of ICMP and IGMP is deferred to Chapter 5. (The Internet standards describe these protocols as part of IP, although they have protocol identifiers the same as the transport layer protocols UDP and TCP.) Dynamic routing protocols are also described in Chapter 5.

This chapter begins with an explanation of the IP datagram and the IP protocol. The addressing of IP datagrams, which is determined from the source and destination address fields in the IP header, is covered next. The chapter concludes with a description of various routing techniques and an introduction to IP Version 6.

3.1 IP Operation

IP provides a connectionless delivery system that is unreliable and that works on a best-effort basis. IP specifies the basic unit of data transfer in a TCP/IP internet as the datagram. Datagrams may be delayed, lost, duplicated, delivered out of sequence, or intentionally fragmented to permit a node with limited buffer space to handle the IP datagram. It is the responsibility of IP to reassemble any fragmented datagrams and deliver them from the source host to the destination host. In some error situations, datagrams are *silently discarded,* while in other situations, error messages are sent to the originators (via the ICMP, a utility protocol described in Chapter 5). IP specifications also define how to choose the initial path over which data will be sent and defines a set of rules governing an implementation of IP [1].

3.2 IP Datagram Format

The IP datagram format, which is illustrated in Figure 3.1, consists of a header and data. Each field in the header is described, as follows.

1. Header Length: The value of the header length field represents the number of octets in the header divided by four, which makes it the number of 4-octet groups in the header. For example, the version, header length and type of service occupies 32 bits, which is equal to one, 4-octet group and causes the header length field to be increased in magnitude by +1. The header length is used as a pointer to the beginning of data and is usually equal to 5, which defines the normal, 20-octet header without options. When options are used, padding may be required to make the total size of the header an even multiple of 4-octet groups. The valid range of values for the header length is 5 to 15.

2. Version: The value of the 4-bit version field identifies the version of IP. The normal IP described in this chapter has a value equal to 4. By means of this field, different versions of IP could operate in the Internet. An experimental stream mode IP used the value 5 and the latest version of IP uses the value 6 (See Section 3.5). All other values are reserved or unassigned.

3. Type of service: The single octet type of service field contains both the precedence and priority fields of the IP datagram. See Figure 3.2.

   Bits +5, +6, and +7 make up the precedence field with a range of zero to 7. Zero is the normal precedence, and seven is reserved for network control. Most routers presently ignore this field but will eventually use it to bypass data with network control messages. (This will exempt control messages from being impacted by congestion.)

   The four bits (+1, +2, +3 and +4) define the priority field, which has the field range of zero to 15. The four priorities presently assigned are value 0, the default, normal service; value 1, minimize monetary cost; value 2, maximize reliability; value 4, maximize throughput; and

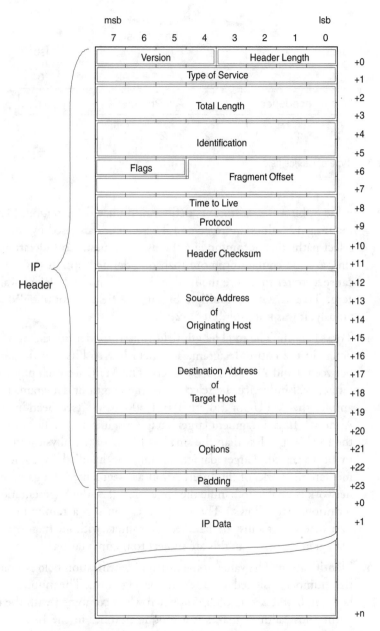

Figure 3.1 IP version 4 datagram format.

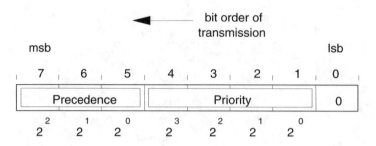

Figure 3.2 Type of service field.

value 8, bit +4 equal to one, defines minimize delay option. (The remaining 12 values are reserved.) These values are used by routers to select paths that accommodate the user's request. Consideration: Setting a value containing more than one defined option (e.g., 3) may cause a router to ignore the field and use the normal, default value of zero. That is, setting multiple bits in the field is not allowed. (Previously, it was not prohibited) [2].

4. Total Length: The total length field is used to identify the number of octets in the entire datagram. The field has 16 bits and the range is between 0 and 2^{16}-1 (65,535) octets. The MTU for each physical interface will influence the selection of the length of datagram. For example, the MTU for Ethernet is 1,500 octets (see Section 2.1.4). Although IP can fragment larger datagrams, it is more efficient to have the total length less than the smallest MTU of the physical interfaces to be traversed. Larger datagrams may be handled by some intermediate networks of the Internet but are segmented if a gateway of a network is unable to handle the larger size. (See the fragmentation description that follows.) The IP specification sets a minimum size of 576 octets that must be handled by routers without fragmentation. Datagrams larger than this are subject to fragmentation.

5. Identification: The value of the 2-octet identification field is a sequential number assigned by the originating host. The numbers cycle between 0 and 2^{16}-1 (65,535), which when combined with the originating host address makes it a unique number in the Internet. The number is used to aid in the assembling of a fragmented datagram.

6. Fragment offset: When the size of a datagram exceeds the maximum of an intermediate network, it is segmented by that network. The fragment offset represents the displacement (in increments of eight octets) of this segment from the beginning of the entire datagram. The 13-bit

fragment offset field provides an offset to the proper location within the original datagram of this fragmented segment. Since the value represents groups of eight octets, the effective range of the offset is between zero and 65,535 (($2^{13} * 2^3$)-1)) octets. The resulting fragments are treated as complete datagrams and remain that way until they reach the destination host where they are reassembled into the original datagram. Each fragment has the same header as the original header except for the fragment offset field, identification field and the flags field. Since the resulting datagrams may arrive out of order, these fields are used to assemble the collection of fragments into the original datagram (form before fragmentation).

7. Flags: The flags field occupies bits +5 through +7 of the +6 octet and contains two flags. The low-order bit (bit +5) of the flags field is used to denote the last fragmented datagram when set to zero. That is, intermediate (nonlast) datagrams have the bit set equal to one to denote more datagrams are to follow. (Similar to the *more* data bit in the packet layer of X.25.) The high-order bit (bit +7) of the flag field is set by an originating host to prevent fragmentation of the datagram. When this bit is set and the length of the datagram exceeds that of an intermediate network, the datagram is discarded by the intermediate network and an error message is returned to the originating host via the ICMP. See Section 5.1.2. Bit +6 of the field is not used.

8. *Time to live* (TTL): The single octet TTL represents a count (in seconds) set by the originator that the datagram can exist in the Internet before being discarded. Hence, a datagram may loop around an internet for a maximum of 2^8-1 or 255 seconds before being discarded. The current recommended default TTL for IP is 64. Since each gateway handling a datagram decrements the TTL by a minimum of one, the TTL can also represent a hop count. However, if the gateway holds the datagram more than one second, it decrements the TTL by the number of seconds held. The originator of the datagram is sent an error message via the ICMP when the datagram is discarded. See Section 5.1.2.

9. Protocol: The single octet protocol field is used to identify the next higher layer protocol using IP. It will normally identify either TCP (value equal to 6) or UDP (value equal to 17) transport layer but may identify up to 255 different transport layer protocols. An upper layer protocol using IP must have a unique protocol number [3]. Protocol numbers are listed in Appendix D.

10. Checksum: The 16-bit checksum field provides assurance that the header has not been corrupted during transmission. The least significant bit of the checksum is bit +0 of octet +11. The checksum includes all fields in the IP header, starting with the version number and ending with the octet immediately preceding the IP data field, which may be a pad field if the option field is present. The checksum includes the checksum field itself, which is set to zero for the calculation. The checksum represents the 16-bit, ones complement of the ones complement sum of all 16-bit groups (double octet pairs) in the header. An intermediate network (node or gateway) that changes a field in the IP header (e.g., TTL) must recompute the checksum before forwarding it.[4] Programs that use IP must provide their own data integrity, since the IP checksum only protects the IP header.

11. Source address: The 32-bit source address field contains both the network and host identifiers of the originator. The source address may be class A, B, or C, which is described in Section 3.3.

12. Destination address: The 32-bit destination address field contains both the network and host identifiers of the destination. The IP address may be of class A, B, C, or D. (See Section 3.3.)

13. Options: The presence of the options field is determined from the value of the header length field. If the header length is greater than five, at least one option is present.

    Although it is not required that a host set options, it must be able to accept and process options received in a datagram. The options field is variable in length. Each option declared begins with a single octet that defines the format of the remainder of the option. Each code used is followed by a length field and data. The format of the single octet is illustrated in Figure 3.3, and a description of each field follows.

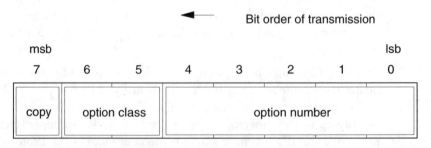

Figure 3.3 Options octet format.

- The copy flag is used by intermediate networks (node or gateway) when a datagram is being fragmented. If the bit is set equal to one, each fragment generated must contain the same option. If the bit is set equal to zero, the option is purged.

- Option class zero denotes datagram or network control, and option class two denotes debugging and measurement. Option classes one and three are reserved for future use. (All assigned options are class zero except option four which is class two.)

- The option number identifies a specific action. It occupies bits +0 through +4, with bit +4 being the most significant. Table 3.1 below summarizes the option numbers and meanings, and Table 3.2 provides detail of the options.

- IP datagram options: Options 0 and 1, which have a single octet equal to zero following the option, are used for alignment. The remainder of options (2, 3, 4, 7, and 9) are followed by a single octet defining the length of the remainder of the option. The value of the single octet (length) is equal to 11 for option number 2 and variable for all others. The remainder of the option following the length octet is defined in Table 3.2.

(Pointers in Table 3.2 are relative to the options octet, and the field number is relative to the pointer.)

- Timestamp option: The timestamp option provides the user with a technique of recording the precise route taken by a datagram and the

Table 3.1
IP Options

Option Class	Option Number	Octets	Description
0	0	1	Ending for alignment (if required)
0	1	1	For alignment within list of options
0	2	11	Security (military applications)
0	3	Variable	Loose source routing
0	7	Variable	Record/trace route
0	9	Variable	Strict source routing
2	4	Variable	Timestamp

Table 3.2
Other IP Options

Option 2 (Security): Four fields
    Field 1 (16 bits): As defined by Defense Intelligence Agency

Option 3 (Loose Source Routing): Variable fields
    Field 1 (8 bits): Pointer (in octets) to field
        of IP address (Initially 4)
    Field 2 (32 bits): First IP address
    Field 3 (32 bits): Next IP address
    -
    Field n (32 bits): Last IP address

Option 4 (Timestamp): Variable fields
    Field 1(8bits): Pointer (in octets) to field
        after field 3 (Initially 5)
    Field 2 (4 bits): Number of gateways unable to
        stamp because of no space.
    Field 3 (4 bits): Format control
        Field 3 value equal to 0
    Field 4 (32 bits): First timestamp
    Field 5 (32 bits): Next timestamp
    -
    Field n (32 bits) Last timestamp
        Field 3 value equal to 1
    Field 4 (64 bits): First IP address and timestamp
    Field 5 (64 bits) Next IP address and timestamp
    -
    Field n (64 bits) Last IP address and timestamp
        Field 3 value equal to 3
    Fields same as for field 3 value equal to 1

Option 7 (Record/Trace Route): Variable fields
    Field 1 (8 bits): Pointer (in octets) to field
        of IP address (Initially 4)
    Field 2 (32 bits): First IP address
    Field 3 (32 bits): Next IP address
    Field n (32 bits): Last IP address

Option 9 (Strict Source Route): Variable fields
    Field 1 (8 bits): Pointer (in octets) to field
        of IP address (Initially 4)
    Field 2 (32 bits): First IP address
    Field 3 (32 bits): Next IP address
    -
    Field n (32 bits): Last IP address

time that each element (node or gateway) handling the datagram processed it. The length field is the number of octets including the four-octet header and eight octets for each node expected to traverse. The maximum size length field is equal to 40. Hence, the maximum number of nodes that may be timestamped is four. (Five nodes would require 44 octets.) See Figure 3.4.

The pointer field is the number of octets from the beginning of the header to the next option to be processed. The minimum value is equal to 5, which points to the first timestamp option.

The flags field determines the precise handling by gateways. If the flag field is equal to one, (bit +0 equal to one) it is as illustrated. That is, the IP address precedes the time stamp. If the value is zero, the IP address is omitted. If the value is three, the originator supplies the IP addresses and a particular intermediate node only timestamps if its address is pointed to by the value in the pointer field.

If the originator does not allocate sufficient space for all the nodes to timestamp, the field identified by LOW (bits +4 through +7 of octet +0) in the Figure 3.4 above is incremented by one for each node unable to timestamp. The information obtained by the timestamp option is

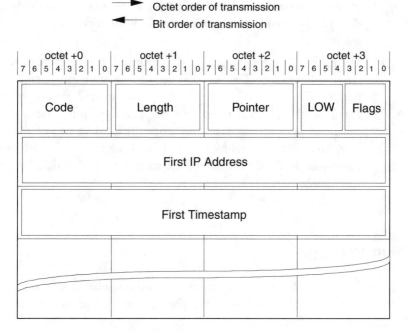

Figure 3.4 Timestamp option format.

only useful if the destination uses it. That is, there is no central location where such statistics are collected.

- Record/strict/loose source routes: The record, strict source and loose source route options are similar in coding to the timestamp option, except the flags field and the LOW field are not present. With record route the originator constructs an empty list with the pointer set to the first entry. Each node of the Internet that handles the datagram then stores its IP address and increments the pointer to the next entry. This provides a routing trace of the datagram. The strict source route option permits the originator to specify precisely the path through the Internet that the datagram must take. This can be useful to force all traffic over a particular path for testing. The strict source routing option is coded with the precise successive IP addresses with a pointer set to the first hop. Each node handling the datagram increments the pointer to the next IP address. Loose source routing is similar, except only the major IP addresses are entered in the list of IP addresses. The Internet may take any desired intermediate path so long as the datagram visits the IP nodes identified.

14. Padding: The pad field, when present, consists of one to three octets of zero, as required, to make the total number of octets in the header divisible by four. (The header length is in increments of 32-bit words.)

15. Data: The data field consists of a string of octets, each with unrestricted contents. That is, each octet has a value in the range of 0 to 255. The string size may have minimum and maximum sizes based on the physical interface type. The maximum size is determined from the total datagram length field. The size of the data field (in octets) is equal to: [(total datagram length) - (header length times four)]. With no options, the maximum size of data length is (65,535-20) = 65,515.

3.3 IP Addressing

Networks are formed by connecting hosts with routers. The connection is with a physical link as described in Chapter 2. Each connection point must have an IP address, except when using unnumbered addressing with a serial link (described later). Elements of a simple network are illustrated in Figure 3.5. For the

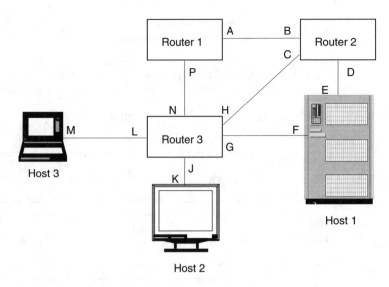

Figure 3.5 IP address assignment.

purpose of addressing, the network (and Internet) consists of hosts, which includes routers. (Some hosts do have router functionality [5].)

Each of the characters A through P represent an IP address. Host 3 has the IP address M that when used by any other host as the destination address will result in the message being sent to host 3. Host 1 has two interfaces (called multihomed) and each end of these interfaces has an IP address (E and F). If host 1 wants to send a message to host 3, it uses IP address M for a destination address. Then it must choose whether the next hop is IP address D or IP address G. It would normally choose IP address G because this path has the least number of hops between host 1 and host 3. Hosts and routers determine the path from a database of routing paths. The database is typically static for hosts and dynamically maintained with a routing protocol for routers. The routing process is described in more detail in Section 3.4.

An IP address consists of four octets (32 bits), which include both a network and a host address. The high-order bits (illustrated in Figure 3.6 as bits 5, 6, and 7) of the most significant octet are used to define how the bits within the IP address are partitioned. This provides the flexibility for small, medium, and large network configurations.

When bit 7 of the first octet is equal to zero, a class A address is defined. Class A addresses use the remainder of bits in the first octet to define the network and the remaining three octets to define the host. When bit 7 of the first octet is equal to one and bit 6 is equal to zero, a class B address is defined. Class B addresses use the remainder of the first octet and all bits from the second octet

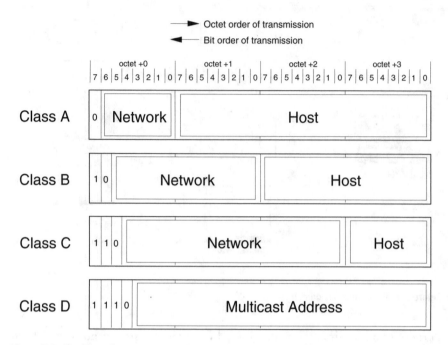

Figure 3.6 IP address format.

to define the network and the remaining two octets to define the host. When bits 6 and 7 of the first octet are each equal to one and bit 5 is equal to zero, a class C address is defined. Class C addresses use the remainder of bits in the first octet and all bits from the second and third octets to define the network and the last (fourth) octet to define the host. When bits 5, 6, and 7 are each equal to 1 and bit 4 is equal to zero, a class D address is defined. Class D addresses are used for multicasting. That is, a single address from this space is used to direct a datagram to multiple unicast addresses. There is also a class E address space, which is reserved for experimental work and not used. The experimental address space begins just past the end of the class D address space.

To define the number of networks and hosts for each class is easy when it is done in either a binary or hexadecimal manner. Table 3.3 illustrates the ranges for all classes of IP addresses, in decimal.

For a class A address the first network number is one (zero is not used), and the last is equal to $(2^7-2) = (7F-1)$ hex $= 126_{10}$. The remainder of bits define the host and are in the range 1 to $(2^{24}-2) = (FFFFFF -1)$ hex $= 16,777,214_{10}$.

For a class B address the first network number is one, and the last is equal to $(2^{14}-2) = (3FFF-1)$ hex $= 16,382_{10}$. The remainder of bits define the host and are in the range 1 to $(2^{16}-2) = (FFFF-1)$ hex $= 65,534_{10}$.

Table 3.3
IP Address Ranges by Class

Class	Valid Network Numbers	Valid Host Number
A	1–126	1–255.255.254
B	128.1–191.254	1–255.254
C	192.0.1–223.255.254	1–254
D	224.0.0.1–239.255.255.255	Not applicable
E	240–255.255.255.255	Reserved

For a class C address the first network number is one, and the last is equal to $(2^{21}-2)$ = (1FFFFF-1) hex = $2,097,150_{10}$. The remainder of bits define the host and are in the range 1 to (2^8-2) = (FFFF-1) hex = 254_{10}.

3.3.1 Reserved IP Addresses

Why was the number -2 used in the above calculations? Some early implementations of IP addressing used all the bits in the host field, for example, to maximize the possible addresses. IP addresses are not permitted to use the value zero or all bits set to one for either the host portion or the network portion of the address, because these addresses are reserved for special situations. This is why the value -2 (e.g. 2^8-2) was used in the above calculations.

IP addresses may be specified as <network host>, meaning the first number represents the network address and the second represents the host address. For example, <64,32> indicates a class A network and the 32nd host on network number 64. The number -1 is used to denote that all bits in the field are each equal to one. (In binary arithmetic, one less than zero is all bits in the field each equal to one, or -1.) The addresses <0,0>, <0,n>, <1,1>, <n,-1>, and <127,n> are special cases and described as follows.

- <0,0>: Indicates this host on this network. It may not be used as a source address except in special circumstances (e.g., BOOTP explained later).

- <0,n>: Specifies host "n" on this network, when n is not equal to all ones (-1). This is also only used in special circumstances.

- <1,1>: This is described as a limited broadcast. It is directed to every host connected to the physical network (e.g., Ethernet segment) but will not be forwarded outside the physical network.

- <n,-1>: This is referred to as a directed broadcast because it is sent to all stations connected to a particular network identified by n.
- <127,n>: This is a class A address specifically reserved for loopback testing. It is never forwarded except to station n on this network.

Unless you have eight fingers on each hand and think in hexadecimal, or read Section 1.6, binary arithmetic, it may be easier to work with decimal numbers, as in Table 3.1 where the range of both network and host addresses for each class of address is illustrated.

3.3.2   IP Address Notation

The convention described earlier for an IP address (<m,n>) is awkward for normal use. Consider an IP address <12625921,101>. It would be difficult to remember, for sure. The notation developed to identify an IP address (both the network and host portion) in one number is called dotted decimal notation. This is essentially four decimal numbers, each representing one octet of the IP address and separated by a period (.). The awkward address cited earlier becomes much easier to use. Namely, the IP address <12625921.101> is represented in dotted decimal format as 192.168.1.101. It is only one digit less but much easier to handle. The value 192 translates to the first octet (11000000_2), the value 168 translates to the second octet (00000001_2), the value 1 translates to the third octet (00000001_2), and the value 101 translates to the last octet (01100101_2).

IP addresses were, in the past, assigned by a central authority, the Hostmaster at the *Defense Data Network* (DDN) *Network Information Center* (NIC). This situation, which has been changed drastically, is described in Section 3.4.4.

3.3.3   Subnet Addressing

The purpose of subnet addressing is to extend the possible network numbers of a particular class of IP address [6]. It does this by robbing the necessary bits from the host portion of the address. The easiest example of subnetting can be illustrated with a Class B address. See Figure 3.7 for an illustration of a Class B address and two subnet masks. Each subnet mask causes the boundary between the network and host portions of the address to be interpreted differently.

In Figure 3.7, there are 14 bits for the network address and 16 bits for the host address. Up to 14 bits of the host address may be used to extend the network address. For example, to make it simple, take an entire octet, leaving only

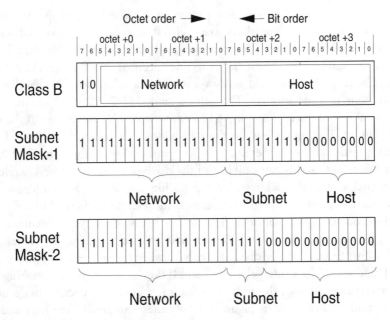

Figure 3.7 Class B subnet address example.

one octet for the host address. There are now 14 bits for the network address, eight bits for a subnetwork address, and eight bits for the host address. Remember that although there are eight bits to define subnets, only 254 of the 256 are available for use. The cost of doing this is to decrease the possible number of hosts that can be addressed.

Identifying the bits that are treated as the network address and the bits that are treated as the host address is done with a template, called a *subnet mask.* A subnet mask (template) is a 32-bit number that corresponds bit-for-bit with the 32-bit IP address. For each bit set to 1 in the subnet mask, the corresponding bit position in the IP address is interpreted as part of the network and subnet address. For each bit set to 0 in the subnet mask, the corresponding bit position in the IP address is interpreted as part of the host address. For example, if the subnet mask bit position is part of the host field and is set to 1, the corresponding bit in the IP address is interpreted as part of the subnet address. However, if the subnet mask bit position is part of the host field and is set to 0, the corresponding bit in the IP address is interpreted as part of the host address. The subnet mask usually has significance to the final (destination) router only. Intermediate routers use only the network portion of the IP address for routing. For example, for a class B address the default subnet mask is 255.255.0.0.

Routers that use route aggregation (supernetting) use less bits in the subnet mask than required by the address class. (This is described in Section 3.4.4.)

For any IP address one has been assigned, each bit in the first (left-most) field of the subnet mask must be set to one (decimal 255, binary 11111111). This is because the first field is always a network address field, regardless of whether one is using subnets. The subnet mask 255.255.255.255 is not used because all bits of the corresponding IP address would be interpreted as the network address, leaving no bits for the host address. (See the exception of host-specific route defined in Section 3.4.3.) Similarly, the subnet mask 0.0.0.0 (binary 0) is not used because all bits of the corresponding IP address would be interpreted as the host address, leaving no bits for the network address. For those who can afford the luxury of a class B address, the entire eight-bit field of the third octet is set to all ones. This makes it easier for system managers to distinguish among the network, subnet, and host fields. Using the values 255 and 0 allows the fields to be divided evenly between the subnet and host parts of the IP address. Values other than 255 and 0 do not divide evenly, making the host and subnet parts of the address difficult to visualize, especially in decimal. The second subnet mask in Figure 3.7 only uses four bits of the host address portion. It represents a typical use of subnetting with a class B address and is more difficult to visualize in decimal.

The overall impact on addressing needs to be mentioned again. Without subnetting, the class B addressing capability is 16,382 networks and 65,534 hosts per network. With subnet addressing, the addressing capability is 16,382 networks, 254 subnets per network, and 254 hosts per subnetwork. The effect is to allow a variable class of IP address that extends the number of networks in smaller increments than going from class A to class B, or from class B to class C.

3.3.3.1 Class B Subnetting Exercise

Your router has a class B IP network address equal to 172.16.x.x, and it has three subnetworks with four stations per subnet and one serial port configured with an unnumbered (discussed later) IP, PPP connection. Design a subnet mask for the router using eight bits from the host address bits for the subnet mask [7].

The solution is as follows: The number of subnets possible with eight bits from the host address bits is equal to 2^8 - 2, or 254. With such a small network, this would be a waste of address space, but it is acceptable for a basic exercise. The addresses selected to satisfy the requirement are illustrated in Figure 3.8. They are described as follows:

• The router address in dotted decimal format is 172.16.0.0.

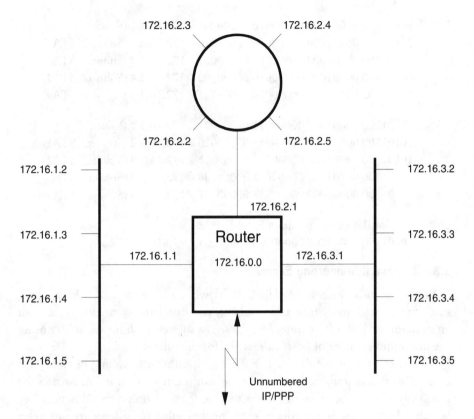

Figure 3.8 Class B IP subnetting exercise.

- In binary the address is 10101100 00001000 00000000 00000000.
- The subnet mask in binary is 11111111 11111111 11111111 00000000.
- The subnet mask in hex is ff.ff.ff.00.
- The subnet mask in decimal is 255.255.255.0.

Now calculate the individual address for each of four hosts on each of 3 LANs. The address of each station on each subnet is as follows.

LAN1: 10101100 00001000 00000001 00000001 = 172.16.1.1 (Router 1)
      10101100 00001000 00000001 00000010 = 172.16.1.2 (Subnet1, STA1)
      10101100 00001000 00000001 00000011 = 172.16.1.3 (Subnet1, STA2)
      10101100 00001000 00000001 00000100 = 172.16.1.4 (Subnet1, STA3)
      10101100 00001000 00000001 00000101 = 172.16.1.5 (Subnet1, STA4)

LAN2: 10101100 00001000 00000010 00000001 = 172.16.2.1 (Router 2)
        10101100 00001000 00000010 00000010 = 172.16.2.2 (Subnet2, STA1)
        10101100 00001000 00000010 00000011 = 172.16.2.3 (Subnet2, STA2)
        10101100 00001000 00000010 00000100 = 172.16.2.4 (Subnet2, STA3)
        10101100 00001000 00000010 00000101 = 172.16.2.5 (Subnet2, STA4)

LAN3: 10101100 00001000 00000011 00000001 = 172.16.3.1 (Router 3)
        10101100 00001000 00000011 00000010 = 172.16.3.2 (Subnet3, STA1)
        10101100 00001000 00000011 00000011 = 172.16.3.3 (Subnet3, STA2)
        10101100 00001000 00000011 00000100 = 172.16.3.4 (Subnet3, STA3)
        10101100 00001000 00000011 00000101 = 172.16.3.5 (Subnet3, STA4)

PPP   (Optional if not using unnumbered IP PPP.)
        10101100 00001000 00000100 00000001 = 172.16.4.1 (Router)

3.3.3.2 Class C Subnetting Exercise

Your private stub router has a class C IP network address of 192.168.1.x, three
subnetworks with four stations per subnet, and one serial port configured with
an unnumbered IP, PPP connection. Design a subnet mask for the router using
the minimum number of host address bits for the subnet mask.

The solution is as follows: The minimum number of subnet mask bits for
three subnets is determined by finding the minimum value of n that satisfies the
inequality $2^n - 2$ is equal to or greater than 3. If $n = 2$, only two subnets can be
handled. Hence, use $n = 3$, which is the smallest value and allows expansion up
to six subnets. The addresses selected to satisfy the requirement are illustrated in
Figure 3.9. They are described as follows.

- The router address in dotted decimal format is 192.168.1.0.
- In binary the address is 11000000 10101000 00000001 00000000.
- Subnet mask in binary is 11111111 11111111 11111111 11100000.
- Subnet mask in hex is ff.ff.ff.e0.
- Subnet mask in decimal is 255.255.255.224.

Now calculate the individual address for each of four hosts on each of
three LANs. The address of each station on each subnet is as follows.

LAN1: 11000000 10101000 00000001 00100001 = 192.168.1.33 (Router 1)
                            21 hex = 33 b10
        11000000 10101000 00000001 00100010 = 192.168.1.34 (Subnet1, STA1)
        11000000 10101000 00000001 00100011 = 192.168.1.35 (Subnet1, STA2)

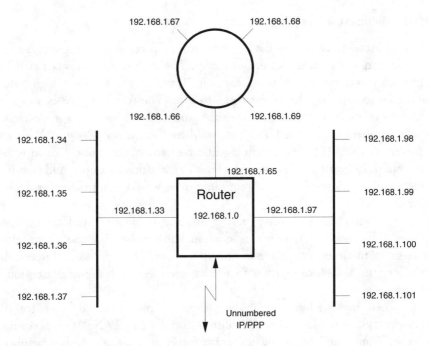

Figure 3.9 Class C IP subnetting exercise.

11000000 10101000 00000001 00100100 = 192.168.1.36 (Subnet1, STA3)
11000000 10101000 00000001 00100101 = 192.168.1.37 (Subnet1, STA4)

LAN2: 11000000 10101000 00000001 01000001 = 192.168.1.65 (Router 2)
41 hex = 65 b10
11000000 10101000 00000001 01000010 = 192.168.1.66 (Subnet2, STA1)
11000000 10101000 00000001 01000011 = 192.168.1.67 (Subnet2, STA2)
11000000 10101000 00000001 01000100 = 192.168.1.68 (Subnet2, STA3)
11000000 10101000 00000001 01000101 = 192.168.1.69 (Subnet2, STA4)

LAN3: 11000000 10101000 00000001 01100001 = 192.168.1.97 (Router 3)
61 hex = 97 b10
11000000 10101000 00000001 01100010 = 192.168.1.98 Subnet3, STA1)
11000000 10101000 00000001 01100011 = 192.168.1.99 Subnet3, STA2)
11000000 10101000 00000001 01100100 = 192.168.1.100 Subnet3, STA3)
11000000 10101000 00000001 01100101 = 192.168.1.101 Subnet3, STA4)

PPP: (Optional if not using unnumbered IP PPP.)
11000000 10101000 00000001 10000001 = 192.168.1.129 (Router)
81 hex = 129 b10

3.3.4   Multicast Addressing

IP multicast addressing provides a mechanism to direct an IP datagram to a set of hosts that form a single multicast group. The multicast address is a class D address and is distinguished from classes A, B, and C with the three high-order bits equal to one and the +4 bit equal to zero. The remaining 28 bits of the address designate a particular multicast group. Some *well-known* multicast group numbers are reserved and controlled by the *Internet Assigned Numbers Authority* (IANA) [8]. The remaining numbers are available and termed transient multicast groups. The maximum limit of a multicast group would amount to a broadcast address (all). The minimum limit would be a unicast, or single, address.

An individual host may belong to one or more multicast groups. However, it is not necessary for an individual host to be a member of a multicast group in order to send a multicast message. It is necessary for a host to belong to a multicast group to receive messages with a multicast group address.

When the hardware technology supports multipoint delivery (called multicasting), it is used. The capability is common in LANs. This saves the software from duplicating the datagram for each multicast group member, which is common in switches with point-to-point connections. When, for example, Ethernet is used, multicasting is designated by setting the low-order bit of the high-order octet of the Ethernet destination address. The IP multicast address is mapped by placing the low-order 23 bits into the low-order 23 bits of the Ethernet multicast address. Note that this restricts the IP multicast address to only 23 of the possible 28 bits available [9]. In Internet standard "dotted decimal" notation, host group addresses range from 224.0.0.0 to 239.255.255.255. See Section D.8 for a list of multicast addresses.

3.3.5   Unnumbered IP Addressing

With the scarcity of IP addresses it seems a shame to waste two IP addresses for a single point-to-point link. To minimize the exhaustion of IP addresses, an exception to the rule that every interface must have an IP address was made for point-to-point links. With the complete absence of an IP address for an interface, problems are encountered, such as what to place in the options field for the IP address when using the record route feature of IP. To avoid this problem (and others) the unnumbered IP address feature actually uses an IP address from another interface in the same router. (Every router must have at least one IP address.)

3.4 Internet Protocol Routing

There are several forms of routing used to direct datagrams. The most basic are static routing and variations such as default routing and host-specific routing. These routing mechanisms operate from a fixed table that is maintained by the network administrator. In larger networks and the Internet, dynamic routing protocols are used to update routing tables. Dynamic routing protocols are described in Chapter 5.

3.4.1 Static Routing

Static routing is effected by the network administrator who manually constructs a host table with valid IP addresses and the associated next hop. Network administrators are also required to manually update the table when a new network address is added or an old one is out of service. In small *autonomous systems* (ASs) the manager can keep up with the activity, but in large networks, a dynamic routing should be used.

The originating host (or in some cases a router) performs the first routing function for an IP datagram. That is, it determines where to send the datagram. If the network portion of the IP address is the same as the originating host, (i.e., the destination resides on the same network as the source) direct routing is performed by converting the IP address to a physical address (e.g., Ethernet) and delivering the datagram. The datagram is never sent to another host for routing back to the originating host. If the network portion of the IP address is not the same as the originating host, the *route table* (RT) is analyzed to determine the next hop. The RT is not the same table used for transparent bridging. If the originating host has a single connection to the Internet, the routing decision tree is simple—direct or the attached router.

A host receiving a datagram from a router handles it as though it has been received from a direct user, except when the destination IP address is for a different host. Hosts do not normally perform routing functions of received datagrams if not addressed to a directly connected workstation. This would take away resources from users of that host, hide routing problems from the internet administrator, and possibly result in misrouting since the hosts do not typically participate in the routing table update process (dynamic routing protocols), as routers do. A host receiving an IP datagram from a router that does not contain an IP address that is local to that host is typically discarded. Some hosts may provide a "network not reachable" error message (via ICMP) when an IP datagram is discarded. See Section 5.1.2.

3.4.2 Default Routing

Default routing minimizes the size of the memory resident RT. The order of search is for direct routing, network addresses for a host-specific route, and all others (default route). Host programs update their routing tables with information obtained from routers via the ICMP redirect mechanism. See Section 5.1.4.

Default routing is accomplished by setting up a default route at the end of the RT that will match any IP address. After entering all the static routes, the default (all others) address is set at the end of the routing table in the same manner as the valid addresses, except this IP address is equal to 0.0.0.0, and the subnet mask is equal to 0.0.0.0. Since all bits in the subnet mask are equal to zero, any address being compared will match. This is actually setting the address to be compared to zero, then comparing it to zero—a sure match. In the process, the router is passing the datagram on to the next router for the address to be properly resolved. (Compare this to specifying a subnet mask equal to 255.255.255.255 in Chapter 4.)

Default routing is very useful when a site has a small set of local addresses and only one connection to the rest of the Internet. For example, default routes work well in hosts that attach to a single physical network and can reach only one router leading to the remainder of the Internet. This is called a stub network or one without pass through. It may define its hosts on that network statically and use a default route for all others. This is a reasonable ploy for a stub network that is connected to the Internet by a single router. It would eliminate all the automatic routing table updates and minimize the size of the routing table.

3.4.3 Host-Specific Routing

Although IP routing is normally based on the network portion of the IP address, this is the exception. Most IP routing software will allow specific-host routes to be used. Specific-host routing gives the network administrator more control over the network use and permits testing. It can also be used to control access for security purposes. When debugging network connections or routing tables, the ability to specify a special route to one individual host turns is very useful.

A host-specific route is defined by any IP address that has the subnet mask equal to 255.255.255.255. Since the RT is constructed with the longest mask first, this will be the first match found. The router attempts to mask out the host portion of the address with the zero bits in the mask, but there are none.

Therefore, the complete address (network and host) is compared to the address in the datagram. Only datagrams containing the precise network and host address will match a routing table host-specific entry.

3.4.4 Classless Interdomain Routing (CIDR)

From the early 1990s, the inadequacy of the IP address space to accommodate the Internet growth rate has been well publicized and documented. Not only was the number of available IP addresses being exhausted, the size of routing tables required to perform the routing function grew even more quickly. This is because IP addresses were assigned sequentially to individuals as they were requested. Anyone with the token fee and patience to complete a request form was issued an IP address without reasonable justification of need. Many assigned IP addresses have never been used, and many small organizations with only limited need for access to the Internet were issued IP addresses. This compounded the problem of the Internet's growth to proportions beyond the wildest dreams of the inventors and designers.

Many proposals [10] were made to slow the exhaustion rate until a long-term solution could be invented and then implemented. The long-term solution focused on IP Version 6, which has 128 bits allocated for the IP address. The short-term solution to slow the rate of exhaustion focused on *classless interdomain routing* (CIDR). The IAB formally recommended the supernetting approach with RFC 1481, and RFC 1338 was revised as RFC 1519. RFC 1518 and RFC 1519 provide the latest description of both policy and function of CIDR.

CIDR, by name, would seem to indicate that five bits of the 32-bit IP address used to define the type of address were converted to address space. The name is a bit of a misnomer since class A, class B, class C and class D addresses are alive and well in the Internet. Since CIDR is assigning all new addresses from the class C address space, it is moving towards a single address space that can be described as classless. An appeal to the Internet community has been made to subscribers with old IP addresses to exchange them for new addresses allocated from the new space. However, some form of the old addresses will likely be around for years (See RFC 1917). There are functional changes required for CIDR, but the most significant change needed is in the policy of IP address allocation.

The policy change involves the assignment of IP addresses in blocks, and the functional change involves the identification technique of these blocks, rather than individual addresses. The scheme is described as *supernetting*. (By policy, service providers are issued large blocks of contiguous IP addresses and individual users [subscribers] rent them.) Supernetting makes it possible for a single IP address to be used in routing tables for an entire block of IP addresses.[10] Within a large block of IP addresses, a subblock may be defined. This permits a large *service provider* (SP) to supply smaller ISPs with subblocks of the larger block. First, let us describe the policy changes and then the workings of supernetting will be evident.

The overall hierarchical scheme of the Internet registry (IR) is to allocate large blocks of IP addresses to regional registries, which will then assign smaller blocks of address space to SPs, which, in turn, will assign yet smaller blocks of address space to ISPs. Finally, individual users will rent IP addresses from the ISPs.

The class A portion of the number space represents 50% of the total IP host addresses; class B is 25% of the total; class C is approximately 12% of the total; and the remainder is used for multicast addresses and classfull overhead (five bits). Future requests for IP addresses will be treated in the following manner.

1. Class A: Don't even ask;
2. Class B: Ask but expect to receive block(s) of class C addresses;
3. Class C : Blocks assigned based on need and justification.

The class C network number space is divided into allocatable blocks, which are reserved by the IANA and IR for allocation to distributed regional registries, as illustrated in Table 3.4. In the absence of designated regional regis-

Table 3.4
Regional Assignments

Area	Address
Multiregional	192.0.0.0–193.255.255.255
Europe	194.0.0.0–195.255.255.255
Others	196.0.0.0–197.255.255.255
North America	198.0.0.0–199.255.255.255
Central/South America	200.0.0.0–201.255.255.255
Pacific Rim	202.0.0.0–203.255.255.255
Others	204.0.0.0–205.255.255.255
Others	206.0.0.0–207.255.255.255

tries in geographic areas, the IR will assign addresses to networks within those geographic areas according to the class C allocation divisions.

The IANA and the IR reserved the upper half of this space, which corresponds to the IP address range of 208.0.0.0 through 223.255.255.255. Network numbers from this portion of the class C space will remain unallocated and unassigned until further notice.

The remaining class C network number space will be allocated in a manner that accommodates address aggregation techniques (supernetting). The intent is to divide this address range into eight equally sized address blocks, as follows. The Reseaux IP Europeans Network Coordination Center (RIPE NCC) had been previously allocated a block of class C addresses (193.0.0–193.255.255).

CIDR is being deployed in the Internet backbone routers as the primary mechanism to improve scaling properties of the Internet routing system. It is also being implemented in all organizations that operate as ISPs. These organizations are expected to support CIDR. Since CIDR is only supported by new versions of dynamic routing protocols (described in Chapter 5), ISPs must support the latest versions of the dynamic routing protocols.

Variable length subnet masking (VLSM) is the generalization of the older subnet mask concept (originally defined in RFC 917) and an integral part of CIDR. The terminology can be confusing, because the length of the subnet mask defined in RFC 917 is also variable. The goal is to eliminate the usage of network numbers A, B, and C by assigning only multiples (or blocks) of original class C addresses, now considered classless.

The concept of supernetting is not difficult to visualize if it is examined in binary or hexadecimal (as the designers saw it). It looks as bad as a tax return in decimal, so the following examples are done in binary with decimal conversions as necessary.

From the earlier section on subnet addressing, recall the example of a class C subnet mask equal to 255.255.255.224, or 11111111 11111111 11111111 11100000_2. This essentially robbed three bits from the host portion of the IP address to define eight possible subnetworks, of which only six could be used (all zeros or all ones not valid). The three bits robbed represent the high-order bits of the host field.

Each network number represents a total of 256 possible host addresses (although only 254 are available for usage). Supernetting combines multiple, contiguous class C network numbers, thereby identifying n x 256 possible host addresses, where n is the number of networks defined by the supernet mask. The key difference in the definition of a subnet mask and a supernet mask is that the reservation bits are set to one for the subnet mask and set to zero for the supernet mask. A visual way to imagine this is to picture the division between the network and host portions of a class C address as a decimal point. Then, the whole numbers to the left of the decimal point can be supernetted (or grouped). In addition, the numbers to the right of the decimal point, which represent fractions of each whole number, may be subnetted. The following example illustrates a subnet and a simple, small supernet mask. See Figure 3.10.

This is not a practical or accurate example, because if it represented a real subnet mask, all bits in the network portion of the mask would each be equal to 1. Both were included to illustrate the difference. This supernet mask defines eight class C networks, each containing $2^8 - 2 = 254$ usable host addresses. An example of an IP address for this supernet mask is 199.255.252.99/255.255.248.0. Notice that the mask consists of the first 21, high-order bits of the 32-bit field. The CIDR notation for this address, 199.255.252.99/21, would appear in the IP RT of an intermediate router as 199.255.252.0/21. The mask for the same address in the destination router (where subnetting is accomplished) is 255.255.255.224 (or /27). Most router manufacturers now support the use of CIDR notation.

The previous example of $8(254) = 2,032$ IP addresses could be subdivided by the subscriber (now an SP if reselling). That is, the SP could delegate (lease or rent) a multiple of the class C addresses to an ISP. For example, if the SP wanted to allocate four of the eight class C network numbers to an ISP, the IP address in intermediate routers would not change. However, the address in the first router of the SP would be 199.255.252.0/22. Since CIDR requires that IP addresses with the longest mask be ordered in the routing table first, the address with a /22 mask (ISP) would be found first before the address with a /21 mask (SP) and properly forwarded to the ISP.

From the nonpractical example in Figure 3.10, it should be evident that the more zeros in the supernet mask, the larger the potential address space being allocated. The subnet mask commences with the highest order bit of the host address and the supernet mask commences with the lowest order bit of the network address. The entire address space for the network addresses 198 through

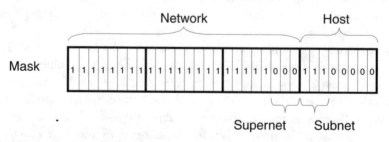

Figure 3.10 Comparison of supernet to subnet.

199.255.255.255 belongs to "North America" and has the CIDR mask 198.0.0.0/8. This address space contains $2^{16}(2^8-2) > 16M$ host addresses. These addresses may be allocated in a hierarchical manner just as illustrated with the previous example.

Consider a new North American SP given 2,048 class C network numbers beginning with 198.24.0.0 and ending with 198.31.255.0. The supernetted route in intermediate routers to this block of addresses is described as 198.24.0.0/13. From these decimal numbers it is difficult to visualize the size and portion of the total address space. This is illustrated in Figure 3.11.

Subtracting the first address from the last address yields a difference of 11111111112 (equal to $7F_{16}$, or 2,048). The supernet mask has the same number of binary zeros as does the difference. That is, the supernet mask has the low order 11 bits each equal to zero and all other bits of the network portion of the address each equal to 1. Notice the similarity to subnet masks (zeros instead of ones and in ascending order instead of descending order).

An example of a subblock of addresses from the above example that could be provided to a smaller ISP needing only 4,096 host addresses (16 class C networks) is the range 198.24.16 through 198.24.31. This block of addresses is represented by 198.24.16.0/20 in intermediate routers. Note that the mask is longer than the mask for the general block belonging to the SP, which would cause the message to be routed to the ISP.

An example of a subblock of addresses from the above example that could be provided to a smaller ISP needing only 4,096 host addresses (16 class C

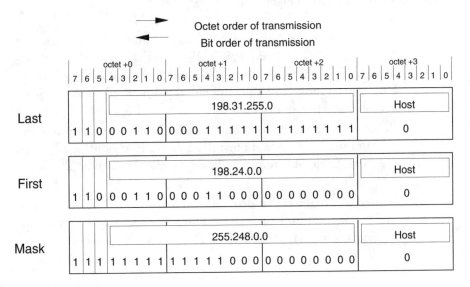

Figure 3.11 Example of CIDR block of addresses.

networks) is the range 198.24.16 through 198.24.31. This block of addresses is represented by 198.24.16.0/20 in intermediate routers. Note that the mask is longer than the mask for the general block belonging to the SP, which would cause the message to be routed to the ISP.

When working with only decimal numbers it is difficult to visualize the concept of either supernetting or subnetting. The book titled *TCP/IP Addressing*, written by Buck Graham, contains a detailed example of supernetting in decimal format.

3.4.5 Network Address Translator (NAT)

A scheme used to reduce the exhaustion rate of IP address space is called *network address translator* (NAT). It is designed for intranets with IP addresses that are invalid to use with the Internet because they are already assigned and in use. NAT permits multiple stub domains to use the same IP addresses by translating the outgoing addresses to globally unique CIDR addresses. Inbound messages with the new CIDR addresses are translated back to the private IP addresses that are not globally unique [11].

Figure 3.12 illustrates an example of two private stub networks that are attached to the Internet, while their hosts continue to use private class A host addresses that are not globally unique. Assume that workstation A sends a message to workstation B that contains the destination IP address 198.76.28.4 and the source address 10.33.96.5. The numbered description that follows coincides with the circled numbers in Figure 3.12.

1. Router A recognizes the message from workstation A addressed to workstation B (IP address 198.76.28.4) and changes the source address from 10.33.96.5 to the globally unique IP address 198.76.29.7. Stub routers must have a pool of globally unique IP addresses to accomplish this function.

2. Router A has a static route configured for an Internet regional router and forwards the message to that IP address. The regional router sends the message to stub router B, based on the destination IP address 198.76.28.4.

3. Router B recognizes the destination address of 198.76.28.4 as one that it translates to the private IP address 10.81.13.22 and sends the message to workstation B.

The process is the same when workstation B sends to workstation A. In this manner, any stub network may use local IP addresses that are the same as

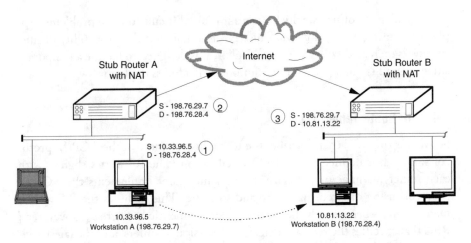

Figure 3.12 NAT example.

other stub networks. The globally unique IP addresses required for routers with NAT that are CIDR-compliant may be obtained from an ISP.

Does this sound too easy? Unfortunately, there are some drawbacks. Several programs use IP addresses, and changing either the destination or source IP address causes problems for them. To accomplish the NAT function in routers, the following issues must be addressed.

1. Both UDP and TCP have a pseudo header for calculating a checksum that includes the source and destination IP addresses from the IP header. The checksum must be recalculated when changing either IP address.

2. The FTP port command has an IP address in ASCII characters. If the number of characters in the local IP address is different from the number of characters in the globally unique IP address, the size of the packet may be changed, which will affect the TCP sequence numbers. Should the FTP port command be encrypted, it will fail.

3. In addition to the checksum recalculation, ICMP messages may have an IP address in the data portion of the message.

4. Any upper layer application program that uses IP addresses must be evaluated to determine if it needs to be changed.

While NAT should work well with simple stub networks, it will not work for more complex applications. Simple static routes are desired. Otherwise, the more robust OSPF routing should be used in preference to RIP. NAT tends to hide the identity of users, which makes it more difficult to trace problems. Applications using encryption will cause NAT to fail, and the probability of misrouting is increased with the use of NAT. It was designed to handle a temporary situation and is not a permanent solution to IP address depletion.

3.4.6 IP Route Table (RT)

So what is the RFC that describes the RT? This key piece of the TCP/IP protocol suite is not defined by an RFC specification. The RT varies slightly with each router manufacturer. Each router manufacturer implements the TCP/IP protocols using different hardware and software. This includes the RT, which manufacturers believe to be better than their competition. There is, however, an RFC that specifies the requirements for IP Version 4 routers, and the latest revision is RFC1812. It is the second longest in the RFC library and is more of a recommendation document than a specification. That is, it provides ideals as compared to the absolute (*must, required,* etc.) statements found in other RFC specifications. It does this by allowing that some features are more important than others, or critical in some applications while irrelevant in others. RFC1812 is the glue that ties together the loose ends between various protocols operating in routers. Since it touches on facets of many TCP/IP protocols, it is treated as the "router handbook" by router manufacturers.

Of interest here is the tie between dynamic routing protocols (RIP, OSPF, and BGP described in Chapter 5) with the RT and the content of the RT. Entries in the RT are either manually configured, the result of an ICMP redirect message (described in Section 5.1.4), or originated from one of the dynamic routing protocols. But none of the dynamic protocols place entries in the IP forwarder's RT. The RT entries are transferred from the private databases maintained by the dynamic routing protocols to the RT by a proprietary piece of software developed by each router manufacturer. This proprietary piece of software is similar (but different) in each manufacturer's router. The RT is also similar (but different) in each manufacturer's router. The private databases are only private to the dynamic routing protocols, which are described in each dynamic routing protocol specification (RIP, OSPF, and BGP.) Each route entry in an RT contains the following minimum information.

- Destination: IP address of the destination network;
- Subnet mask: Identification of the network portion (prefix) of the IP address;
- Next hop: The IP address of the neighboring router that is closer to the destination;

- Port: Internal identification of the physical port attached to the next hop;

- Metric: A weight assigned to the route for determining preference to other routes (with static and RIP routes this is a hop count, or the number of routers necessary to traverse between this router and the destination router);

- Type: Identification of either a directly connected destination (zero routers crossed) or a remotely connected destination, which has one or more routers to be traversed between this router and the destination;

- Source: Identification of the route source (may reflect manually entered routes [static], local [directly connected], ICMP, or one of the dynamic routing protocols [RIP, OSPF, or BGP]);

- Vendor-specific: A weighted value that affords, when desired, the overriding of other information in selecting a route.

In addition, entries originated by one of the dynamic routing protocols may contain the following information.

- TOS: The type of service for this route (may be the least cost, maximum throughput, highest reliability, or minimum delay);

- *Autonomous system* (AS): Identification of the AS containing the destination IP address;

- Source IP address: The source IP address of the route originator;

- Precedence: A weighted value for different dynamic routing protocols;

- Preference: A weighted value determined by the route originator.

These values may be configurable by the network administrator.

3.4.7 IP Forwarder

So how does the IP datagram get from the local host through the possible maze of routers to the final router that is local to the addressed remote host? It happens one hop at a time. Forwarding IP datagrams require choosing the IP address and physical interface of the next router independently from other datagrams—that is, one datagram at a time without state information regarding the previous datagram (a stateless operation). Remember from Section 3.3 that intermediate routers are concerned with only the network number and that the destination router works with the host number, including the subnet bits.

The routing function in routers is handled much the same way as routing at the originating host. Typically, this amounts to evaluating the network portion of the IP address to determine if this is the last hop (router attached to the addressed host). If not, the IP forwarder (router) uses the RT (memory resident and possibly cached) to select the best path to the destination IP address. An exception to this is when the strict source routing option is used by the originator to dictate the path through the Internet. (See Tables 3.1 and 3.2.) The datagram is encapsulated again and sent a hop closer to the final destination. This continues until the final router is reached.

Routers have a skeleton default RT for startup, but manually updating it would be labor intensive and not feasible. Also, the dynamic nature of the Internet (and large intranets) prohibits a manual update. Routers that participate in a dynamic update process have an RT that is current with an optimum path for destinations in their AS. There may be multiple paths (entries in the RT) to the final destination, but only the optimum path is selected by the router. The optimum path may be based on more than just the minimum number of hops to the destination. It may take into account, for example, the time delay in getting messages to a destination and the cost of sending messages to a destination. The process of selecting the optimum path is described as follows. (At any stage of the process if the number of route entries in the short list becomes equal to 1, the process ends and that route entry, possibly the default route, is used.)

1. Filtering: The process begins with the validation of both the source and destination IP addresses in the datagram header. If either fail the validation tests the datagram is silently discarded. Validation includes testing for a source IP address equal to zero or one not legal to be originated on the particular physical interface. Validation for the destination IP address includes testing for a loopback address and an address on network 0. The router may also implement include and exclude lists of source and destination IP address pairs. A pair of addresses in the include list would permit communication between the pair. A pair of addresses in the exclude list would not permit communication between the pair. Filtering is usually configurable and is controlled by the network administrator.

2. Pruning: Selecting the optimum route begins with a process of elimination referred to as pruning. That is, the entries in the RT that do not match the destination IP address (except the default route, which is a special case) are removed. The subnet mask is used to isolate the network portion of both the destination IP address from the IP datagram and the IP address from the RT before comparing.

3. Subnet mask: The next elimination of the short list is determined by the length of the subnet mask. That is, all but the route(s) with the longest subnet mask are discarded. Hence, a host-specific route would always qualify for the short list.

4. TOS: Next, the TOS from the IP datagram is compared to the TOS of the RT entries remaining. Any entry containing a TOS that does not match the requested TOS field in the IP datagram is discarded. (See Section 3.2 and Figure 3.2.)

5. Best metric: Assuming that there is more than a single entry in the short list, the metric of each entry is compared to determine the best route(s) remaining. As a subprocess, the origination of the route may have a preference, which is used for pruning before comparing the best metric. For example, local and static routes may be used over all others, while OSPF internal routes may be used over RIP, BGP, and OSPF externals. After this pruning, only routes with the same type metric should remain. For example, if the remaining two entries are RIP routes and one has a higher hop count, it is discarded.

6. Administrative preference (vendor specific): The final process is to allow the vendor and network administrator to supply the last criteria for determining the route selected. This may be accomplished by assigning a weighted value (0 to 254) to the most and least desired routes. The value zero is the highest value, and if present it would cause those with a value greater than zero to be discarded.

3.5 IP Version 6

IP Version 6 (IPv6) was designed to be the replacement for IPv4. The most compelling reason for IPv6 is expanded address space. It is, by design, the long-term solution to the limited, 32-bit address space. (CIDR is the short-term solution.) The address space of IPv6 is defined by 128 bits. To attempt to put a number of this magnitude in perspective, consider the present address space of 32 bits. This is more than 4 billion addresses. The 128-bit address space is approximately 8×10^{28} times bigger than the entire present, 32-bit address space. (This should accommodate some loose accounting.)

Since it not an easy effort to migrate to a new version of IP (see Section 3.6.5), IPv6 was simplified in some areas (fields dropped) and enhanced with new features. The coding scheme of options, which was made extensible, allows greater flexibility in the length. Special handling of selected traffic and authentication has also been added.

3.5.1 IPv6 Header Format

The format of the IPv6 header is simple compared to that of IPv4. Each field of this format, which is illustrated in Figure 3.13, is described as follows.

1. Version: IPv6, a new version of the IP, is identified in the version field with the value equal to 6 (versus the value equal to 4 for IPv4). It continues to be a four-bit field and occupies the same relative position in the IPv6 header as in the IPv4 header.

2. Priority: The four-bit priority field allows a host to specify the delivery order of its packets. There are two categories of traffic, those that yield to congestion in a router and those that cannot yield because of the time-sensitive nature of the traffic (constant rate). Priorities 8 and above represent the nonyielding traffic. The priority values for yielding traffic recommendations are illustrated in Table 3.5.

3. Flow label: This is a 24-bit field that aims to provide type of service handling. Type of service is experimental, and the field should be set

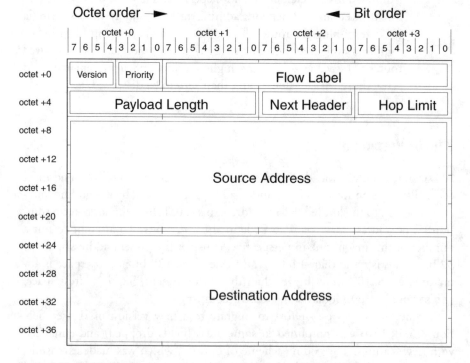

Figure 3.13 IPv6 header format.

Table 3.5
Priority Values for Yielding Traffic

Value	Type Traffic
0	Uncharacterized traffic
1	Filler traffic (e.g., Netnews)
2	Unattended data transfer (e.g., email)
3	Reserved
4	Attended bulk transfer (e.g., FTP, NFS)
5	Reserved
6	Interactive traffic (e.g., telnet, X)
7	Internet control traffic (e.g., routing protocols, SNMP)

to zero by the originator and ignored by intermediate routers and the destination host.

4. Payload length: This 16-bit field identifies the number of octets (1 to 65,535) in the packet following the IPv6 header. When the value is equal to zero, it indicates the jumbo hop-by-hop option, which is described in the hop-by-hop header option.

5. Next header: The IPv6 header size is minimized by eliminating fixed fields for features or options whether they are used or not. For example, although a fragment offset is not used, the space is allocated in an IPv4 header. With IPv6 several features (e.g., fragment offset) are handled with an extension header. If the feature is not used in IPv6, the extension header is absent. The extension headers are stacked, in a prescribed order, one after the other. The eight-bit next header field identifies the next header to be processed; the first extension header immediately follows the destination address field. If special extension headers are not used, the next header field contains the same value as the protocol field used with IPv4 and defined in the latest Assigned Numbers RFC (presently RFC1700). For example, the value is equal to 6 for TCP and equal to 17 for UDP. When special extension headers are present, the last one must point to the transport layer function (for example, TCP). The defined extension headers are summarized as follows.

   • Value 0: Hop-by-hop header is used to define options such as the jumbo payload size packet greater than 65K in length.

- Value 43: Routing header used to perform the equivalent of strict or loose directed routing with a maximum of 23 hops.

- Value 44: The fragment header is used to define the fragment off-set, similar to IPv4's fragment offset, except IPv4's offset occupies a fixed 12 bits of the header.

- Value 59: No-next header is used to end extension headers.

- Value 60: Destination option header is used for variable processing by destination only.

6. Hop limit: This is an eight-bit field and decremented by one each time it is handled by a node. If the hop limit reaches the value of zero, the packet is discarded.

7. Source and destination addresses: These are 128-bit fields and get defined in the following sections.

3.5.2   IPv6 Addressing Modes

The addressing modes with IPV6 have been simplified. There are only two types of addressing, unicast and multicast. The unicast mode is the same as with IPv4, single address results with a single delivery. The multicast mode is also the same as with IPV4, except broadcast is defined as a subset of multicast and a special application of multicast called *anycast* is defined that results with only the closest destination defined by the multicast group receiving the datagram.

3.5.3   IPv6 Address Format

The address format for IPv6, similar to IPv4, is dictated by the high-order, eight bits of the 128-bit address, which is called the address prefix. Table 3.6 provides an identification of the categories allocated space, the associated prefixes, and the percent of total space used by the allocation.

The value zero for the address prefix (noted with an asterisk) is reserved and never assigned as a destination IP address. It may be used (for example) as the source IP address during initialization of a station that does not yet know its IP address.

When the address prefix is equal to zero and the remainder of the address is equal to one (only the low-order bit of the entire 128 bit field is set), a loop-back address is identified. This is equivalent to the class A loopback address with each of the network bits set equal to one (<127,n>).

There are two methods of imbedding IPv4 addresses in an IPv6 address. This permits mixing of IPv4 and IPv6 routers or hosts in the same network,

Table 3.6
IPv6 Address Allocation

Allocation	Prefix	% of Space
Reserved*	0000 0000	1/256
Unassigned	0000 0001	1/256
NSAP	0000 001	1/128
IPX	0000 010	1/128
Unassigned	0000 011	1/128
Unassigned	0000 1	1/32
Unassigned	0001	1/16
Unassigned	001	1/8
Unicast	010	1/8
Unassigned	011	1/8
Reserved	100	1/8
Unassigned	101	1/8
Unassigned	110	1/8
Unassigned	1110	1/16
Unassigned	1111 0	1/32
Unassigned	1111 10	1/64
Unassigned	1111 110	1/128
Unassigned	1111 1110 0	1/512
Link local	1111 1110 10	1/1024
Site local	1111 1110 11	1/1024
Multicast	1111 1111	1/256

which is the technique used for migration. When the first 96 bits of an IPv6 address is equal to zero, the remaining 32 bits is an IPv4 address. This type of address is used for an IPv6 compliant host or router that routes some traffic over an IPv4 network. The value of the address is the same for 32 bits as it is for 128 bits. The second type of imbedded IPv4 address in an IPv6 address is preceded by 16 bits each equal to one. This type of an address is used in an IPv6 network to identify a router or host that supports only IPv4 addressing.

3.5.4 IPv6 Address Representation

The problem of representing the IPv6 address is no different than that of representing an IPv4 address, just bigger (four times bigger). New ideas have been introduced to minimize the foot print of the big address. The basic representation of an IPv6 address is, x:x:x:x:x:x:x:x, where each x represents a 16-bit number coded in hexadecimal, and each is separated by a colon (:). (Call it coloned hexadecimal notation!) A sample IPv6 address is, 0:0:0:0:0:ffff:c001: 165. Actually, this is the address used in Section 3.3.2 (192.168.1.101) fit into the space reserved for IPv4 only routers or hosts.

For brevity, strings of zeros may be replaced with double colons (::), once only per address. With this space saver, the previous address can be specified as, ::ffff:c001:165. To further ease specifying addresses, the low-order 32 bits may be specified in dotted decimal format, just as is done with IPv4 (usual). Hence, with this gimmick, the address becomes recognizable as ::ffff.192.168.1.101. Moreover, if it is being used in an environment that is IPv6-compliant, the address becomes, ::192.168.1.101. Even simpler, the IPv6 loopback address can be represented as, ::1. Hence, most users will not observe a change when they commence using IPv6. That is, the configuration program will accept the normal IPv4 address (class C for example) and convert it as required.

3.5.5 Migration from IPv4 to IPv6

It would appear from the format of the IPv4 datagram that migration to IPv6 would be an easy task. Simply change the version number, and a router receiving a datagram would process either format. For several reasons, it is not that simple. For example, IPv4 and the transport layer are integrated to an extent that a simple replacement of IPv4 is not possible. (See Sections 4.1 and 4.2.4.8 for further discussion.) Also, upper layer programs are written to handle 32-bit IP addresses. Furthermore, IPv6 does not enforce the maximum packet lifetime (requires upper layer programs to handle this), and the list goes on: Algorithms that use the maximum IP header size must be changed to use 60 octets instead of the 40 octets for IPv4. A subtle difference that even the designers did not remember is probably still lurking in the dark. Suffice it to say that IPv6 is not backward-compatible with IPv4, and only small, isolated implementations have been made on the Internet backbone.

3.5.6 IPv6 Specifications

Although there is a specification for IPv6 (RFC 1883), it only provides the general features of the protocol and references other RFCs for specifics. Table 3.7 provides a list of the current RFCs describing specifics of IPv6.

Table 3.7
IPv6 References

RFC	Title
1881	IPv6 Address Allocation Management
1883	Internet Protocol Version 6 Specification
1884	IP Version 6 Addressing Architecture
1885	Internet Control Message Protocol (ICMPv6) for the IPv6
1897	IPv6 Testing Address Allocation
1924	A Compact Representation of IPv6 Addresses
1933	Transition Mechanisms for IPv6 Hosts and Routers
1970	Neighbor Discovery for IP Version 6 (IPv6)
1971	IPv6 Stateless Address Autoconfiguration
1972	A Method for the Transmission of IPv6 Packets Over Ethernet Networks
2019	Transmission of IPv6 Packets Over FDDI
2073	An IPv6 Provider-Based Unicast Address Format
2080	RIPng for IPv6
2133	Basic Socket Interface Extensions for Ipv6
2147	TCP and UDP Over IPv6 Jumbograms

Endnotes

[1] IP is part of the IAB official Standard Number 5. The IP portion of Standard #5 is defined in RFC 791. Standard #5 also includes: RFC 792 (Internet Control Message Protocol), RFC 919 (IP Broadcast Datagrams), RFC 922 (IP Broadcast Datagrams with Subnets), RFC 950 (IP Subnet Extension) and RFC 1112 (Internet Group Multicast Protocol).

[2] This definition of type of service field is consistent with RFC 1349, which updates many others.

[3] For the official list of protocol numbers, see RFC 1700, Assigned Numbers.

[4] Coding examples, including "C," of checksum calculation using different machines are provided in RFC 1071.

[5] See Router Requirements, RFC 1812.

[6] The Internet standard subnetting procedure is described in RFC 950. RFC 1219 describes techniques of assigning subnet numbers.

[7] Note that this IP address is taken from the private Class B IP address space per RFC 1597.

[8] The latest is published in RFC 1700.

[9]   The specification for multicasting is contained in RFC #1112, Host Extensions for IP Multicasting.

[10]   TCP and UDP with Bigger Addresses (TUBA) was described in RFC 1347. This was a Trojan horse to get OSI IP into the Internet. Dual network assignment was described in RFC 1335. Other RFCs either providing analysis or proposing schemes include 1365, 1366, 1367, 1375, and 1380. A scheme involving block assignment of Class C addresses (Supernetting) is described in RFC 1338.

[11]   NAT is described in RFC 1631, Network Address Translator.

4

Transport Layer Protocols

This chapter provides a description of the transport layer protocols UDP and TCP. The selection by an upper layer program to use either UDP or TCP is based primarily on the requirement for reliability. Some upper layer protocols were designed to operate with either UDP or TCP, while others use one or the other exclusively.

The selection by IP of either UDP or TCP is based on the protocol number in the IP header. That is, going up the protocol stack, IP passes its datagram to the transport layer program identified in the protocol field of the IP header. (See Section 3.2.) The TCP/IP designers used the protocol number in the IP header to demultiplex distinct services. Not all programs receiving datagrams from IP are transport layer programs—only UDP and TCP. For example, ICMP, IGMP, and OSPF are identified by protocol numbers in the IP header, but they function as a utility. The key distinction is that these programs are not called by upper layer programs. Hence, the ICMP, IGMP, and OSPF descriptions are deferred to Chapter 5.

4.1 User Datagram Protocol (UDP)

UDP provides upper layer programs with a transaction-oriented datagram type service. The service is similar to IP in that it is connectionless and unreliable. It is simple, efficient, and ideal for application programs such as TFTP and DNS. UDP provides a "no frills" datagram multiplexing service. An IP address is used to direct the user datagram to a particular machine, and the destination port number in the UDP header is used to direct the UDP datagram (or user datagram) to a specific application process (queue) located at the IP address. The

163

UDP header also contains a source port number that allows the receiving process to know how to respond to the user datagram. It effectively loads up one round with the entire fixed-length message, aims it at the intended receiver (IP address and destination port number), and, finally, fires one shot into the network. There is no acknowledgment, flow control, message continuation, or other sophistication offered by TCP (described in Section 4.2).

UDP operates at the transport layer and has a unique protocol number in the IP header (number 17). This enables the network layer IP software to pass the data portion of the IP datagram to the UDP software. UDP uses the destination port number to direct the data from the IP datagram (user datagram) to the appropriate process queue. The format of the UDP datagram is illustrated in Figure 4.1 [1]. Since there is no sequence number or flow control mechanism, users of UDP must either not need reliability or provide their own. The reliability of UDP is characterized by the phrase "send and pray."

- Source/destination port numbers: The source and destination port numbers, in conjunction with the IP addresses, define the end points of the single shot communication. The source port number may be equal to zero if not used. The destination port number is only meaningful within the context of a particular UDP datagram and IP address.

  There are some fixed, preassigned port numbers used for services on the Internet. For example, number 7 is used for the UDP echo

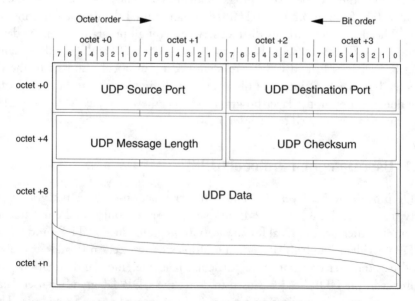

Figure 4.1 UDP datagram format.

server and number 69 is used for TFTP. These fixed, preassigned port numbers are referred to as, *well-known* ports and controlled by the IANA. There is also a group of *registered* port numbers, which the IANA does not control. As a service to the Internet community, the IANA publishes a list of *registered* port numbers [2].

- Length Field: The 16-bit length field contains a count of the total number of octets in the user datagram, including the header. The minimum value of the length field is 8, which is the number of octets in the UDP header.

- Checksum: Usage of the UDP checksum is optional. However, the field must be set to zero when not used. Since the IP layer does not include the data portion of the IP datagram in its checksum (it protects the IP header only), UDP has its own checksum to provide data integrity. The UDP checksum is the 16-bit one's complement of the one's complement sum of the UDP header, UDP data, and some fields from the IP header, as illustrated in Figure 4.2. If the computed checksum is zero, it is transmitted as all ones (the one's complement of zero).

  To include fields from the IP header, a pseudo UDP header that contains the source IP address, destination IP address, protocol number, and the length of the UDP user datagram (length field is redundant) is constructed. This is followed by the UDP header, UDP data, and a pad octet equal to zero if necessary to force 16-bit alignment.

  The 12-octet pseudo UDP header is used only during the calculation of the checksum. The receiver of a UDP datagram must reconstruct the same pseudo UDP header in order to verify the checksum.

  Including fields from the IP datagram header in the checksum protects against misrouted UDP segments but makes UDP dependent on IP. That is, a different network layer protocol, such as CLNS or IPX, could not be substituted for IP. This is a clear violation of the separation of functionality between layers. The designers made the compromise for efficiency in the dialog between UDP and IP to obtain the correct source IP address. (UDP could not know the correct source IP address when the host has multiple interfaces.)

4.2 Transmission Control Protocol (TCP)

TCP provides traditional circuit-oriented data communications service to programs. For those familiar with CCITT's X.25, TCP provides a virtual circuit for programs, which is called a connection. The communication on a connection is

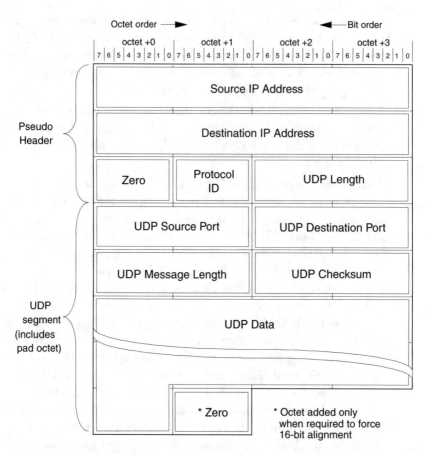

Figure 4.2 UDP checksum area.

asynchronous in that a segment sent does not have to be acknowledged before sending the next segment. Unlike upper layer applications using UDP, those using TCP enjoy a connection service between the called and calling programs, error checking, flow control, and interrupt capability. Unlike X.25, a connection can be initiated simultaneously at both ends and have the window size for flow control dynamically adjusted during the connection.

TCP uses the unreliable, connectionless IP in the same manner as UDP. That is, it has a unique protocol number (#6) in the IP datagram that signals IP to pass the data from the datagram (TCP header and data) to the TCP processing layer. The transmission unit for IP is called *datagram;* for UDP it is called *user datagram;* and for TCP it is called *segment,* or sometimes *packet.* The term *packet* was coined by Donald Davies and originally referred to the program *data unit* of the packet layer of an X.25 network. It is now used to describe most anything. The OSI reference model defined these terms in an unambigous manner (N-PDU for network and T-PDU for transport), but with the passing of the OSI stack the terms are not used.

4.2.1 TCP Transmission Control Block (TCB)

There are two interfaces between TCP and all other programs. The first is down the stack to the network layer. Since there is only one network layer program, IP, the interface is dictated by IP. The second is the interface to the user programs going up the stack. This interface varies with different operating systems; the general characteristics are described here.

Generally, the interface involves the user program calling a system routine that makes entries in a data structure called the *transmission control block* (TCB). The entries may initially be made into a hardware stack and transferred to the TCB by the system routine, which is implementation-specific. The entries in the TCB enable TCP to associate an application process with a particular connection in order to accept continuing commands from that application process and send them to the application process at the other end of the connection. TCP uses unique identifiers from each end of the connection to remember the association between two users. The user is provided with a connection name to use in making future entries in the TCB for the connection. The unique identifiers for each end of the connection are called sockets. The local socket is constructed by concatenating the source IP address with the source port number. The remote socket is constructed by concatenating the destination IP address with the destination port number. A socket for a well-known service such as TELNET or FTP may have many connections. However, the pair of sockets for one connection forms a unique number within the Internet. Note that UDP has the same set of sockets as TCP but they are not "remembered" by UDP. This is the difference between connection-oriented and connectionless. Each UDP user datagram transmitted is the first and last. It does not make a setup with the destination. An elobrate setup is made between a TCP sender and receiver before the first data carrying segment is transmitted. Furthermore, transmission of continuing segments after the first segment is based on the success of the previous segment transmitted. The following are generalized user commands specified in a system call.

1. Open: A passive open command causes the operating system to commence listening on a socket. An active open command causes a connection to be initiated. The variables associated with an open

command are local port, foreign socket, active/passive status, timeout, precedence, security, and options. The user is provided with a local connection name, which acts as a pointer into the TCB for issuing future commands.

2. Send: The send command causes the data in the specified buffer to be sent. A send command issued to a passive connection causes the connection to become active. The variables associated with a send command are local connection name, buffer address, octet count, push flag, urgent flag, and timeout value.

3. Receive: The receive command causes the operating system to assign a buffer for incoming data and normally to return control when the buffer is full or an error condition prevails. An error message is returned if the local name supplied has not been used with an open command. The variables associated with the receive command are local connection name, buffer address, and octet count. Control is given back to the user when the supplied buffer is full or an error return is made.

4. Close: The close command causes the connection specified to be closed. An error return can occur if the connection is not open or the user program is not authorized to close the connection. The only variable associated with the close command is the local connection name. The user may close the connection at any time, or in response to various prompts from TCP (timeout exceeded, remote close executed). When the connection is closed, all buffers associated with the connection are released, and the entry in the TCB for the connection is deleted.

5. Status: The only variable associated with the status command is the local connection name. The response message includes local connection name, foreign socket, receive window, send window, connection state, number of buffers awaiting acknowledgment, number of buffers pending receipt, urgent state, precedence, security, and transmission timeout. Some of this information may not be provided (implementation-specific) or may not be available depending on the state of the connection.

6. Abort: The abort command causes all pending send commands and receive commands associated with the local connection name to be aborted. Also, the user entry into the TCB is removed, and a special reset message is sent to the TCP entity at the other side of the connection. The only variable associated with the abort command is the local connection name.

During a connections lifetime, TCP remembers the state of each connection from information stored in the TCB. When a connection is opened, an entry in the TCB unique to the connection is made. A connection name is supplied to the user to enable directing commands to the connection. As conditions change for a connection, the status and other variables are changed in the entry for that connection in the TCB. When the connection is closed, the entry for the connection is deleted from the TCB. The states of TCP are described in more detail in Section 4.2.5.

4.2.2 TCP Connections

With all the added capability over UDP, one might expect the overhead for TCP to be greater than UDP. It is much greater. Not only is the header larger, connections must be established and terminated. The format of the TCP header is covered in Section 4.2.4, but a cursory review of Figure 4.3 will give the

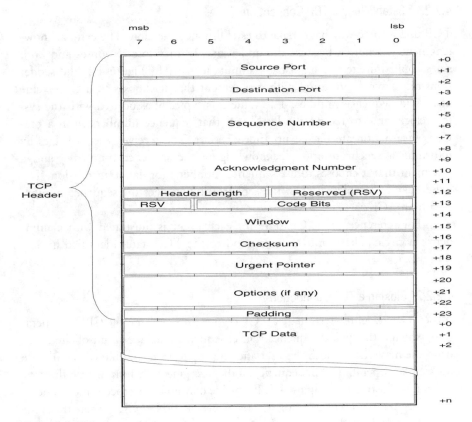

Figure 4.3 TCP segment format.

reader a general idea of the TCP header contents. This will aid in the understanding of TCP connections.

The source and destination port numbers in the TCP header identify
the application programs at each end of the TCP connection. As defined
above, a pair of sockets, one from each end of a connection, uniquely
identifies that connection. Another way of thinking of it—to illustrate the
multiplexing capability of TCP—is the pair of sockets identifies a unique connection thread through the Internet. The IP address in the IP datagram is
used to deliver the TCP segment to the correct machine. The protocol
number in the IP datagram directs the segment to TCP. The source and destination port numbers in the TCP header are used to direct the segment data
to the appropriate application layer entity (software program). Since the port
number in the TCP header is a 16-bit field, there could be, theoretically, up
to $65K^2$ connections between two peer TCP Layers using the same set of IP
addresses.

4.2.2.1 Establishing a TCP Connection

TCP uses a three-way handshake to establish a connection. The extra acknowledgment is required because the underlying IP may lose responses and both
sides could initiate the handshake simultaneously. In Figure 4.4, the sender
transmits a synchronization with a statement that it chooses to use the value
of x as its sequence number. The value x is typically associated with the system clock time to reduce the likelihood that sequence numbers from a previous incarnation are picked up after a restart. The receiver acknowledges by
returning a synchronization, acknowledging the sequence number x and a
statement that it chooses to use sequence number y for data transmission. The
handshake is completed when the sender acknowledges the sequence number
y with an ACK.

For reference, the TCP state for each event is illustrated with a number in a circle. The numbers correspond to the TCP states identified in Section 4.2.5.

4.2.2.2 Closing a TCP Connection

A user may gracefully terminate a TCP connection by using the CLOSE operation, which is the typical sequence. Other sequences include an unsolicited FIN
from the network, or both the user and the application simultaneously initiate
the CLOSE operation. The sequence of the user gracefully terminating the connection is illustrated in Figure 4.5. Each side continues to process and acknowledge segments received prior to receiving a FIN ACK. After this, the
connection is deleted.

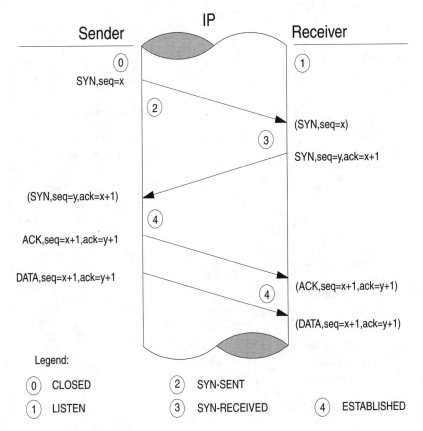

Figure 4.4 TCP connection sequence.

In Figure 4.5, the TCP state for each event is illustrated with a number in a circle. The numbers correspond to the TCP states identified in Section 4.2.5.

The modified three-way handshake on connection close is necessary to close both sides of the *full duplex* (FDX) connection. The sender has indicated that one side of the FDX connection will no longer be used (i.e., done sending) and waits for a similar statement from the other side. However, the other side only acknowledges the sequence number x and advises the application program of the desire to close the connection. Since this may involve an operator intervention, the receiver allows a large timeout before sending a FIN. Upon receiving confirmation from the application program, the receiver of the first FIN sends a FIN segment with sequence y and again acknowledges the sender's sequence number x (with $x + 1$). Note that the receiver's sequence number y did not change since the previous ACK did not occupy sequence

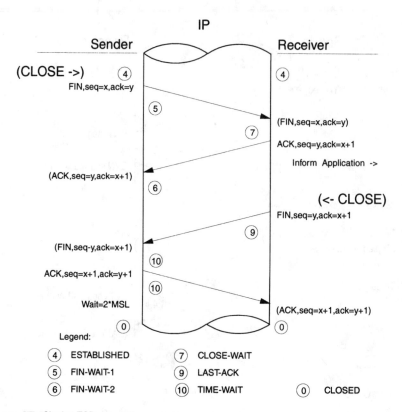

Figure 4.5 Closing TCP connection.

number space. The original sender receives this and acknowledges with an ACK, sequence number $x + 1$ and acknowledges the receiver's sequence number y (with $y + 1$). To guarantee that old segments from a previous incarnation are discarded, the original sender waits for four minutes (twice the maximum segment life of 120 seconds) and closes the connection [3]. Upon receipt by the receiver of the segment with sequence number $x + 1$, the connection has been gracefully closed.

4.2.2.3 Resetting TCP Connections

When it is necessary to abort a connection, the RST bit is set in the code field of a segment. The receiver of a segment with the RST bit set immediately aborts the connection and advises the application program, after the fact. This means that both sides of the FDX connection are aborted, and any buffers associated with the connection are released.

4.2.3 Flow Control

Side A of a TCP connection can control the amount of data that side B may send by sending the window field in the TCP header equal to the maximum number of octets that it has the capacity to receive. Side B uses this value to determine if side A has the capacity to continue receiving data, or if it must hold off until side A recovers buffer space. The window field tells side B that this is the maximum number of octets that may be sent before receiving further permission.

The TCP window is dynamically adjusted throughout a connection as compared to the window of the packet layer of CCITT's X.25, which may be negotiated only during connection setup with a facility code. (This should not be confused with the TCP option to set the maximum segment size, which also can only be set during connection establishment, described in Section 4.2.2.1.) Since congestion conditions vary during a connection, it is advantageous to be able to dynamically adjust the window.

If side B receives a segment containing a window equal to zero (closed), it cannot transmit data segments to side B (ACK and probe segments continue). A probe is a segment containing a single octet that is used to detect an unreachable host or network. The ACK segment is used to acknowledge received data. Thus, it is possible for a TCP peer to advertise a zero window while transmitting data and receiving ACKs. However, even when advertising a zero window, a TCP peer must process the RST and URG fields of all incoming segments.

The practice of advertising a large window and quickly shrinking it results in poor performance. The receiver of a window classifies it as the advertised window of the sender. The receiver then calculates a usable window by subtracting from the advertised window the number of octets already sent. The difference (equal to or less than the advertised window) may be much less than the sender can actually handle due to the asynchronous nature of the connection. This condition, called *silly window syndrome,* can reduce the throughput extremely. The situation is corrected by either or both of the following:

1. The sender advertising a window of zero instead of a small value;
2. The receiver compares the ratio of the advertised and usable windows and if less than some number, say 50%, treats the advertised window as zero [4].

Both large and small segments have, in the past, caused network congestion problems. A retransmission caused by a segment too large should not change the congestion window. To assure this, a *slow-start* algorithm, which permits only one segment to be retransmitted until acknowledgments begin to arrive, is used.

When retransmissions are frequent, the throughput of the network may be greatly reduced to the point called congestion collapse. Once reached, this condition is constant, being refueled by more retransmissions. The condition is typically caused by small (41 octet) packets made up of a single octet of user data and a 40-octet TCP header (TELNET hot-key input). To prevent this condition, TCP inhibits the sending of new segments if new data arrives from the user and previously transmitted data remains unacknowledged [5].

Should a TCP segment get lost, the intended receiver will either acknowledge a number less than has been sent, or nothing. Eventually the sender times out and retransmits the entire stream. The retransmission may be done synchronously—a positive acknowledgment for each segment sent. This is a cumulative acknowledgment mechanism. See Section 4.2.4.6 for a description of a new option that permits selective acknowledgment.

4.2.4   TCP Segment Format

The TCP segment consists of a TCP header and data. The header portion of the TCP segment is relatively fixed in size. The only optional field is the options field, which may necessitate a pad field to assure that the overall header length is a multiple of four-octet groups. The format of the TCP segment is illustrated in Figure 4.3. Sections 4.2.4.1–4.2.4.8 describe each of the fields in the TCP segment and general characteristics of TCP associated with each field description.

4.2.4.1   Source/Destination Port Numbers

Just as with UDP, the TCP segment has some fixed port numbers (typically the same for TCP and UDP). Each port number is a 16-bit unsigned integer; the fixed port numbers are identified in Appendix D. (See the discussion of the port mapper protocol in Section 6.4.)

4.2.4.2   Sequence Numbers

There are two sequence numbers in the TCP header. The first is the *send sequence number* (SSN). The SSN is a 32-bit unsigned integer and occupies bits +0 through +7 of octets +4, +5, +6, and +7 of the TCP header. The entire data to be sent from one program to another is called a stream. If a stream is too large for a single TCP path segment (this is typical), it is broken up into smaller segments. The SSN of the first TCP segment identifies the first octet of the entire stream. Assume this value is n, which was established when the TCP connection was made (see Section 4.2.2.1). Then, the value of the SSN of the second TCP segment is equal to $n + m$, where m is the octet displacement within the total stream to the beginning of the second TCP segment. In general, the SSN is an octet pointer within the total stream to the first octet of the particular TCP segment. The value of the field cycles between 1 and (2^{32} - 1). It provides each octet of a stream with a sequence number.

The second sequence number is called the expected *receive sequence number* (RSN)—also called the acknowledgment number. The RSN is a 32-bit field and occupies bits +0 through +7 of octets +8, +9, +10, and +11 of the TCP header. The RSN acknowledges the receipt of octets up to and including octet m - 1 by stating the next expected SSN of m.

For the reader familiar with either the frame or packet layers of the CCITT X.25 Recommendation, the SSN and RSN serve a similar function to the N(s) and N(r) numbers in HDLC or the P(s) and P(r) numbers in the packet layer, except that there is no implied window size associated with SSN and RSN. The unit number for the SSN and RSN is an octet, while the unit number for the N(s)/N(r) is a frame and the unit number for the P(s)/P(r) is a packet.

From the scenario above with the SSN of n for the first segment and $n + m$ for the second segment, the receiver of the first segment would send an ACK with the RSN equal to $n + m$, which acknowledges the receipt of octets n through $n + m$ - 1 by advising the next expected SSN is equal to $n + m$. Note that the acknowledgment process is accumulative, not selective. For example, an ACK can only acknowledge a group of octets, not an individual octet, excepting the case in which there is only one octet, or for the first octet of a group.

The TCP connection is FDX, and variables are maintained in the TCB for each direction of the connection to enable proper assignment of SSNs and to determine the validity of received RSNs—much the same as the variables V(S) and V(R) used by the X.25 frame layer. The variable SND.NXT is used to assign the next SSN. The variable RCV.NXT is used to keep track of the next expected SSN. The variable SND.UNA is used to keep track of the oldest un-acknowledged sequence number. The variable SND.WND is used to determine the range of new SSNs that may be sent, which is between SND.NXT and (SND.UNA+SND.WND1). There are several other variables maintained in the TCB for management of the connection, but they are not required for the simple example that follows. Considering only a one-way transmission from side A to side B, the variables SNDa.NXT, SNDa.UNA and RCVb.NXT will become equal if the connection is idle for a short period. This is illustrated in Figure 4.6. In Figure 4.6, the following assumptions (numbered as comments below) are made for ease of illustration:

1.   Only side A has data to send. ACK segments are used to acknowledge SSNs. ACKs would normally be piggybacked with data segments but would be more difficult to illustrate (and to follow).

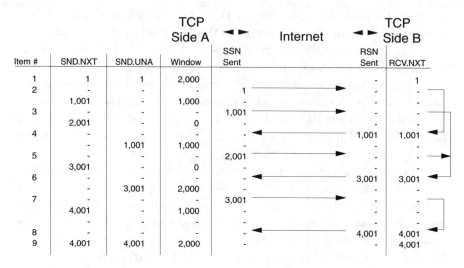

Figure 4.6 TCP sequence number example.

2. The randomly selected SSN is equal to 1.

3. The window size advertised by side B is 2,000.

4. Side A has a stream of 4,000 octets to send and divides the stream into equal 1,000-octet segments.

When a segment is sent, the sender advances SNDa.NXT by the length of the segment. When the segment is received, the receiver advances RCVb.NXT by the length of the segment and sends an acknowledgment. Upon receipt of the ACK, the sender advances the SNDa.UNA by the length of the segment. For the simple case with only a single segment, the three variables are equal.

The following comments are numbered the same as the item numbers in Figure 4.6.

1. Initial, conditions resulting from assumptions above.

2. Side A sends octets numbers 1 through 1,000 (first segment). The SSN of the segment is obtained from the variable SND.NXT. The variable SND.NXT is updated by adding to the present SND.NXT the length of the segment sent. The effective window is reduced by the segment length sent (2,000 - 1,000 = 1,000).

3. Since the window is still open, side A sends the second segment with SSN equal to 1,001. The SND.NXT is updated to 2,001 (by the length of the segment) and the effective window is set to zero (closed).

4. Side B removes the first segment (with SSN = 1) from its queue. It adds the number of sequence numbers in the segment to the RCV.NXT variable and sends an ACK segment with the RSN = RCV.NXT (1,001). Side B updates its oldest unacknowledged sequence number (SND.UNA) to the first sequence of the second segment and adds the acknowledged sequence numbers to the effective window (0 + 1,000).

5. With the effective window now open, Side A sends the third segment with the SSN = SND.NXT (2,001). The SND.NXT variable is updated to 3,001.

6. Side B processes both segments 2 and 3 and updates the RCV.NXT variable by the number of sequence numbers in both segments (adds 2,000). It sends an ACK segment with the RSN = RCV.NXT (3,001). Side A updates its SND.UNA and effective window by the number of acknowledged sequence numbers (2,000).

7. With the window full open again, Side A sends the last segment with the SSN = SND.NXT (3,001) and updates the SND.NXT variable (4,001) and the effective window (decreased by 1,000).

8. Side B processes the final segment. It updates the RCV.NXT and sends an ACK segment with RSN = RCV.NXT (4,001).

9. Side A updates its SND.UNA and effective window (again full open). The SNDa.NXT, SNDa.UNA and RCVb.NXT variables are again equal.

With the TCP variables identified above and the window field described in Section 4.2.6, each end of a connection is able to determine if a received RSN refers to a SSN sent but not yet acknowledged, if a received SSN plus the segment length are in the window, and if all sequence numbers of a segment have been acknowledged.

An acceptable received RSN is one passing the following inequality test:

SND.UNA < RSN   SSN to be sent next

The sequence numbers of a received TCP segment are in a valid range if the following inequality is true:

RSN expected   (SSN received + segment length - 1) < (RSN + window)

4.2.4.3 Header Length

The header length is a four-bit field and occupies bits +4 through +7 of octet +12. It contains an integer equal to the total number of octets in the TCP header, divided by four. That is, it represents the number of four-octet groups in the header. The value of the header length field is typically equal to five unless there are options. Since there may be options in the TCP header, the pad field is used to force the number of octets in the header equal to a multiple of four. There may be up to three octets in the pad field, each containing the value zero.

4.2.4.4 Code Bits

The purpose and content of the TCP segment is determined by the settings of the bits in the code bit field. The six bits are identified in Table 4.1 and are described as follows.

1. URG bit (bit +5): The URG code bit and the urgent pointer (octets +18 and +19) provide the sending program with the ability to bypass the normal stream with urgent data. Actually, the urgent data is delivered as received (in-band) but identified to the receiver as urgent data by setting the URG bit in the segment containing the urgent data. When the URG bit is set to one, the urgent pointer is used an octet displacement from the sequence number of the segment to the last octet of urgent data. That is, the data following the urgent pointer is the routine or nonurgent data. Depending on the *urgent mode* handling by the receiver, the urgent data could appear to the user as out-of-band data. When the URG bit is not equal to one, the urgent pointer is not used.

Table 4.1
Code Bit Field Definitions

Mnemonic Name	Bit Number	Meaning (when set)
URG	F5	Urgent pointer field is valid
ACK	4	Acknowledgment field is valid
PSH	3	This segment requests a push
RST	2	Reset connection
SYN	1	Synchronize sequence numbers
FIN	0	Sender reached end of octet stream

2. ACK bit (bit +4): When the ACK bit is equal to one, the ACK number in octets +8 through +11 is valid. This should always be the case after the connection is established. (See Section 4.2.3.)

3. PSH bit (bit +3): Although a transmit buffer may not be full, the sender may force it to be delivered. This procedure is often used with TELNET hot-key entry to create single octet segments, which is required by some applications.

4. RST bit (bit +2): Setting the RST bit in a segment causes the connection to be aborted. All buffers associated with the connection are released, and the entry in the TCB is deleted.

5. SYN bit (bit +1): The SYN bit is set during connection establishment only to synchronize the sequence numbers. (See Section 4.2.3.)

6. FIN bit (bit +0): The FIN bit is set during connection closing only. (See Section 4.2.4.)

4.2.4.5 Window

The window field is a 32-bit unsigned integer contained in bits +0 through +7 of octets +14 and +15 of the TCP segment. The window field is used to advertise the available buffer size (in octets) of the sender to receive data.

4.2.4.6 Options

The option field permits the application program to negotiate, during connection setup, characteristics such as the maximum TCP segment size able to receive. (Some future options may not be limited to connection setup.) Ideally, the TCP segment size would be the maximum possible without causing fragmenting. If the option is not used, any segment size is allowed. The TCP maximum segment size option is illustrated in Figure 4.7.

A type octet equal to zero signals the end of options. A type octet equal to one is used as a *NO OP*, or filler. The option for maximum segment size commences with an octet defining the type (equal to two), followed by an octet identifying the overall length of the option. (Including the type octet and length octet, the length field is equal to four.) The two octets that follow define the maximum segment length in number of octets.

New option codes are being added for scaled windows, timestamps, and *selective acknowledgment* (SACK). The scaled window option permits using a 32-bit window for calculation by shifting the transmitted window. The timestamps option permits the sender of a SYN segment to include a timestamp of the sending time and the remote TCP to indicate the sending time of the corresponding ACK. There is a SACK option being developed but details are still

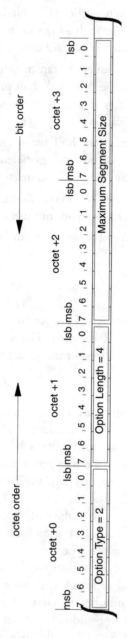

Figure 4.7 TCP maximum segment size option.

being investigated for its combination with the scaled window option [6]. The SACK mechanism permits the TCP receiver to advise the TCP sender of data received. From this, the TCP sender is able to retransmit only the missing data segments. (The normal accumulative acknowledgment mechanism would result in retransmission of all data segments commencing with the first missing.) Both the TCP sender and receiver agree to provide the mechanism with an option type 4 during connection setup. After the connection is established, the receiving side may send an option type 5 to initiate the retransmission of missing data segments [7].

Options have not typically been needed, or used, with TCP connections. However, with the increased speed of links, more new options are sure to follow.

4.2.4.7 Padding

The padding field, when present, consists of one to three octets, each equal to zero, to force the length of the TCP header to be in multiples of four octets. If options are not used, padding is not required. If options are used, padding may or may not be required.

4.2.4.8 Checksum

Since the IP Layer does not include the data portion of the datagram in its checksum (protects the IP header only), TCP has its own checksum to provide data integrity. The TCP checksum field occupies bits +0 through +7 of both octets +16 and +17, as illustrated in Figure 4.3. The checksum field of the TCP header is set to zero before the checksum calculation. The TCP checksum is the 16-bit one's complement of the one's complement sum of the TCP header, TCP data, and the pseudo header illustrated in Figure 4.8. Including fields from the IP datagram header in the checksum protects against misrouted TCP segments. If the computed checksum is zero, it is transmitted as all ones (the one's complement of zero).

To include fields from the IP datagram, a pseudo TCP header is constructed that contains the source IP address, destination IP address, protocol number, and the length of the TCP segment (header and data), which is followed by the TCP header, TCP data, and a pad octet equal to zero if necessary to force 16-bit alignment. (This should not be confused with padding for TCP options alignment.)

The pad used for check summing, when required, along with the entire pseudo header is used only during the calculation of the checksum and is not transmitted as part of the segment. The receiver of a TCP segment must, of course, extract the same information from the IP header in order to reconstruct the pseudo header, and possibly the pad octet, in order to verify the checksum received in octets +16 and +17. The received checksum is saved and the field

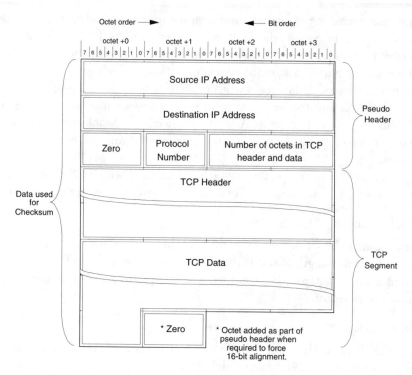

Figure 4.8 TCP checksum format.

replaced with zero prior to the recalculation of the checksum. Obtaining the source address, destination address, and protocol number is straightforward. The TCP segment length may be obtained by subtracting four times the datagram header length (bits +0 through +3 of octet +0) from the total length of the IP datagram (octets +2 and +3).

4.25   TCP States

The establishment, maintenance, and closing of a TCP connection requires TCP to remember state information, timers, and variables associated with each connection. As described in Section 4.2.1, this information for each connection is stored in a unique entry (one per connection) of the TCB. When a connection is opened, an entry in the TCB is made with all the variables, timers, and state information initialized. During the connection, the information in the entry of the TCB is updated to reflect the current status.

A TCP connection state changes from one state to another in response to events. The events may be user commands such as SEND, RECEIVE, and

STATUS. The events may also be expiring timers (e.g., user timeout or re-transmission timeout) or received code bits in a segment, such as RST, SYN, and FIN.

Since an entry does not exist in the TCB for nonexistent connections, the state used to describe no connection is *closed* and it is referred to as State 0. The following is a brief description of each of the TCP states.

0. CLOSED: Does not exist; for reference only;

1. LISTEN: Represents waiting for a connection request from any remote TCP;

2. SYN-SENT: Represents waiting for a matching connection request after having sent a connection request;

3. SYNRECEIVED: Represents waiting for a confirming connection request ACK after having both received and sent a connection request;

4. ESTABLISHED: Represents an open connection; data received can be delivered to the upper layer protocol (the normal state for the data transfer phase of the connection);

5. FINWAIT1: Represents waiting for a connection termination request from the remote TCP, or an ACK of the connection termination request previously sent;

6. FINWAIT2: Represents waiting for a connection termination request from the remote TCP;

7. CLOSEWAIT: Represents waiting for a connection termination request from the upper layer protocol;

8. CLOSING: Represents waiting for a connection termination request ACK from the remote TCP;

9. LASTACK: Represents waiting for an ACK of the connection termination request previously sent to the remote TCP (which includes an ACK of its connection termination request);

10. TIMEWAIT: Represents waiting for enough time to pass to be sure the remote TCP received the ACK of its connection termination request.

Progressing from one state to another is straight forward for the normal case. However, when error conditions prevail, the process becomes complex. Straightforward examples are illustrated in Sections 4.2.2.1 and 4.2.2.2. In these examples, the state number is illustrated in the associated figures (Figures 4.3 and 4.4) as circled numbers. The numbers correspond to the state numbers

1 through 10 described above. A states diagram is offered in the TCP specification. However, trying to determine the state changes by looking at a state table is like looking at a world map to find a street address [8].

Endnotes

[1]　The specification for UDP is contained in RFC 768, User Datagram Protocol.

[2]　*Well-known* port numbers in most cases are the same for UDP and TCP. The *well-known* port numbers are identified in Appendix C and defined in RFC 1340, Assigned Numbers.

[3]　The three-way handshake was designed for more than protection from an unreliable IP service. This was intended to protect against old duplicate segments from a previous incarnation. Recent analysis shows that old duplicate segments can still be mishandled when a RST segment is found during the wait period designed for old segments to die—the waiting period is truncated. The short-term fix recommended is to ignore RST segments during this time period (four minutes). The long-term fix may be to use a 64-bit sequence number. See RFC 1337, TCP TIMEWAIT Hazards.

[4]　The *silly window syndrome* was first explained by David Clark—see RFC 813. The RFC offered simple solutions to the problem, which are referenced in RFC 1122, Requirements for Internet Hosts Communication Layers.

[5]　See RFC 896, Congestion Control in IP/TCP Internetworks.

[6]　See RFC 1323, TCP Extensions for High Performance, with the status of *elective*. RFC 1323 obsoletes RFCs 1072 and 1185 (which defined other new options such as *echo* and *echo reply*), and updates RFC 793, the original specification for TCP.

[7]　See RFC 2018, TCP Selective Acknowledgment Options.

[8]　Fortunately, there is also a very detailed description of states versus events contained in the TCP Specification, RFC 793. Corrections are contained in the Requirements for Internet hosts Communication Layers Requirements, RFC 1122.

5

Internal Utility Protocols

This chapter provides a description of the internal utility protocols used in the Internet. These protocols operate from the Internet layer to the upper layer and are not directly accessible by the user. These protocols are normally used to aid the upper layer protocols and operate behind the scene of the typical user session. They are important but may only be used for special events such as for a particular interface type or error condition.

The ICMP is an integral part of IP but is only used in error or other special cases. The IGMP is only used for broadcast addressing, and the ARPs are only used to translate addresses. BOOTP describes the initial load process of a diskless workstation, and DHCP is an extension to BOOTP that dynamically assigns IP addresses. DNS provides a domain name to IP address translation service.

The dynamic routing protocols—RIP, OSPF, and BGP—automate the tedious task of building the RT for each router. The network administrator may manually construct the RT for a small intranet. This is unacceptable (near impossible) for the Internet, or even larger intranets. The dynamic routing protocols fall in two categories, as illustrated in Figure 5.1.

The first category of the dynamic routing categories is an interior routing protocol, called *interior gateway protocol* (IGP) that functions within a single AS—that is, within a single administrative domain using the same IGP(s). Within the IGPs there are two major types. The first is called a vector distance because it works in terms of the distance (in hops) between the source and destination. The *routing information protocol* (RIP) is a vector distance type IGP that is simple to implement and has been around for a long time. Although it continues to be enhanced with new features, the design limits its capability in large

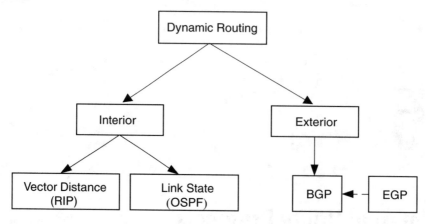

Figure 5.1 Dynamic routing categories.

networks. The second type is called a link state protocol because its information exchange with other routers contains status of links—that is, whether the links are up or down. Each router using a link state IGP solicits the status of its adjacent router(s) and broadcasts the status of each link (link state up or down) to all other routers. From the status of each link received, each router is able to update its RT of reachable destinations. The first link state type protocol was the *shortest path first* (SPF) protocol, which was developed by BBN and used in ARPANET for 10 years. The latest link state protocol is the OSPF protocol, which was designed as the replacement for RIP. The design weaknesses of RIP were used as a requirements document for OSPF. RIP is an Internet Standard (STD 34) with *recommended* status. The OSPF protocol is the dominant IGP in current large networks, and it will probably replace RIP in the future. It is presently a Draft Standard with *elective* status.

The second category of dynamic routing protocols, an exterior protocol that operates between IGPs, is called *exterior gateway protocols* (EGPs). The first EGP was called EGP, which was not a good choice of names for explanation. (The protocol name is the same as the name of the category.) EGP (the protocol, not the category) has been replaced with a new, more robust EGP called *border gateway protocol* (BGP). BGP Version 4 and OSPF Version 2 are both compliant with the requirements of CIDR, described earlier in Chapter 3.

SNMP provides a network management for the Internet. The echo protocol provides the diagnostic service of returning messages, and the *network time protocol* (NTP) provides a synchronized time service.

The architectural level of the utility protocols vary from the Internet layer to upper layer applications. Control and data flow for each will be identified.

5.1 Internet Control Message Protocol (ICMP)

The Internet is an autonomous system without central control. The ICMP provides a vehicle for software programs in intermediate routers and hosts to communicate with one another. It is not a protocol directly accessible by users, as upper layer programs are. For example, there are no ICMP commands. The communication is used to regulate traffic, correct routing tables, and check the availability of a host. Just as the IP datagram is encapsulated in an Ethernet frame, the ICMP message is encapsulated in an IP datagram. In that sense ICMP functions as a transport layer protocol, above IP. On the other hand, it does not provide end-to-end functionality as a transport layer protocol generally does. It is an integral part of IP, and its status is *required.* It provides a utility function with direct access to IP [1]. The ICMP has its own unique protocol number (number 1) enabling it to use IP directly. Each ICMP message has a type field, code field, ICMP checksum field, and other variable information depending on the type and code fields. Table 5.1 summarizes the ICMP message types. The type field value in an ICMP message identifies the function and format.

The echo request and reply messages, illustrated in Figure 5.2, use an identifier field and sequence number field that are not used by the other ICMP messages. It further has a variable amount of data (supplied by the user) that is

Table 5.1
ICMP Type Codes

Type Field	ICMP Message Type
0	Echo reply
3	Destination unreachable
4	Source quench
5	Redirect (change route)
8	Echo request
11	Time exceeded for a datagram
12	Parameter problem on a datagram
13	Timestamp request
14	Ttimestamp response
17	Address mask request
18	Address mask response

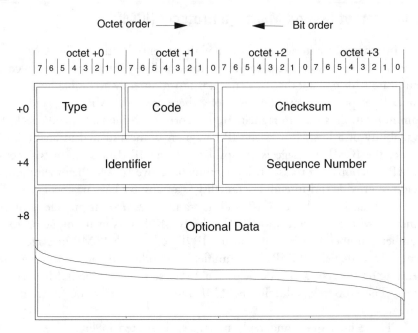

Figure 5.2 ICMP echo request/response format.

returned. The remainder of the ICMP messages are nearly identical in format and illustrated in Figure 5.3.

5.1.1 Echo Request and Reply Message

One host may verify that another host is operational by sending it an echo request. A recipient of an echo request returns it to the originator. The name given to this service application is packet Internet groper (referred to as Ping). The Ping application encapsulates the ICMP echo request (type = 8) in an IP datagram and sends it to the IP address. Intermediate routers forward it to the final destination. The recipient of the echo request switches the source and destination addresses in the IP datagram (see Figure 3.1), changes the type code to response (type = 0) and sends it back to the originator. Most implementations of Ping permit the user to specify a series of echo requests with variable data and a pause between echo requests. The format of the ICMP echo request is illustrated in Figure 5.2.

The type field in a request message is equal to 8 and equal to 0 in the response message. The code field is always zero. The checksum is calculated in the same manner as for the IP datagram, and the field value during the calculation is zero. The identifier and sequence number are used by the sender to

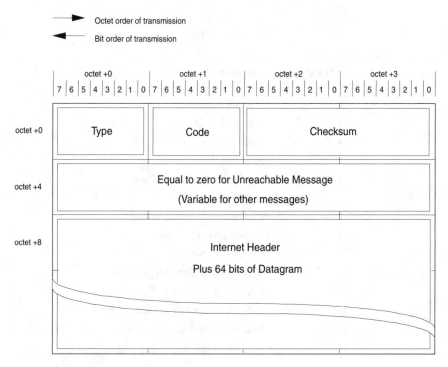

Figure 5.3 Format of unreachable message.

match replies with requests. The optional data field typically contains user-specified data (e.g., "Now is the time for , , , ") and is variable in length.

5.1.2 Reports of Unreachable Destinations

If a router cannot deliver a datagram to the destination address, it sends an ICMP error message to the originator. The format of the unreachable message is illustrated in Figure 5.3. The value of the type field is three, and the type of error is identified by the code field. The possible values of the code field, with the associated error condition, are illustrated in Table 5.2. Octets +4, +5, +6, and +7 are not used and are set equal to zero.

5.1.3 Flow Control

To handle a surge of IP datagrams, an intermediate router relies on a memory buffer pool. If the surge of traffic is prolonged and buffers become saturated, the router simply discards received datagrams until a safe buffer threshold is obtained. Each discarded datagram results in the router sending an ICMP quench

Table 5.2
Unreachable Codes

Code Value	Description
0	Network unreachable
1	Host unreachable
2	Protocol unreachable
3	Port unreachable
4	Fragment needed with DF set
5	Source route failed
6	Destination network unknown
7	Destination host unknown
8	Source host failed
9	Network administratively prohibited
10	Host administratively prohibited
11	Network service type not reachable
12	Host service type not reachable

message to the originator. This is notification that a message has been discarded. Note that IP, considered unreliable, is being used to send ICMP messages. When an ICMP message is discarded, an ICMP message is not generated.

Originally, the ICMP quench message was not sent until it was necessary to discard a message. By this time the system was overloaded and could ill afford to handle a retransmission. The algorithm was changed to send the ICMP quench message when the receiver used 50% of its buffer capacity [2].

The format of the quench message is identical to the ICMP unreachable message illustrated in Figure 5.3 except that the type field is equal to four and the code field is equal to zero.

5.1.4 Route Change

Routers contain the most current routing tables, and hosts must learn this from them. This permits hosts to be initialized with only a default route. When the default route is not the most efficient route, the router will send the host an ICMP redirect message containing the correct routing for the least number of hops to reach the desired destination. The most efficient route (e.g., least number of hops away) may not necessarily be the best route. A particular route, although it entails fewer hops, may not accommodate the MTU for some traffic. This is referred to as the *path MTU,* or PMTU. A host uses this information to update its routing tables [3]. The format of the ICMP redirect message is the same as the unreachable message format illustrated in Figure 5.3, except the type is equal to 5, the code is variable from 1 to 3, and octets +4 through +7 contain the correct IP address (instead of zero as in the unreachable message). The reasons for redirection and the associated code values are illustrated in Table 5.3.

5.1.5 Time-to-Live Exceeded

To prevent routing loops, the IP datagram contains a TTL field that is set by the originator. As each router processes the datagram, it decrements the field and checks the value for zero. When zero is detected, the gateway sends an ICMP error message to the originator and discards the datagram. The format of the error message is the same as the unreachable message illustrated in Figure 5.3 except that the type is equal to 11, the code is equal to zero (count exceeded) or one (fragment reassembly time exceeded), and the +4 through +7 octets are each equal to zero.

5.1.6 Parameter Errors

A parameter error may occur because the originator constructed the datagram incorrectly or because the datagram was corrupted. If a gateway finds an error in a datagram, it sends an ICMP parameter error message to the originator and discards the datagram. The format of the ICMP parameter error message is the same as the unreachable message illustrated in Figure 5.3 except that the type is equal to 12 and the code is equal to zero (pointer used) or one (pointer not used). The pointer is contained in the +4 octet of the ICMP header. Its value is an octet increment from the beginning of the datagram header to the offending octet. For example, the value 20 would indicate a problem with an option (see

Table 5.3
Codes for Redirection

Code Value	Reason for Redirection
1	For the host
2	For type of service and network
3	For type of service and host

Figure 3.1). When the error is a missing field, the code is set equal to one and the pointer is not used (octet +4 is equal to zero).

5.1.7 ICMP Timestamp Message

The ICMP timestamp message is a useful tool for diagnosing Internet problems and collecting performance measurements. The *network time protocol* (NTP) may be used for the source time in the timestamps and will achieve a milli-second clock synchronization. However, this is not a requirement. See Section 6.4 for other sources, such as the UDP time server. Figure 5.4 illustrates the format of an ICMP timestamp message.

The type field is equal to 13 for the originator and equal to 14 for the remote host (responder). The code is equal to 0. The identifier and sequence number are used to identify the reply. The originate timestamp is the time the sender initiated the transfer; the receive timestamp is the initial time of receipt by the receiver and the transmit timestamp is the time the receiver initiated the return of the message.

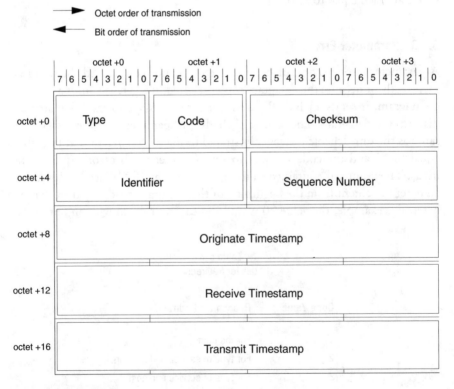

Figure 5.4 ICMP timestamp format.

5.1.8 Subnet Mask

Several examples of using a subnet mask were discussed in Section 3.3.3, but the manner in which a host obtains its subnet mask was not. The subnet mask is typically defined when the IP address is defined. Some hosts (for example, a diskless workstation) rely on a server to provide both the IP address and the subnet mask. (This is discussed later in Sections 5.4 and 5.6.) ICMP was enhanced (RFC 950) with the capability to provide the subnet mask for hosts that have only an IP address. The feature is not often used with current host and routers.

When a host needs to know the subnet mask for a physical LAN, it may send an ICMP subnet mask request to the attached router. If the host is connected to multiple routers, a request must be sent to each router because the subnet mask could be different. The request format is similar to the first eight octets of the ICMP timestamp request. The type value for the subnet mask request is 17, and it is 18 for the subnet mask response. The code value is zero, and the identifier and sequence numbers are used to identify the reply (i.e., associate requests with replies). The originator sets the four-octet field following the identifier and sequence number to zero, and the responder overlays the zeros with the subnet mask. The returning mechanism is similar to that of the echo request. This feature is not often used with current host and routers.

5.2 Internet Group Management Protocol (IGMP)

A multicast address is a Class D address as described in Section 3.3.4. This section describes the IGMP, which is a mechanism of maintaining the members of a multicast group address.

The IGMP is a *recommended* protocol that functions at a layer above IP, although it is considered an integral part (or extension) of IP. It has a protocol number (number 2) for using IP layer 3, in much the same manner as the ICMP uses IP layer 3. IGMP is used exclusively by members of multicast groups to maintain their status as members and to propagate routing information. A multicast router sends queries (type field in IGMP message header equal to one) once per minute (maximum). A receiving host responds with an IGMP message (type field in the header equal to two), which marks the host as an active member. A host not responding to the query is marked inactive in the multicast group routing tables [4].

The IGMP mechanism (four-byte header) with a type field equal to three is used to dynamically maintain source information of multicast members. This amounts to a subprotocol of IGMP since it uses ICMP's protocol number to receive data from IP. The subprotocol is called *distance vector multicast routing protocol* (DVMRP) [5]. DVMRP uses a request (subtype code equal to two) to solicit information, and the response uses a subtype code equal to one. The actual DVMRP commands (following the IGMP header) are in tagged data format and aligned on 16-bit boundaries. Information conveyed by DVMRP include the source addresses, subnet mask, metric, split horizon, and other route characteristics.

A more recent development of multicast addressing is with the dynamic routing protocol OSPF, described in Section 5.8. This protocol, called multicast extensions to OSPF (MOSPF) defines forwarding of multicast and unicast addressed datagrams within an autonomous system.

5.3 Address Resolution Protocol (ARP) and Proxy ARP

As described earlier, the link layer protocols provide an interface between IP and the physical layer interfaces. In addition, IP sends to and receives from the link layer protocols an IP datagram. One problem with this is that the destination IP address provided by the IP forwarder cannot be used by the physical medium. It requires a specific address format for that medium and in the case of Ethernet and token ring, that is a 48-bit MAC address. Since an IP address is only 32 bits in length, the mapping is not one-to-one. That is, one address can not be derived from the other.

The ARP was designed to handle the translation of a 32-bit Internet address to a 48-bit MAC address. After the IP forwarder supplies the destination IP address and it is determined that the destination is directly connected on an Ethernet or token ring network, a broadcast message is given to the link layer protocol containing a special type code that identifies it as an ARP message. The message is placed in a MAC frame with a MAC broadcast address. All stations will analyze the ARP message, and if one recognizes its own IP address, it will respond back to the originator with the correct 48-bit MAC address to associate with the particular IP address. The response message is also sent in a MAC broadcast frame and contains the same special type code to enable the ARP function at the original source to receive the information. This process would be very time consuming if performed for every message being sent. To reduce this type of activity, the originator stores the binding (IP address and associated MAC address) in an ARP cache. The ARP cache is always checked first before engaging in the ARP process of broadcasting the request. Since all stations received the ARP reply containing the binding, they may also make an entry in their ARP cache to preclude using the broadcast method. The ARP cache entries must be deleted periodically and relearned to allow for failed or removed stations. The ARP mechanism can be used for other conversions since the format is general and contains fields easily applied to other uses [6].

The ARP message is identified in an Ethernet frame by the type code field being equal to 0806 hex. As viewed by a machine interpreter, the ARP is an independent layer 3 protocol that gains control and data from layer 2 in the same manner as IP.

Proxy ARP is a variation of ARP whereby a gateway responds to a broadcast ARP request containing an address other than its own. It is also called promiscuous ARP. The proxy ARP does this when it recognizes the destination IP address in an ARP request that it can access, but the originator of the ARP request cannot. The proxy ARP knows this because the address in the ARP request is for a different network than the originating network. The originating ARP then sends datagrams to the proxy ARP, which must then forward them to the final destination [7].

Proxy ARP, which will only work on networks that support ARP, requires the use of subnet addressing [8]. The proxy ARP knows the originator of the ARP request is on a different physical network because the subnet numbers have been set up to correspond to the physical networks. In summary, a proxy ARP fakes out multiple hosts on different LANs (all attached to the proxy ARP) by responding to ARP requests with its own address. When a host addresses another host not on the same physical network, the proxy ARP takes the message and places it on the proper physical network, thus providing a transparent subnet gateway service.

5.3.1 ARP Message Format

The ARP message, which consists of 28 octets, is illustrated in Figure 5.5. The following field descriptions (and those in Figure 5.5) assume an Ethernet interface type and assume that bit +0 of the +0 octet (or +0 word if preferred) is the least significant bit, which is the first bit transmitted.

- Hardware type: The 16-bit hardware field identifies the hardware interface type. For example, the value for an Ethernet interface is one. The value in this field is used to determine the format of the hardware address field. The possible values are noted in Table 5.4. For example, if the value is 11, a LocalTalk address is identified with a 16-bit network

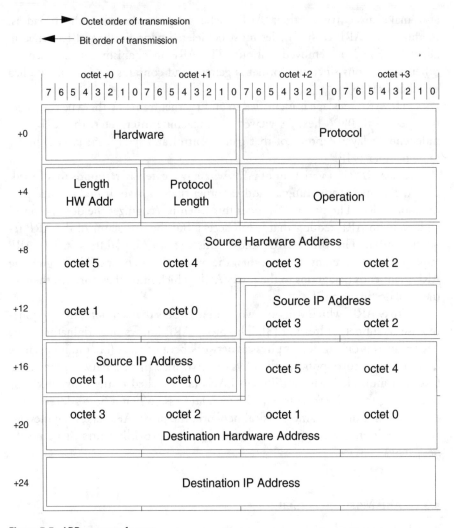

Figure 5.5 ARP message format.

address and an eight-bit node ID, or if the value is 7, an ARCnet address is identified with a field length of eight bits.

- Protocol Numbers: The 16-bit protocol field identifies the ether protocol used. For example, the value for an Ethernet interface is 0800 hex. The value is the same as the type code in the Ethernet frame. See Appendix D for a list of possible protocol numbers. Although all Ethernet protocol numbers are listed as possible values for the field, only those representing a hardware interface with specific address formats

Table 5.4
Hardware Interface Type

Type	Description
1	Ethernet (10 Mbps)
2	Experimental Ethernet (1 Mbps)
3	Amateur radio AX.25
4	Proteon ProNET token ring
5	Chaos
6	IEEE 802 Networks
7	ARCNET
8	Hyperchannel
9	Lanstar
10	Autonet short address
11	LocalTalk
12	LocalNet (IBM PCNet/SYTEK LocalNET)

for conversion will be found. For example, the Ethernet type code for ARP (0806 hex) will not be encountered because it does not have physical addresses.

- Length of hardware address: This eight-bit field identifies the length of the hardware address. The value is the number of octets. The typical value for Ethernet is 6, which provides for a 48-bit address, but it could be equal to 2 for a 16-bit address.

- Protocol length: The eight-bit protocol length field is used to define the length of the network address. For an IP network, this is a count of the octets required to contain the IP address; it is equal to four. Since both the hardware address length and the network address length fields are relatively large, the mechanism is reasonably unlimited.

- Operation: The 16-bit operation code field has the value of one (bit +0 of octet +7 equal to one) for a request and the value of two for a reply.

- Source hardware address: The source hardware address, source IP address, and destination IP address are completed by the sender, if known. The receiver adds the proper *destination hardware address* and returns the message to the sender with the *operation* field set equal to two.

The octets of the source hardware address are numbered in Figure 5.5 as they are transmitted, which is most significant octet first (octet 5) and least significant octet last (octet 0). Assembled in this order, they represent a 48-bit MAC address.

- Source IP address: The 32-bit source IP address field may contain a class A, B, or C IP address. The octets of the source IP address illustrated in Figure 5.5 are labeled in the order transmitted, with octet 3 being the most significant and octet 0 being the least significant.

- Destination hardware address: The interpretation of this field is the same as that for the source hardware address above. Note that the octet labeled *octet 5* is the most significant and the octet labeled *octet 0* is the least significant.

- Destination IP address: The interpretation of the 32-bit destination IP address is the same as the source IP address described above. Note that octet +24 of the ARP message header is the most significant of the destination IP address.

5.4 Reverse Address Resolution Protocol (RARP)

When a diskless host attached to an internet is initialized, it must learn its IP address. The diskless host sends a broadcast message encapsulated in an Ethernet frame, which is nearly identical to an ARP request. RARP interfaces directly with the link layer protocol in the same manner as ARP and the process and format of the RARP is similar. The delegated RARP server on the Internet responds to the RARP request with the corresponding IP address associated with the physical address supplied [9]. A host may use this mechanism to glean the IP address of any other host if it knows the MAC address (sender address different from target address). If a backup server is configured, it simply saves the time of the request, and should the backup server receive the same request again in a short time frame, it responds with the IP address. (It assumes the primary server is not in service.)

RARP servers providing responses that frequently require repeating for other clients may hold the responses in a cache to minimize the work and time to respond to future requests. If a cache is used, the entries are periodically deleted to force obtaining the current addresses from a RARP server.

The RARP message is identified in an Ethernet frame by the type code field being equal to 8035 hex. As viewed by a machine interpreter, the RARP is an independent layer 3 protocol that directly interfaces the link layer in the same manner as IP.

The format of the RARP message is identical to the ARP message. (See Figure 5.5 above.) The value of the operation field for a request is three, and the value for the reply is four.

5.5 Inverse Address Resolution Protocol (IARP)

Inverse ARP (IARP, also called InARP) is an extension to the basic ARP. Hence, the format is identical to that defined and illustrated for ARP (see Figure 5.5). It was designed primarily (although not exclusively) for frame relay networks [10].

When a new *data link connection identifier* (DLCI) is configured for a frame relay network in a router, it must somehow obtain an IP address associated with the DLCI to be sent messages. This may be done by manually configuring an address, or using IARP. When IARP is used, as soon as the DLCI is configured, the router sends an IARP to the DLCI (treated as a hardware address) and requests that the IP address of the recipient be returned. Note that the hardware type for frame relay is 15, the IARP request operation code is equal to 8, and the reply operation code is equal to 9. Since this is a nonbroadcast network, only the single station defined by the DLCI receives the IARP. If the receiver supports IARP, it replaces the zero filled IP address field in the request with its own IP address, changes the operation code from a request (8) to a response (9), flips the source and destination DLCIs, and returns the message to the requester. Most modern routers support IARP. If the remote neighbor router does not support IARP, it is necessary to manually associate each DLCI with an IP address.

When a router has more than one IP address (multihomed), it must return the IP address associated with the subnet containing the DLCI.

5.6 BOOTP and Dynamic Host Configuration Protocol (DHCP)

BOOTP and DHCP are upper layer programs that automate the initial loading and configuration of hosts. Instead of using the RARP server to obtain an IP address, a newly booted machine may use these programs to obtain an IP address, a bootable file address, and configuration information. There are other advantages to using these programs than obtaining bootable information. For example, an application program supplies the information instead of a network program, which makes it easier to revise the software if necessary. These programs may be used on networks that dynamically assign hardware addresses (which precludes using RARP). They also provide a centralized management of IP addresses and configuration files, which eliminates the need of per-host information files.

BOOTP and DHCP are actually the same program with different options. BOOTP was implemented in 1985, and new options were introduced in 1993. BOOTP enhanced with the new options is called DHCP. This provides an easy distinction between clients and servers with and without the new options. That is, those with the new options are called DHCP clients and servers [11]. DHCP offers the superset of options, and BOOTP has only a subset of the total options. Hence, a DHCP server can perform as a BOOTP server, but a BOOTP server cannot perform as a DHCP server. The architecture of BOOTP/DHCP is illustrated in Figure 5.6. For ease of description, the names BOOTP and DHCP will be used interchangeably—unless the description is specifically for one or the other.

BOOTP is an upper layer program that uses the transport layer UDP. The BOOTP server listens on *well-known* port number 67, and the BOOTP client listens on *well-known* port number 68. These are referred to as BOOTPS and BOOTPC respectively. Had only a single *well-known* port been used for both, a broadcast response from the server could unnecessarily initiate handling by other servers, only to find it is a response message. By using a different *well-known* port, the responses may be filtered (discarded silently).

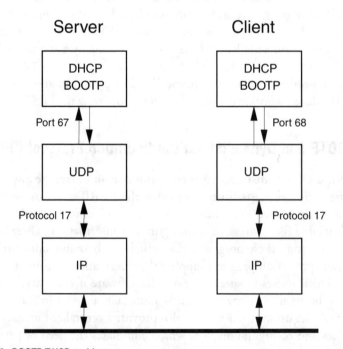

Figure 5.6 BOOTP/DHCP architecture.

To protect against corruption from the lack of robustness of UDP, BOOTP may set up the UDP datagram with *checksum required*. If being done in PROM, it may be desirable to omit the checksum to save PROM. All implementations must support a minimum length of 300 octets in the UDP data field. As with RARP, the request is sent with a limited broadcast if the BOOTP server is on the same network, or with a directed broadcast address if the BOOTP server is not local [12]. The directed broadcast IP address is equal to 255.255.255.255 in dotted decimal format and also noted as (-1,-1). (Minus one indicates ones compliment of zero, or each bit of the field is equal to one.) The directed broadcast IP address, which has the form (<network address>, -1), causes the message to be sent to each host on the specified network [13].

The client may include its IP address in the BOOTP request message if known. Otherwise, it is set equal to zero as a flag to the BOOTP server that replacement with the client's IP address is required. The BOOTP server returns a message containing the client's IP address and bootstrap information to the requester using a broadcast address. The broadcast address in the response message is used to preclude another program on the network from addressing the requester before it knows its IP address.

The boot process is a two-step operation, because the response message from BOOTP only contains the file name for the boot load file. Hence, to obtain the memory image required for the boot load, the client must use another protocol (typically TFTP, which listens on *well-known* port number 69). TFTP, which is more reliable for the transfer of a memory image, is described in Section 6.3.

5.6.1 BOOTP Sequence

The BOOTP sequence commences with a client, typically a diskless workstation, being booted. It initiates a BOOTP request with a broadcast address to all stations on the local segment with a minimum of its hardware (MAC) address. If the station knows its IP address, this is included. If a particular server is desired, a unicast address can be specified, if known, instead of the broadcast address. (See field descriptions in Section 3.3.) If the client is unable to receive a unicast message, it must set the broadcast flag in the BOOTP request message to make BOOTP servers aware of the limitation.

The BOOTP server monitors for BOOTP requests. The same format is used in the response message as the request message. The request type code sent by the client is changed to a response type code by the server. The server looks up the assigned IP address and puts it in the response message. It also adds the IP address of the BOOTP server and the name of the appropriate load file that may be executed. Depending on the implementation, it may also add other configuration parameters such as the local subnet mask, default routers, and Internet server addresses. The server either sends the response message with a broadcast address if the client station did not know its IP address, or as a unicast message if the IP address was known, as dictated by the broadcast flag in the BOOTP request message.

When the client diskless workstation receives the BOOTP reply, it uses the information supplied by the server to initiate a TFTP get message to the server specified. The response to the TFTP get message is an executable load file.

In Figure 5.7, a diskless workstation initiates a BOOTP request, which is noted as item 1 (number circled). The BOOTP server could also be located on a different network—colocated with the TFTP server, for example. If so, the BOOTP request is either sent there with a directed broadcast or with a BOOTP helper, called a relay agent, located in the attached router. A relay agent is an application layer program that listens on BOOTPS and relays the request to the appropriate BOOTP server.

Item 2 represents the response from the BOOTP server, which contains the information necessary to initiate a TFTP request for the actual load file. Item 3 represents the TFTP get command, which is recognized by the attached router and forwarded (item 4) to the router where it is sent to the TFTP server. The TFTP server sends a response message to the client (item 5), which contains the executable load and configuration file.

5.6.2   BOOTP/DHCP Field Descriptions

The format of the BOOTP message is illustrated in Figure 5.8. A description of each field follows. (Some refinements have been made for DHCP and noted.

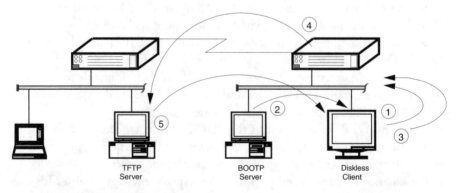

Figure 5.7 BOOTP/TFTP sequence.

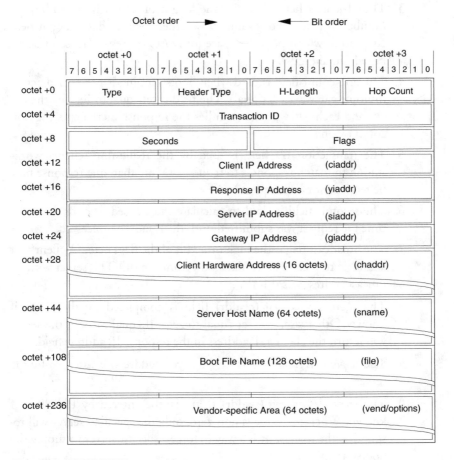

Octet order ⟶   ◀— Bit order

Figure 5.8 BOOTP/DHCP message format.

The power of the vendor extension field, or options field as defined by DHCP, is discussed in Section 1.1.3 [14].)

1. The type field identifies whether the message is a request or a response. The value equal to 1 indicates a request, and the value equal to 2 indicates a response.

2. The header type field identifies the hardware address type. This is the same type identified for ARP; a list of types is located in the Assigned Numbers RFC, currently RFC1700. For example, the hardware address type for Ethernet is equal to 1.

3. The H-length field identifies the length of its hardware address in number of octets. For example, Ethernet (type 1) has a length field equal to 6, which provides space for MAC addresses (12 semi-octet, hexadecimal digits).

4. The hop count field is used when bootstrapping is performed across multiple gateways (optional). The field is set equal to zero by the client and each gateway that handles the response increments the hop count by one.

5. The transaction ID field is used by the requesting (client) workstation to match responses to requests. (More than one response may be received for one request.)

6. The seconds field is used to calculate the elapsed seconds from the time the request was submitted until the response is received.

7. The broadcast flag is the high order bit of the flags octet. A client that is unable to receive an IP datagram from the BOOTP server with an unicast address sets this flag.

8. The *client IP address* (ciaddr) field is completed by the client if known. Otherwise, it is set equal to zero. If equal to zero, the server will return the client's IP address in the response IP address field.

9. The *response IP address* (yiaddr—for your address) is assigned by the server.

10. The *server IP address* (siaddr) field may be entered by the client if known. When the value is non-zero, only the specified server will respond to the request. It is a way to force a specific server to supply the bootstrap information.

11. The *gateway IP address* (giaddr) field is set to zero by the client (0.0.0.0), and if the request is handled by a gateway, it records its address in the field. This is not a client default router address. It is for BOOTP relay agents.

12. The client hardware address field is used by the client for its MAC address.

13. The server host name field is optional and may be set to zero by the client and server.

14. The boot file name field may be set to zero by the client or optionally set to a generic filename to be booted. For example, "UNIX" or "ETHERTIP." The server will replace the field with a fully qualified path and file name of the appropriate boot file.

15. The Vendor-specific (options) area field may have a key entered by the client. The server will use the key to access information on its database that can be used during the download of the memory image. Other possible information returned by the server may include configuration options and codes for the remote file access of the memory image [15].

5.6.3 Options

The real power of DHCP lies in the options field, called the vendor-specific field by the BOOTP specification. The options field is used to pass tagged data between the client and server. Each option is preceded by a single octet tag code (up to 256 options) and followed by a single octet length field. The length field permits the options to be of variable length. A zero tag code is reserved for padding the entire options field to attain 32-bit alignment, and the tag code 255 identifies the end of the options field.

The options field commences with a four-octet magic cookie number, which is 99.130.83.99 (in dotted-decimal format) [16]. The magic cookie number is followed by one or more options. Each option is identified by an option number (tag) and length field and followed by the desired parameter. For example, option number 26, which sets the interface MTU size, is illustrated in Figure 5.9. The length for this option is equal to two, which defines the next two octets as an unsigned, 16-bit integer. In this example the MTU size is 1,500 octets (or 5DC16, or 101110111002).

The first 128 options are nonproprietary, and the second 128 options are used for vendor-specific applications. While DHCP may use any of the options, BOOTP only uses a subset of the total options. In time BOOTP clients and servers will be phased out. There are only a few options that are used exclusively by DHCP clients and servers, as identified in Table 5.5 with an asterisk.

DHCP extends the number of type codes in the +0 octet (request and reply only) with eight new commands without actually changing the type code

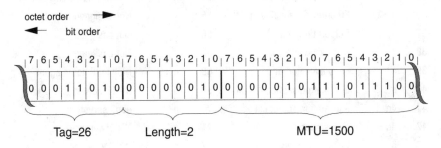

Figure 5.9 Option coding example.

Table 5.5
DHCP/BOOTP Options

Code	Length	Name	Code	Length	Name
1	4	Subnet mask	12	1-n	Host name
2	4	Time offset	13	2	Boot file
3	4n	Router (IP addresses)	14	n	Path to dump file
4	4n	Time server	15	n	Domain name
5	4n	Name server	16	4	Swap server
6	4n	Domain name server	17	n	Path to root disk
7	4n	Log server	18	1-n	Path name extension
8	4n	Cookie server	19	1	IP forwarding
9	4n	Lpr server	20	1	Nonlocal source routing
10	4n	Impress server	21	8-8n1	Policy filter
11	4n	Resource relocator	22	2	Reassembly size
23	1	Default time to live	*52	1	Option overhead
24	4	MTU aging timeout	*53	1	DHCP message type
25	n	MTU plateau	*54	4	Server identifier
26	2	Interface MTU	*55	n	Parameter equest list
27	1	Subnets are local	*56	n	NAK message
28	4	Broadcast address	*57	2	Max dhpc message size
29	1	Mask discovery	*58	4	IP address renewal time
30	1	Mask icmp response	*59	4	Rebinding time
31	1	Router discovery	*60	n	Vendor class identifier
32	4	Router solicit address	*61	n	Type client identifier
33	8n	Static route list	62	-	Not assigned
34	1	Trailer encapsulation	63	-	Not assigned
35	1	ARP cache timeout	64	n	Inform service and domain
36	1	Ethernet encapsulation	65	-	Not assigned
37	1	Default time to live	66	-	Not assigned
38	4	Keep alive interval	67	-	Not assigned
39	1	Keep alive garbage	68	n	Mobile IP home agent
40	n	Information service domain	69	n	SMTP server
41	4n	Information server list	70	n	pop3 server
42	4n	NTP server	71	n	NNTP server
43	n	Vendor-specific	72	n	WWW server

Table 5.5 (continued)
DHCP/BOOTP Options

Code	Length	Name	Code	Length	Name
44	n	netbois-tcp/ip name server	73	n	Finger server
45	n	netbios-tcp/ip distr server	74	n	Internet relay server
46	1	netbios-tcp/ip node type	75	n	Streetalk server
47	n	netbios-tcp/ip scope	76	n	Streetalk dir assist server
48	n	x window font server	77	-	Not assigned
49	n	x window display manager	78	-	Not assigned
*50	4	Requested ip address	79	-	Not assigned
*51	4	Resource relocator	80	-	Not assigned

*DHCP only.

field. The new type codes, called message types, are defined with an option code equal to 53, which is identified in Table 1.1. This was a convenient way to enhance the operation and maintain backward compatibility with older BOOTP clients and servers. BOOTP clients are not aware of the option, and a BOOTP server receiving an option code equal to 53 treats it as an unknown option code. A BOOTP request is identified by a DHCP server by the absence of option code 53. The DHCP server simply provides BOOTP service to the client. The new message codes provided by option code 53 are identified in Table 5.6, and a description of each follows.

1. Discover: A broadcast sent by a client to locate servers and solicit (or confirm an existing) an IP address and/or configuration parameters. It may be necessary for a relay agent to forward the message to the DHCP server.

2. Offer: Response message from the server to the requesting client. Since the solicitation was sent as a broadcast, there may be multiple offer messages. The client must evaluate the offers and select a server.

3. Request: The client sending a request message to the server concludes the shopping. This binds the client and a specific server.

4. Inform: The shopping may also be concluded by the client sending an inform message, which is used to request only configuration parameters because an IP address is not needed.

Table 5.6
DHCP Supplemental Message Types

Value	Message
1	Discover
2	Offer
3	Request
4	Decline
5	ACK
6	NAK
7	Release
8	Inform

5. ACK: The server assigns an IP address (or other required action such as extending the lease) and provides the configuration parameters with an ACK message. If for an IP address request, it is effective upon receipt by the client for the length of the lease. Whether the request is for a new IP address or confirmation of a supplied IP address, the server confirms the uniqueness of the IP address with an ICMP echo request.

6. NAK: The server could send a NAK message in response to the request if the format of the request message was invalid.

7. Decline: The server may send a decline message if, for example, the requested IP address is already in use.

8. Release: The client checks the validity of an assigned IP address from the server by using the ARP. Assuming it is valid, the client will use the IP address for up to the lease period and then send a release message to the server. The server will void any remaining part of the lease period and make the IP address available to other clients by returning it to a pool.

The additional option codes and message types give DHCP the ability to assign an IP address for a finite lease and reassign it later to another client. DHCP is a one-stop source for all IP configuration information needed for operation.

5.7 Routing Information Protocol (RIP)

The first IGP used by early routers, called *Interface message processors* (IMPs), was called *gateway-to-gateway protocol* (GGP). An IMP would periodically broadcast to all other gateways in its autonomous group an RT containing the destinations that it was able to address with the associated vector distance for each (vector distance in the form of number of hops). Gateways able to receive the routing table assumed that it could reach any IMP that the sender could reach. The hop count to the sender was added to the hop count of the advertised route for the new metric. The process could take minutes before all IMPs were able to obtain an accurate RT. This is called the convergence time. The mechanism is based on the Bellman Ford algorithm, which shows convergence on the optimum path with the limited update between neighbors only. Because of the amount of time required for convergence, the maximum number of hops that can be supported with the vector distance protocols is 15. (The amount of time is not linear with regard to the number of hops.)

RIP, the replacement for GGP, is an IGP developed several years after GGP. It is based on Xerox's *Palo Alto Routing Information Protocol* (PRIP), which was generalized to cover multiple network families and called RIP. The UNIX implementation of RIP, which is called *routed*, was originally part of the Berkeley UNIX distribution. This made it a de facto standard for many universities [17].

5.7.1 RIP Operation

The basic operation of RIP has not changed, although several enhancements have been made. RIP sends a routing update message every 30 seconds (at most) that contains pairs of IP network addresses and an integer distance vector (hop count). The hop count ranges from one to a maximum of 15. A hop count equal to 16 is considered not reachable and referred to infinity. To sense topology changes, a router invalidates any route in its route table that has not been reaffirmed for more than 180 seconds. Hence, it can take several minutes for routers to correct their route tables when either an interface or other router fails.

A major problem with RIP is its limitation to a maximum of 16 hops. The limitation is caused by the slow convergence rate of the vector distance algorithm and not a field size restriction. Novell IPX running RIP is impaired for the same reason.

Possible routing loops that may be formed during the convergence period are handled by the TTL option of the IP datagram (see Section 3.2). To minimize possible loops and reduce the convergence time, routers never send RT information to an interface if that interface was the source of the information. This is called split horizon—suggesting above and below the horizon as legal or illegal destinations for routing information. On point-to-point links (e.g., PPP) and nonbroadcast multi-access NBMA networks (such as frame relay) the restriction is to a configured neighbor. A neighbor is no more than one hop away, as in a PPP link. When RIP neighbors are configured, RIP updates are only accepted from a neighbor—not the interface. A broadcast network (e.g., Ethernet) permits RIP to accept updates from any RIP router attached to the same network. Updates on broadcast networks may be sent as directed broadcasts, multicast (RIP Version 2 only), or unicast. (See Section 3.3.)

A variation of split horizon is to send a route entry for a router that it was received from, except with the hop count equal to 16. Since a hop count of 16 is considered not reachable, it reduces the convergence time. The specifications refer to this as split horizon with poison (hop count equal to 16).

The frequent RIP updates (every 30 seconds) has been a problem with RIP from the beginning. It used too much of the available bandwidth of WAN links and inhibited on-demand services like ISDN from going idle. An enhancement to RIP, called *on demand,* limited the sending of routing updates for WAN links to only when there was an actual change in the route table or the status of an interface. To further improve the operation of RIP, a new feature called *triggered RIP* limits the amount of data in the routing update to only the changed data. Routes learned for WAN links are permanent until an actual change, while routes learned for LAN continue to be timed out if not refreshed.

5.7.2   RIP Architecture

RIP functions two layers above IP since its messages are encapsulated in UDP datagrams. Unlike the other IGPs that use IP directly with the protocol number 9 in the IP header, RIP uses a UDP header with the destination port number equal to 520 for requests. (See Figure 5.10.) The response has the source port number equal to 520 to identify it as a RIP response. Unsolicited routing update messages have both the source and destination port numbers equal to 520 [18].

5.7.3   RIP Message Format

The format of RIP messages is illustrated in Figure 5.11. A description of each field follows.

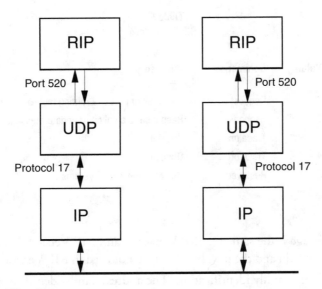

Figure 5.10 RIP architecture.

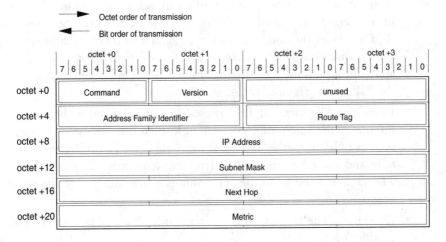

Figure 5.11 RIP message format.

- Command: The command field identifies the purpose of the RIP message. There are four commands used by RIP and one reserved for vendor-specific purposes. The commands with the associated field value of each are illustrated in Table 5.7.

- Version: The value of the version field is equal to one for RIP version 1 and equal to two for RIP version 2. Note that the size of the RIP

Table 5.7
RIP Commands

Value	Commands	Purpose
1	Request	Solicit all or part of a routing table
2	Response	Response to a solicit or update message
3	Trace on	Obsolete
4	Trace off	Obsolete
5	Reserved	Used by Sun Microsystems

message is the same for RIP Version 1 and RIP Version 2. The subnet mask field and the next hop field are not used in RIP Version 1.

- Address family identification: The address family identification field is equal to two for both RIP versions 1 and 2. RIP version 2 identifies an authentication with one entire entry in the RIP message. When authentication is used, it is identified by setting the address family identification field equal to FFFF hex. The value of the next 16-bit field identifies the authentication type (equal to two for simple password), followed by 16 octets of a password field. Note that when authentication is used, the maximum number of information groups in the RIP message is equal to 24.

- Route tag: The route tag field is used to identify whether the route is internal to the autonomous system, or learned from an external source such as BGP or another IGP. The route tag is used with RIP version 2 only and must be preserved and propagated when being advertised. The value may be set to the autonomous system number of the route originator.

- IP address: The IP address is the standard, 32-bit address defined in Section 3.3.

- Subnet mask: The subnet mask is only used by RIP version 2. This enhancement enables intermediate routers to know the subnet mask that will be applied by the destination router, if any, and to perform route aggregation.

- Next hop: The next hop field may contain an IP address that provides an optimal path to the destination. If none is provided, the field is equal zero (0.0.0.0). If the field is non-zero and the IP address is not directly connected, it should be treated as zero.

- Metric: The number of routers that must be crossed in order to reach the destination IP address. All routers will add this number to the number of hops to the advertiser of the route in order to determine its metric to the destination IP address. Should the router receive a route to the same destination with a smaller metric (including its own to the source or the advertisement), it will switch paths.

5.7.4 Reference Material

The ISOC Standard #34 only identifies RFC 1058 for RIP. Other related RFCs are identified in Table 5.8.

Another common problem to IGPs using vector distance routing is characterized by a two-stage oscillation. The first stage is when the routers discover a shorter path and expedite traffic onto the new, faster path. The second stage is when they discover that the new path is slower than the old path (because of all the new users) and all return to the old path, making it slow again (much like drivers on the freeway).

5.8 Open Shortest Path First (OSPF)

The newest IGP is OSPF, which is available in published literature (no license fees). OSPF is a link state protocol that was first used during the ARPANET era

Table 5.8
RIP-Related RFCs

RFC	Title
2092	Protocol analysis for triggered RIP
2091	Triggered extensions to RIP to support demand circuits
2082	RIP-2 MD5 authentication
1724	RIP version 2 MIB extension
1723	RIP version 2 carrying additional information
1722	RIP version 2 protocol applicability statement
1721	RIP version 2 protocol analysis
1582	Extensions to RIP to support demand circuits
1581	Protocol analysis for extensions to RIP to support demand circuits
1388	RIP version 2 carrying additional information
1058	Routing information protocol

and is the basis for all other link state protocol, including *NetWare link services protocol* (NLSP). (Link state refers to the status of a link, and not to the state or stateless protocol operation.)

As mentioned in the introduction to Chapter 5, the weaknesses of RIP were used as a blueprint to implement OSPF. It offers a much higher degree of sophistication with many new features, and OSPF version 2 is CIDR-compliant [19]. The features of OSPF are described as follows.

1. Paths based on both types of service and distance;

2. Load balancing of multiple paths with equal metrics;

3. Subnetting within an autonomous group;

4. Authentication schemes to prohibit a bogus advertisement of a non-existent route;

5. Host- and network-specific routes;

6. Designated routers to minimize broadcast messages ($<<n2-n$) and enhance security;

7. Virtual network paths (may cross a transient subnet);

8. Exchange of routing information learned from other autonomous systems;

9. Multicast capabilities for bandwidth reduction;

10. Quick detection of topological changes and a fast convergence time, relative to RIP

5.8.1 OSPF Architecture

OSPF operates at the transport layer and places its messages directly in IP datagrams. It receives messages in the same manner—that is, directly from IP with protocol number 9, as illustrated in Figure 5.12.

OSPF was originally named OSPFIG, and the protocol number 89 was allocated. It may now be recognized on a monitor by the protocol number 9 in the IP header. (See Section 3.2.)

5.8.2 OSPF Operation

Each router in an autonomous system using OSPF maintains a *link state database* (LSD) that describes the entire system topology, and each LSD is identical. A router's usable interfaces and reachable neighbors are reflected in the LSD,

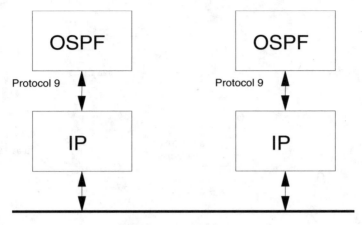

Link and Physical Layers

Figure 5.12 OSPF architecture.

which is flooded throughout the autonomous system. Each router sees itself as the root of the LSD.

When multiple equal-cost routes to a destination exist, traffic is distributed equally among them. The cost of a route is either the same as the interface cost (within the AS) or artificially set high to favor an internal path. It is a dimensionless metric that may be established by the network administrator taking into account the type and speed of the interface. For example, the interface cost may be derived by dividing the link bit rate into an arbitrary number, say 100M. Then for a 10-Mbps Ethernet interface the cost is 10.

Within an OSPF autonomous system, separate areas may be configured to minimize routing table size, minimize broadcasts, and enhance security. When multiple areas are used, the LSDs within each area are identical instead of the AS. There are four types of areas, four types of routers, and seven types of *link state advertisements* (LSAs), as illustrated in Figure 5.13. Each area has a unique 32-bit address, which is specified in dotted-decimal format. (It is not an IP address, although a portion of an IP address may be used for identification.) A description of the four area types follows.

- Backbone area: The backbone consists of those networks not contained in any area, their attached routers, and those routers that belong to multiple areas. The backbone must be contiguous, except for the exception. It is possible to define areas in such a way that the backbone is

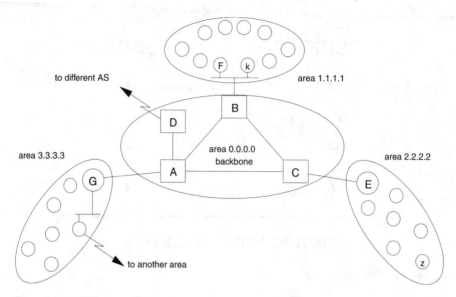

Figure 5.13 OSPF areas and routers.

no longer contiguous. In this case the system administrator must re-store backbone connectivity by configuring a virtual link.

Virtual links can be configured between any two backbone rout-ers that have an interface to a common nonbackbone area. Virtual links belong to the backbone. The protocol treats two routers joined by a virtual link as if they were connected by an unnumbered point-to-point network. On the graph of the backbone, two such routers are joined by arcs whose costs are the intra-area distances between the two routers. The routing protocol traffic that flows along the virtual link uses intra-area routing only.

The backbone is responsible for distributing routing information between areas. The backbone itself has all of the properties of an area and is identified with the OSPF area number 0.0.0.0. The topology of the backbone is invisible to each of the areas, while the backbone itself knows nothing of the topology of the areas.

- Transit area: A transit area is one capable (and consenting) of originat-ing and propagating OSPF external link state advertisements (LSAs). In this case it is an LSA type 5, defined later. The LSAs are originated by an AS border router and propagated throughout the OSPF routing domain. The backbone area is by definition a transit area and will al-ways pass traffic. A transit area will only pass traffic if it has a virtual link or an AS border router. Aggregation of networks within an area may be performed by an *area border router* (ABR).

- Stub area: A stub area is not capable of propagating or originating external LSAs. Routing from these areas is done with a default route, which reduces the size of the LSD for routers internal to the stub area. Stub areas have the same path to and from the backbone and there is only one. Virtual links cannot be configured through stub areas and cannot have an AS boundary router.

- NSSA area (not-so-stubby-area): NSSA areas are the same as stub areas except that they are capable of aggregating and advertising routes similar to a transit area.

Two routers that have interfaces to a common network are neighbors and when they exchange routing information an adjacency is formed. (Not all neighbors have adjacency.) The types of routers in an OSPF AS are the following.

1. Backbone router: A backbone router is one that has an interface to the backbone. These routers have the router ID equal to 0.0.0.0. Routers A, B, C, and D in Figure 5.13 are backbone routers.

2. Internal router: An internal router is one that has all its interfaces in the same area.

3. *Designated router* (DR): Each broadcast and NBMA network must have a DR. The DR is elected by the hello protocol of OSPF and generates network LSAs for all internal routers in the area. The DR becomes adjacent to all routers in the network. (The process of adjacency is explained in the next paragraph.)

4. Backup designated router: Each broadcast and NBMA network also has a backup DR. It forms the same adjacencies as the DR and is essentially a hot-standby for recovery.

5. ABR: An ABR has an interface to a router in at least one other area. Routers E, F, and G in Figure 5.13 have an interface in their own area and the backbone area. Hence, they are both backbone routers and ABRs. Backbone routers can optionally be ABRs.

6. *AS boundary router* (ASBR): An ASBR has an interface to one or more other ASs. It exchanges information with routers belonging to other ASs. Router D in Figure 5.13 is an ASBR. An ASBR may be an ABR and may or may not have access to the backbone.

Routers either auto discover their neighbors or must have their neighbors' IP addresses manually configured. On point-to-multipoint networks the neighbors may be discovered with IARP. An adjacency is formed by the routers exchanging hello packets and synchronizing their databases. Remember that the LSD of routers in the same area must be identical, or synchronized. Hello packets initiate forming the adjacency relationship. A database exchange process is entered after the initial exchange of hello packets. During this process, *database description* (DD) packets are exchanged to determine which router is the master (declares sequence numbers to be used) and which has the most current LSAs. A list of noncurrent LSAs is made by each router and link state request and link state update packets are exchanged until synchronization is attained. The two routers are then fully adjacent. The process is summarized by the neighbor's state. That is down, initialization (hellos), exstart (determining master/slave relationship), exchange (determining most current LSD), loading (LSA requests and updates), and full (adjacency and synchronized).

Each LSA type has a different function, and together they provide the information to construct and maintain the LSD. The seven LSA types, identified in Table 5.9, can be either router, network, summary, or group membership functions; they are described in the following.

1. Type 1 LSAs describe the collected states of the router's interfaces. The LSA is flooded throughout the area only. The LSA identifies whether the router is an ABR or ASBR.

2. Type 2 LSAs describe the set of routers attached to the network. The LSA is initiated by the designated router for transit and NBMA net-

Table 5.9
LSA Types

LSA Type	Function	Description
1	Router	Router interface states
2	Network	Routers attached to the network
3	Summary	Routes to network
4	Summary	Routes to ASBR
5	Summary	Routes external to the AS
6	Group-membership	Multicast group with MOSPF
7	Summary	External routes carried within NSSA

works. It is broadcast throughout the area that contains the transit or NBMA network only.

3. Type 3 LSAs describe inter-area routes to networks. The LSA is initiated by an ABR and flooded throughout a single area only. The destination described is a network external to the area but not to the AS.

4. Type 4 LSAs describe inter-area routes to an AS boundary router. These are the same as type 3 LSAs except that the destination described is an ASBR.

5. Type 5 LSAs describe routes to destinations external to the AS. Type 5 LSAs are initiated by an ASBR and describe destinations external to the AS.

6. Type 6 LSAs are specific to an OSPF area and are never flooded out of the area of origination. They describe the location of multicast destinations [20].

7. Type 7 LSAs describe external routes carried within an NSSA. The function of a type 7 LSA is the same as a type 5 LSA, except that type 7 LSAs are advertised only within a single NSSA, may be initiated by and advertised throughout an NSSA (NSSAs do not receive or originate type 5 LSAs), and have an internal flag (propagate bit) that enables the ABR to translate the type 7 LSA into a type 5 LSA [21].

The type of router sending LSAs, the area being sent from and to, and the format of each are described in Section 5.8.3.

When routing a packet between two areas, the backbone is used. The path that the packet will travel can be broken up into three contiguous pieces: an intra-area path from the source to an area border router, a backbone path between the source and destination areas, and then another intra-area path to the destination. The algorithm finds the set of such paths that have the smallest cost.

Looking at this another way, inter-area routing can be pictured as forcing a star configuration on the AS, with the backbone as hub and each of the areas as spokes.

The topology of the backbone dictates the backbone paths used between areas. The topology of the backbone can be enhanced by adding virtual links. This gives the system administrator some control over the routes taken by inter-area traffic.

The correct area border router to use as the packet exits the source area is chosen in exactly the same way routers advertising external routes are chosen. Each ABR in an area summarizes for the area its cost to all networks external to the area. After the SPF tree is calculated for the area, routes to all other networks are calculated by examining the summaries of the ABRs.

5.8.3 OSPF Packet Formats

The OSPF message format is considerably more complex than the RIP format. It has a fixed 24-octet header plus a variable portion depending on the type of message. There are five different types of messages, each with a header that follows the fixed, 24-octet header. The message types that carry an LSA also have a fixed 20-octet header and a variable header for each of seven LSA types. This format architecture is illustrated in Figure 5.14.

As seen from the OSPF format structure, there are many formats. There are two variables, and both are type codes. The key to finding the format of the desired packet type is to follow the type codes. The formats illustrated in a bold box have detailed illustrations later. (For ease of illustration, the OSPF fixed header type codes are shown out of sequence.)

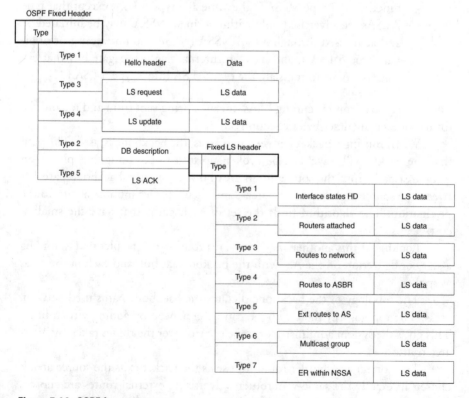

Figure 5.14 OSPF format structure.

The fields of the fixed OSPF packet header, which is illustrated in Figure 5.15, are described as follows.

- Version: The current version number is 2.
- Type code: There are five type codes defined and identified in Table 5.10. The value in the type code field determines the next header format to follow the basic OSPF header.
- Packet length: The field size limits the maximum packet size to 65K octets. OSPF relies on IP to fragment for packets larger than the network MTU. The recommended practice is to limit the size to 576 octets unless MTU discovery is being performed.

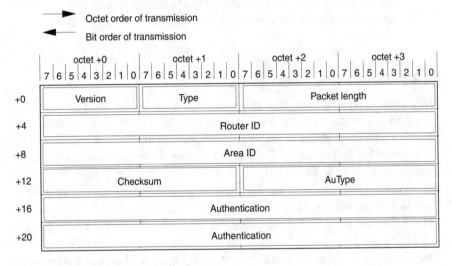

Figure 5.15 OSPF fixed-packet header format.

Table 5.10
OSPF Type Codes

Type Value	Meaning
1	Hello (used to test reachability)
2	Database description (topology)
3	Link status request
4	Link status update
5	Link status ACK

- Router ID: The router ID is the 32-bit identification of the packet's source.

- Area ID: Each packet is associated with a single area, and the area ID is the 32-bit identification of that area.

- Checksum: The checksum is calculated as the 16-bit one's complement of the one's complement sum of all 16-bit words in the packet, excluding the authentication field. (Zero padding is added, as required, for 16-bit alignment.)

- AuType: The AuType field identifies the procedure to be used for authentication. There are three types of authentication defined and identified in Table 5.11.

- Authentication: The 64-bit authentication field may be zero, contain a simple password, or contain a *message digest,* based on the value of AuType.

5.8.3.1 Hello Format

Hello packets are identified by the type code being equal to one. Hellos are sent periodically (default interval is 10 seconds) on all interfaces to establish and maintain neighbor relationships. Items exchanged with the hello packet include the interval for hellos, a dead interval timer, and a network mask. The default dead interval timer is 40 seconds. That is, if I have not heard from you within 40 seconds, you're dead. Other interval timers used include a poll interval (default to 120 seconds for NBMA), retransmit interval, and transit delay interval. (There are also single-shot-timers that, for example, cause an interface to exit a waiting state.) Hellos are also used for auto discovery on broadcast networks. The format of the hello packet is illustrated in Figure 5.16. A description of the fields is as follows.

Table 5.11
Authentication Types

AuType	Description
0	Null authentication
1	Simple password
2	Cryptographic authentication
-	Reserved by IANA

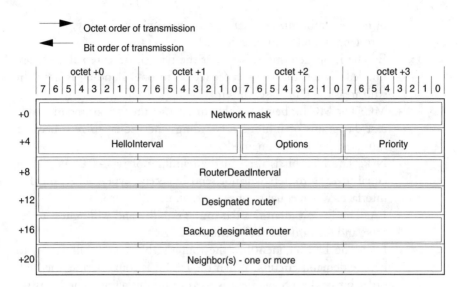

Figure 5.16 Hello packet format.

- Network mask: The network mask field contains the network mask for this interface. For example, for a Class B network the field could be FFFFFF00 hex.

- Options: The options field is contained in the hello packet and all LSA packets. Bits within the option field are used to convey support for optional capabilities. In this manner, routers of differing capabilities can be mixed within the same OSPF routing domain. The options code is illustrated in Figure 5.17.

    The following bits are contained within the option field.

    - T: The T bit is used to determine TOS capability. When the T bit is equal to zero, only TOS-0 is provided, which is normal service. When the T bit is equal to one, different paths are maintained based on the TOS field in the IP header. See Section 3.2 for identification

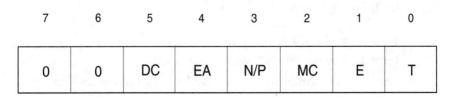

Figure 5.17 OSPF hello options.

of normal, minimize monetary cost, maximize reliability, maximize throughput, and minimize delay.

- E: The E bit determines whether the type-5, AS-external-LSAs are flooded. When the E bit is set (equal to one) flooding is used. In stub areas the E bit is set to zero.

- MC: The MC bit being equal to one causes the forwarding of multi-cast datagrams or inhibits forwarding when equal to zero because multicast is not implemented.

- N/P: The N/P bit is used in the handling of type-7 LSAs. When equal to zero, routers will not send or receive type-7 LSAs on that interface. When equal to one, the E bit must be zero.

- EA: The EA bit identifies the router's ability and willingness to receive and forward external-attributes-LSAs.

- DC: The DC bit identifies the router's ability and willingness to handle demand circuits. When the EA bit is equal to one, the high order bit of the LS age field is used to inhibit aging. The bit is referred to as DoNotAge LSAs.

- Priority: The priority field value is used in the election of the designated router. If set to zero, the router is ineligible to become a designated router.

- RouterDeadInterval: This field contains the number of seconds before declaring a silent router down.

- Designated router: This field contains the IP address of the designated router of the sender and is set to 0.0.0.0 if there is not one assigned.

- Backup designated router: This field contains the IP address of the backup designated router of the sender and is set to 0.0.0.0 if there is not one assigned.

- Neighbor: This field contains one or more IP address(es) of active neighbors.

5.8.3.2 LSA Header Format

All LSA packets have a fixed 20-octet header, as illustrated in Figure 5.18. The LSA packet fields are described as follows.

- LS age: The LS age field is a 16-bit unsigned number representing the seconds since the last LSA was originated.

- Options: Bits within the options field are used to convey support for optional capabilities. This field is described in Section 5.3.3.1 and illustrated in Figure 5.16.

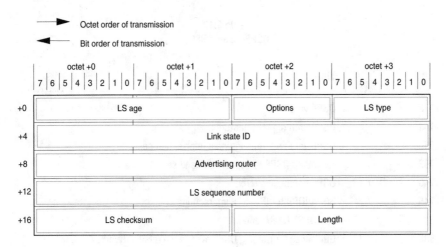

Figure 5.18 LSA fixed header format.

- LS type: The LS type field describes the nonfixed header type to follow. The types are defined in Table 5.9.

- Link state ID: The contents of this field depend on the LS type field value. For example, in type-2 LSAs the field will contain the IP interface address of the DR. Type 3 and 5 LSAs may have bits set in the host portion (unused) of the IP address. This is done to permit routers to originate separate LSAs for two networks with the same IP address but different masks (CIDR-related).

- Advertising router: The advertising router field contains the router ID of the LSA originator. For example, for a type-2 LSA the field is equal to the router ID of the network's DR.

- LS sequence number: This field contains a sequence number used to detect old or duplicate LSAs.

- LS checksum: The checksum field is used to detect data corruption either during flooding or in router storage. The checksum includes the complete LSA, with the exception of the LS age field that is regularly updated. This is not an optional field and a zero checksum is invalid.

5.8.4 Reference Material

The RFC library does not have a summary list of OSPF RFCs. The current specification for OSPF version 2 is contained in RFC 2178. Other related RFCs are identified in Table 5.12.

Table 5.12
OSPF-Related RFCs

RFC	Title
2178	OSPF version 2
2154	OSPF with digital signatures
1850	OSPF version 2 management information base
1793	Extending OSPF to support demand circuits
1765	OSPF database overflow
1745	BGP4/IDRP for IP—SPF interaction
1587	The OSPF NSSA option
1586	Guidelines for running OSPF over frame relay networks
1585	MOSPF: analysis and experience
1584	Multicast extensions to OSPF
1403	BGP OSPF interaction
1370	Internet architecture board applicability statement for OSPF
1246	Experience with the OSPF protocol

5.9 Exterior Gateway Protocol (EGP)

A routing domain is a group of routers that use a common IGP. A concept to reduce the volume of routing information exchange on a peer-to-peer basis is that of a routing domain using a selected gateway to communicate routing information with selected gateways from other domains. The selected gateway is referred to as an exterior gateway, and the protocol between the selected gateways is EGP. An exterior gateway learns about addresses within its AS and communicates this information to the other exterior gateways with messages relayed directly by IP to the immediate EGP neighbor (single hop). EGP uses IP directly; its protocol number is 8.

The EGP has three parts listed as follows:

1. Neighbor acquisition protocol;
2. Neighbor reachability protocol;
3. Network reachability (NR) determination.

The neighbor acquisition is simply a two-way handshake (request, response) to establish communication. The neighbor reachability amounts to a *hello* message (command) and a response message of "I-heard-you" amounts to a keep alive. The *network reachability* (NR) message is used to find out if the immediate neighbor is a suitable path for a particular destination network. Each of the three parts has a different message format [22]. Each message format contains a 16-bit autonomous system number that identifies the particular autonomous system. There are over 600 autonomous systems in the Internet [23].

The key disadvantage of EGP is that it creates a tree structure, or bottleneck. For example, if the exterior gateway fails, there are problems in the Internet since exterior gateways in other autonomous groups only know about the failed exterior gateway. (See RFC 1092 and RFC 1093, which document the key disadvantages of EGP and provide other suggested improvements.)

A UNIX program called *gated* uses routing data from multiple IGPs (RIP or GGP) to form its own routing base of information. It then uses EGP to communicate the routing information between selected gateways from different ASs. Gated also provides the capability of discouraging the use of an AS as a normal path for routing. This is done by advertising the distance vector as the value *n* within the AS, but as *n* + *m* outside the AS. The program was developed at Cornell University.

5.10 Border Gateway Protocol

The EGP shortcomings noted in the Section 5.9 motivated the creation of BGP, which was built on the operational experience gained from using EGP in the Internet. These disadvantages of EGP served as a requirements document for the design of BGP [24].

OSPF with multiple areas can handle very large networks, but all the areas are under a single administrative domain and constitute a single AS. The BGP is an interautonomous system (inter-AS) routing protocol. An AS may contain multiple routing domains, each with its own intra-autonomous routing protocol (IGP) and possibly more than one. Within an AS there may be multiple BGP speakers (routers) that communicate with the speakers of other ASs, or one speaker may be elected to assure a consistent reporting of routing information for the autonomous system. In either case, an AS appears to other ASs consistently with the same NR information. This eliminates the tree structure of EGP. BGP is not restricted to routers—hosts may exchange routing information with a border router in another AS.

A key feature of BGP is path attributes. The optional path attribute (AS-PATH) results in a list of ASs traversed for each route, similar to an IP trace route option. A route here is defined as a unit of information that pairs a destination with the attributes of a path to that destination. Path attributes provide for the straightforward and fast suppression of routing loops. BGP does not place topological restrictions on the interconnection of ASs. There may be stub ASs (no pass through), multihomed ASs (pass through blocked by policy), and transit ASs (designed to permit pass through).

BGP Version 4 (BGP-4) is CIDR-compliant, and routes may be aggregated to reduce the size of RTs. BGP also enforces policy-based route selection as defined by the AS administrator who may, for example, choose to permit transit traffic from one AS but not from another. BGP identifies all paths for each address prefix and assigns an index (non-negative integer) to each. The index values may be administratively set. The extent of interoperation between BGP-4 and BGP-3 and EGP is also controlled by the configuration defined by the AS administrator. The configuration may fully restrict, restrict limited paths, or allow full injection of BGP-4 paths into BGP-3 and EGP ASs. The source of information used in determining a path may be used in establishing the index value for a path.

Each BGP-4 speaker (e.g., AS router) stores routes in three *routing information bases* (RIBs). The RIBs must be logically distinct but need not be physically distinct. The three RIBs are the *in RIB* (IRIB), *out RIB* (ORIB), and *local RIB* (LRIB). The ORIB is used for routes that will be advertised to other BGP speakers. Routes that will be used by the local speaker are contained in the LRIB. The IRIB is used to collect routes from other BGP speakers.

What is the purpose of this elaborate scheme? Keeping multiple RIBs (books) is the tool used to enact policy-based routing. That is, for example, peers report one thing (IRIB) and the router applies a policy based on the configuration file and other learned information and tells other peers a new policy-based story (ORIB).

5.10.1 BGP Architecture

BGP, which is an upper layer program, uses the transport layer TCP, *well-known* port number 179, instead of directly using IP as does EGP. See Figure 5.19. This adds a little overhead but provides reliability and solves the packet loss problem with EGP (attributed with slow or lack of convergence on routing routes). Overall the worst performance of BGP occurs during congestion, which is equivalent to the average performance of EGP.

5.10.2 BGP Operation

BGP is a states-driven program starting in an idle state and progressing to an established state. A states diagram is flashy but never relates the preciseness of a state table. Fortunately, the specification for BGP contains a states table, which is extracted here for definition of the operation of BGP. The six possible states are listed in Table 5.13.

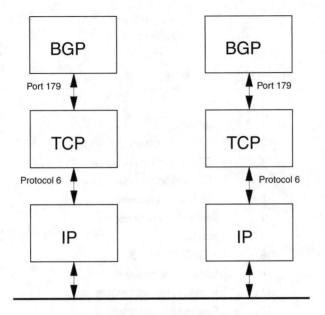

Link and Physical Layers

Figure 5.19 BGP architecture.

Table 5.13
BGP States

State	Name
1	Idle
2	Connect
3	Active
4	OpenSent
5	OpenConfirm
6	Established

An abbreviated set of events that will cause a states change is identified in Table 5.14.

Table 5.15 lists the state transitions of the BGP states machine. From this abbreviated states table and the BGP events in Table 5.14, it is easy to predict the next state based on the present state and an event. In the idle state (S1),

Table 5.14
BGP Events

Event	Description
1	BGP start
2	BGP stop
3	BGP transport connection open
4	BGP transport connection closed
5	BGP transport connection open failed
6	BGP transport fatal error
7	ConnectRetry timer expired
8	Hold timer expired
9	KeepAlive timer expired
10	Receive open message
11	Receive KeepAlive message
12	Receive update message
13	Receive notification message

Table 5.15
BGP States Table

Events	Idle (1)	Connect (2)	Active (3)	OpenSent (4)	OpenConfirm (5)	Established (6)
1	2	2	3	4	5	6
2	1	1	1	1	1	1
3	1	4	4	1	1	1
4	1	1	1	3	1	1
5	1	3	3	1	1	1
6	1	1	1	1	1	1
7	1	2	2	1	1	1
8	1	1	1	1	1	1
9	1	1	1	1	1	5
10	1	1	1	1 or 5	1	2
11	1	1	1	1	6	6
12	1	1	1	1	1	1 or 6
13	1	1	1	1	1	1

when a transport connection (E3) is made the state moves to connect (S2). From the connect state (S2) the state moves to OpenSent (S4) when the transport connection is opened (E3). Note that the active state (S3) is only attained when the connect return timer is restarted, which results from a failed transport connection open (E5). When the open message (E10) is received in the OpenSent (S4) state, the state moves to OpenConfirm (S5). The established state (S6) is reached by receiving a KeepAlive message (E11).

In the established state (S6) several events may occur and the state does not change. For example, restart KeepAlive timer (E11), restart hold timer, and process an update. Other events such as release resources, failed update process, and clear transport will cause the state to move to state 1 (idle).

5.10.3 BGP Message Format

After a TCP connection is established with an external peer, the pair of routers exchange open messages. The maximum message size is 4,096 octets, and each message contains a fixed header of 19 octets, as illustrated in Figure 5.20. A description of each field follows.

- Marker: The marker field, which contains 64 bits, is used to authenticate incoming messages and to detect the loss of synchronization. When authentication is used, the marker field contains a value that may be calculated by the receiver. If authentication is not used, each bit of the field is equal to one.

- Length: The length field contains a 16-bit unsigned number that identifies the total length in octets of the BGP message, including the header.

- Type: The type field contains a single octet, and its value determines the format of the remainder of the BGP message. The possible values and function are illustrated in Table 5.16.

A variable length header may follow the fixed header, depending on the type code in the fixed 19-octet header and (optionally) routing information. When the type code is equal to one representing an open message, another header with a variable portion follows the fixed BGP header. The open variable portion of the header contains a version number, AS number, hold timer, BGP identifier, and possible options. If the open messages are mutually acceptable, update, KeepAlive, and notification type messages may be exchanged. The initial data flow contains the entire BGP RT. Updates are required only for the changed portions of the BGP routing table. KeepAlive type messages are sent to maintain the TCP connection. Update messages containing IP layer reachability

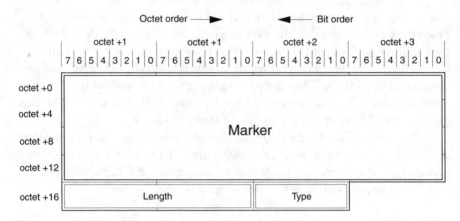

Figure 5.20 BGP fixed header format.

Table 5.16
BGP Type Codes

Value	Function
1	Open
2	Update
3	Notification
4	KeepAlive

information are used to advertise routes. If an error condition is encountered, a notification type message is sent and the TCP connection is closed.

5.10.3.1 Open Message Header Format

The open message header format is illustrated in Figure 5.21. A description of each field follows.

1. Version: The version field contains the BGP version number, which is currently equal to 4.

2. My autonomous system: This field contains a two-octet unsigned integer indicating the AS of the sender.

3. Hold timer: The two-octet hold timer field contains the number of seconds before the next KeepAlive or update message should be received.

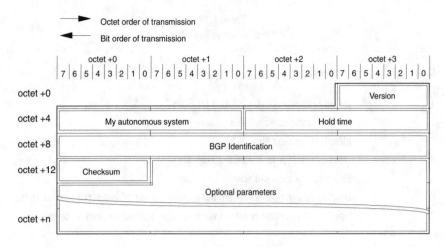

Figure 5.21 BGP open header.

4. BGP identifier: The BGP identifier field contains a 32-bit unsigned integer indicating the BGP identifier of the sender. The value is determined on startup and is the same for every BGP peer. It is an IP address assigned to each BGP speaker.

5. Optional parameters length: When options are specified, this field contains the length of all the options. If the value of this field is equal to zero, no options are contained.

6. Optional parameters: This field may contain optional parameters encoded as a one-octet parameter type and a one-octet parameter length and followed by a variable length parameter. The only parameter type presently defined is an authentication code, which has authentication data.

5.10.3.2 Update Message Header Format

An update message is used to advertise single routes and to withdraw one or more unfeasible routes. Update messages include the BGP fixed header (type code equal to two) plus a variable amount of data to add reachability information for new routes and withdraw unfeasible routes. Path attributes for new routes contain attribute flags and an attribute type code. The flag field (8 bits in length with only four used) defines whether the attribute is optional, if transitive or not, if complete or partial and if the attribute length is one or two octets. The type codes and their purposes are identified in Table 5.17.

Table 5.17
Update Message Type Codes

Code	Purpose
1	Identify route origin (0 for IGP, 1 for EGP)
2	Identify path characteristics (ordered or unordered)
3	Identifies the next hop
4	Used to select neighboring autonomous system
5	Indicates the local preference for a route
6	Used to inform other speakers the route is contained in a more specific route
7	Used to inform other speakers what speaker formed the aggregate route

5.10.3.3 KeepAlive Message Header Format

All the information necessary for a KeepAlive message is contained in the fixed 19-octet BGP header. The rate of sending KeepAlive messages is based on the hold timer (usually equal to one third of the hold timer) but is never sent more frequently than once per second. If the hold timer interval is equal to zero, KeepAlive messages are never sent.

5.10.3.4   Notification Message Header Format

When an error condition is detected, a notification message is sent and the connection is closed. In addition to the fixed 19-octet BGP header, the notification message has a small two-octet header and data. The first octet of the notification header contains an error code, and the second octet contains a subcode for error code. The error codes are identified in Table 5.18.

Table 5.18
Notification Error Codes

Code	Symbolic Name
1	Message header error
2	Open message error
3	Update message error
4	Hold timer expired
5	Finite state machine error
6	Cease

5.10.4 Reference Material

The RFC library does not have a summary list of BGPs. The current specification for BGP version 4 is contained in RFC 1771. Other related RFCs are identified in Table 5.19.

5.11 Domain Name System (DNS)

Users (people) prefer to use a symbolic name that is easy to remember to address a computer rather than numeric digits. For example, acc.com is easier to remember than 129.192.0.0. To allow this flexibility to users, a server is needed to map the symbolic name into the correct network address. This was originally done in the Internet with a single table in a centrally located server where all host names were maintained (a few hundred names in the early 1980s). The increase in network traffic and the number of hosts over the years (several hundred thousands in the late 1980s) necessitated decentralizing the name server function into multiple domain name servers. This reduced the response time for the name server function and distributed the network traffic load. The domain naming structure is similar to the directory of a DOS or UNIX file system. That is, it is a tree structure and files are identified with a path name. However, with

Table 5.19
BGP Reference RFCs

RFC	Title
2042	Registering new BGP attribute types
1998	Community attribute in multihome routing
1997	BGP communities attribute
1966	BGP route reflection: An alternative to full mesh IBG
1965	Autonomous system confederations for BGP
1863	A BGP/IDRP route server alternative to a full mesh routing
1774	BGP-4 protocol analysis
1773	Experience with the BGP-4 protocol
1772	Application of the border gateway protocol in the Internet
1771	A border gateway protocol 4 (BGP-4)
1745	BGP4/IDRP for IP—OSPF interaction
1657	Definitions of managed objects for BGP-4 using SMIv2

the domain name server, the path commences with the named node (least significant information) rather than with the root. (One of these systems is backwards, depending on which was learned first.) See comparison to DOS in Figure 5.22.

In the example above, the domain path for an individual address in the Network Department of Communications (e.g., PC410) is, "PC410.NET. COM.USC.EDU", which might translate to an IP address of "193.44.13.4". Another PC410 could be in the LAN department on the same subnet. Its IP address could be 193.44.13.3 and the name path is stated, "PC410.LAN.COM. USC.EDU". (Note that DNS is case-insensitive, so PC410 and pc410 are the same.) Contrast this to "C:.DOC", which is a DOS name of a channel connect document (CC.DOC) under *SPEC*, which is under *MSW*, and so on to the root "c:.

The *domain name service* (DNS) enabled delegation of the name database (master file) maintenance function to multiple domains, each having responsibility for its own names (names need only to be unique within a domain). Within each of the multiple domains the name space is further divided into

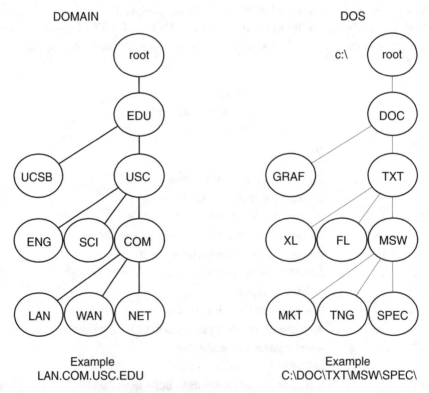

Figure 5.22 Domain versus DOS directory.

zones and each zone has a name server that is responsible for the accuracy of the data. Data found in a record that is located at the name server is called authoritative data. The same zone of data could be in another subdomain for redundancy, but it would not be authoritative. DNS also reduced the number of hops required to provide an address for a name.

For a hypothetical example of zones, consider the server in Figure 5.22 that handles all addresses under the COM (Communication) and be the authority for zones NET, WAN and LAN. A server at SCI could also have information for zones WAN and LAN, but the data would not be authoritative. Should a zone become too large, it may be subdivided.

From the user program's perspective, (e.g., TELNET, FTP, SMTP) the service is simple. It supplies a domain name and is returned an IP address. User programs such as UNIX "dig" or "nslookup" (or equivalent DOS programs) provide a direct translation of domain names to IP addresses by a DNS call. There is a service at UCSF.NET (128.218.1.109) that provides IP addresses for domain names. Simply connect to port number 5555 at IP address 128.218.1.109 and type the domain name. The IP address is returned and the connection closed. DNS also supports host address to host name and general lookup services, but the primary use is host name to IP address, which is the focus of this description.

5.11.1 Domain Name

The user program supplies the domain name as a sequence of labels, each separated by a period. The labels are listed from left to right and the label representing the closest to the user is first. This is the order entered by the user and transmitted by a resolver to a name server. DNS programs manipulating the domain name supplied by the user program format it in a manner that is easily interpreted by other programs. To the programs, each domain name consists of a sequence of labels and each label contains a length octet followed by a string of characters from a subset of the ASCII characters. The subset includes alpha characters (A–Z), digits (0–9), and the minus sign (-). (The original specifications restricted the first character to an alpha character, but this has been relaxed—see RFC #1123, Requirements for Internet Hosts. The maximum length of a label is 63 octets, which means that the high-order two bits of the octet are each equal to zero. When bits +6 and +7 (high order) are each equal to one, the length field is used as a pointer to the beginning octet of a previously defined domain name.) The maximum length of a domain name is 255 octets. All hosts should handle the maximum of 255-octet names, but *must* be able to handle, at a minimum, 63-octet names. The domain name is terminated by a length octet equal to zero. So, a domain name in program format may be reformatted to the user format by deleting the first-length octet and replacing interim-length octets with periods.

5.11.2 DNS Architecture and Program Elements

DNS is an applications layer protocol and classed as a utility, because it is a convenience to users and the system administrator rather than an integral part of user services such as FTP and TELNET. With the ever-increasing size of the Internet, it has become an essential convenience. However, with a small internet, only a limited portion of DNS will be used. (Everyone knows all the addresses.) In smaller internets (e.g., groups of PCs), the resolver (explained in Section 5.11.4) portion of DNS may be used to supply addresses from a table maintained by the local administrator.

In terms of the client/server model described in Section 1.3, DNS consists of a user, client, local name server, and remote name server. In terms of the specification, DNS consists of a user program (e.g., TELNET, FTP, and SMTP), resolver (client), name server (level 1 server), and foreign name server, including foreign resolver, (level 2 server) [25]. A smaller host (a PC) may only have a resolver that works from a local storage or cache. Every host *must* implement a DNS resolver, and it *must* implement a mechanism using the DNS resolver to convert host names to IP addresses and vice versa.

5.11.3 Data Elements of DNS

A DNS node is represented by a label within the domain name and all nodes have *resource records* (RRs) that contain information that enables the DNS programs to find the requested domain name. The basic format of an RR is illustrated in Figure 5.23. A description of each RR field follows.

- Owner name (or SNAME): This is the name of the node to which this RR pertains. It is the name that will be compared to the name supplied by the user program. The name is in DNS form with length octets followed by ASCII strings.
- Type: A 16-bit integer describing the type of RR. Table 5.20 summarizes the types of RRs. The different types of RRs are illustrated in Table 5.8. Values greater than 23 are for QTYPE fields, which appear in the question part of a query. QTYPEs (covered later) are a superset of TYPEs.
- Class: A 16-bit integer defining the class of the RR. An Internet RR has the field equal to one. As with the TYPE field, the QCLASS is a

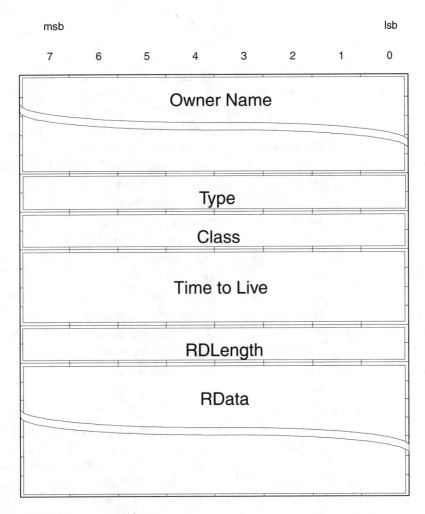

Figure 5.23 Resource record format.

superset of CLASS values. The only additional value defined by QCLASS is 255, which is for any other class.

- TTL: A 32-bit integer that specifies the time interval the resource record may be cached before refreshing the information from the source. Zero values, which are interpreted to mean that the RR can only be used for the transaction in progress, should not be cached. Zero values can also be used for extremely volatile data.

- RDLength: A 16-bit integer that specifies the length in octets of the RDATA field.

Table 5.20
RR Types

Value	Code	Meaing
1	A	A host address
2	NS	An authoritative name server
5	CNAME	The canonical name for an alias
6	SOA	Start of zone of authority
11	WKS	Well-known service description
12	PTR	A domain name pointer
13	HINFO	Host information
14	MINFO	Mailbox or mail list information
15	MX	Mail exchange
16	TXT	Text string
22	NSAP	String to OSI transport service
23	NSAP-PTR	A NSAP domain name pointer
252	AXFR	Request for transfer of entire zone
253	MAILB	Request mailbox records
255	*KS	Request for all records

- RData: A variable length string of octets that describes the resource. The format of this information varies according to the TYPE and CLASS of the resource record. For the TYPE A RR (Internet), the RData field contains a 32-bit IP address. If the host happens to be multihomed, the RData field will contain a 32-bit IP address for each connection.

Another element of data affecting the operation of DNS is the SLIST. The SLIST is a structure describing the name servers and zone where the resolver is currently trying to query. SLIST contains a states counter and time out for each user program request that enables the resolver to know which servers have been queried. SLIST also contains the known addresses for name servers of the zone and an indicator enabling the resolver to determine how close this zone is to the desired SNAME.

Each host's name servers and resolvers are configured by a local system administrator [26]. The configuration data identifies the local master files for name servers, which are in text format making it easy to exchange them between hosts using FTP. The name server uses the master files or copies to load its zones. The configuration data identifies the name servers for resolvers [27].

5.11.4 Mechanics of DNS

A user program (e.g., TELNET, FTP, and SMTP) first presents a request to a resolver that contains a domain name (sequence of labels separated by periods) for which it would like the associated IP address. See Figure 5.24.

The request is typically made with a subroutine or system call containing the domain name, or a pointer to the domain name in a system stack. In the simplest case, the resolver finds the domain address in its cache (or in the shared database if one is available) and returns the associated IP address. This is the millisecond response case.

The domain names in the resolver cache (or name servers) are in a standard format contained in RRs. Type A RRs (type field value equal to one) contain domain names and IP addresses. The format of the domain name in an RR is as described in Sections 4.1 and 4.4—that is a string of labels with each preceded by a length octet. There are four possible responses from a resolver to the calling user program, as follows.

   1. One or more RRs containing the appropriate address(es) in the format appropriate for the calling user program;

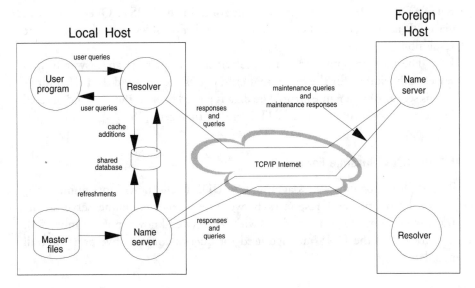

Figure 5.24 Mechanics of DNS.

2. In the event that the name supplied by the user program is an alias, (type field equal to 14—CNAME RR) resolver returns the domain name referenced by the alias—not an IP address of the alias;

3. A name error message, which happens when the supplied name does not exist (for example, a mistyped host name);

4. A data not found error, which happens when the supplied name exists but does not reference an IP address.

Should a resolver be unable to find the requested name in its cache or shared database, it must solicit help from a name server (local or foreign) by sending queries. The format of the DNS message permits multiple queries in a single message. This is particularly useful for a resolver since it may be servicing requests from multiple sources, including itself. The queuing process enables simple requests to be answered promptly and the more difficult requests (cross domain) to take longer.

The conversation between the user program and resolver is recursive in that the resolver continues to look for a find. Although there is provision for the resolver to request recursive service in the DNS message, the conversation between resolver and a name server is typically iterative in that the name server provides whatever it has with recommendations, which concludes the conversation. (We don't carry that, try the store down the street.)

Initially, the resolver directs queries to a name server found in a set of default name servers. As resolver directs and redirects queries, it builds a more reliable set of name servers for different zones in the SLIST. Queries are sent via UDP using port number 53. The mechanism enables cross-domain name resolution.

The ideal response from a name server to a query is an answer from a server authoritative for the query that either gives the required data (IP address in this scenario) or a name error. The data is passed back to the user and entered in the cache for future use if its TTL is greater than zero [28].

5.11.5 DNS Message Format

The DNS protocol uses messages sent by UDP (well-known port number 53) to convey queries and responses between resolvers and name servers. The transfer of complete zones is done with TCP (well-known port number 53). The format of the DNS message used for queries contains five parts, not all mandatory.

1. Header: Defines the format of the remaining parts;

2. Question: The target to resolve;

3. Answer: The resolution of the target;

4. Authority: Reference to an authoritative name server;

5. Additional: Related information but not the answer.

5.11.5.1 Header Format

The header is present in all DNS messages and contains fields as identified in Figure 5.25. These fields are briefly described as follows.

- ID: A 16-bit field used to correlate queries and responses (number assigned by the resolver—unique for a query—and preserved in the response by the name server;

- QR: A one-bit field that identifies the message as a query (value equal to zero) or a response (value equal to one);

- OPcode - A four-bit field that describes the type of message. There are only three types in use, as identified in Table 5.21.

- A: This is a one-bit field that, when equal to one, identifies the response as one made by an authoritative name server.

- T: This is a one-bit field that when equal to the value one, indicates the message has been truncated.

- RD: This is a one-bit field that is set equal to one by the resolver to request recursive service by the name server. Recursive service is not normally provided.

- RA: This is a one-bit field that signals the availability of recursive service by the name server.

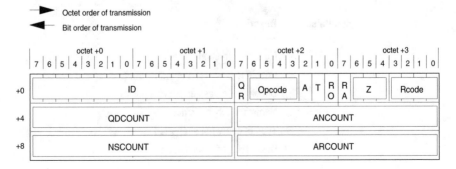

Figure 5.25 DNS message header format.

Table 5.21
Operation Code/Message Type

Code Value	Description
0	A standard query (name to address)
1	An inverse query (address to name)
2	A server status request

- Z: This is a three-bit field that is reserved for future use and must be set equal to zero.
- RCode: This is a four-bit field that is set by the name server to identify the status of the query. Table 5.22 identifies the valid response codes and their meaning.
- QDCount: A 16-bit, unsigned integer that defines the number of entries in the question section.
- ANCount: A 16-bit, unsigned integer that defines the number of resource records in the answer section.
- NSCount: A 16-bit, unsigned integer that defines the number of name server resource records in the authority section.
- ARCount: A 16-bit, unsigned integer that defines the number of resource records in the additional records section.

Table 5.22
Status of Query

RCode Value	Description
0	No error condition
1	Unable to interpret query due format error
2	Unable to process due server failure
3	Name in query does not exist
4	Type of query not supported
5	Query refused (zones set by administrator)

5.11.5.2 Question Section Format

The question section is constructed by the resolver and is always present. It contains the target domain name (sequence of length octets with associated labels), followed by the QType and QClass. The section is the same in length and format as that defined for the CName, type, and class fields in the RR format defined in Section 5.11.3 and illustrated in Figure 5.23.

The answer section contains one (or more) RR(s) identical to the format defined in Section 5.11.3 and illustrated in Figure 5.23. Ideally, there is only one RR, and it provides an authoritative IP address.

The authority section contains one (or more) RR(s) that point to sources of authoritative information. The format is identical to the format defined in Section 5.11.3 and illustrated in Figure 5.23.

The additional section contains one (or more) RR(s) that provide additional sources of information. The format is identical to the format defined in Section 5.11.3 and illustrated in Figure 5.23.

5.11.6 NSAP Address Resolution

The preceding description of DNS has been limited to the IP type (1) of RRs, although other types have been identified (host address to host name and general lookup services, for example). However, the primary focus has been host name to IP address. *Network service access point* (NSAP) is a network layer address used by an OSI stack. To provide multiprotocol operation, DNS handles the address format of the *connectionless network protocol* (CLNP) from the OSI stack, and the DNS format is introduced here [29].

NSAP support by DNS uses two new RR types (numbers 22 and 23) to define and store longer Internet (NSAP) addresses. The type code 22 format and handling is the same as IP (type 1) except the answer portion of the DNS message defines a NSAP address rather than an IP address. The RR type 23 is analogous to the PTR record used for IP addresses (type 12). The NSAP address representation in storage is a character string that is syntactically identical to TEXT and HINFO text (RR types 13 and 15). That is, there are two character strings for each NSAP. The first is a two-character string that identifies the length of the second character string. For example, the ASCII characters "20" would define the length for the full NSAP address including the initial domain part (IDP) and domain-specific part (DSP). The second character string is the characters of the NSAP address. The original design provided sufficient flexibility to add NSAP addressing. This should not be surprising since it also provides general search capability. The inverse conversion from NSAP address to domain name is done by setting the operation code in the query header equal to one (instead of zero).

The NSAP RR allows mapping from DNS names to NSAP addresses, and will contain entries for systems that are able to run Internet applications over TCP, UDP, or OSI CLNP.

5.12 Simple Network Management Protocol (SNMP)

Network management in earlier communications systems (single switch) was performed with a single operator console, which had control over all lines connected to the switch. An operator could enter commands at the operator console to request various status and statistical reports from the link layer programs in the same machine. With the information obtained from the reports, the operator took appropriate corrective actions, such as the following.

1. Suspend or resume operation on a single line (or all);
2. Cause alternate routing;
3. Execute diagnostics and loopback messages on a line.

When the network was expanded to multiple nodes, it was necessary for the operator console program to encapsulate an operator command in a supervisory message and send it to the link layer program of the appropriate node. Further, the remote link layer program returned the response to the operator console by encapsulating it in another supervisory message. Since the entire network (proprietary hardware and software) was provided by a single vendor, interacting with the link layer protocols was both the most efficient and reliable way to accomplish network management. In particular, the link layer protocols require less overhead and even if a routing table is corrupted, a node may still be booted remotely.

So why was SNMP developed? SNMP was developed to manage multiple physical networks provided by different vendors—namely the Internet where there is no link layer protocol that is common to all interfaces, or vendors [30].

5.12.1 SNMP Structure

SNMP operates as an upper layer program to avoid the proprietary link layer protocols. This enables management of the entire Internet because the upper layer (unlike the link layer) has total Internet access. It also permits development of network management programs without the involvement of network programmers. The disadvantages are more overhead, more traffic on the network, slower response time, and the inability to force a boot for a node when only a minor corruption of a routing table has occurred. See Figure 5.26 for an illustration of the architecture of SNMP.

5.12.2 SNMP Model

SNMP follows the basic client/server model, with a little different twist from client/servers like FTP or the model described in Section 1.3. Here the SNMP client is called the *network management station* (NMS). There may be one or more NMSs—typically only one for a small intranet and many for the Internet. The SNMP servers are called agents and every network element with the intelligence to participate in network management (e.g., host or router) has an agent function. Unlike the typical client/server model, there are many servers (agents) and limited clients (NMSs).

Each agent maintains a set of structured tables that contains information pertaining to hardware interfaces and program functions from each layer that are under the purview of that agent. This information is then solicited by the NMS in order to provide network operational reports and management capability. See Figure 5.27 for an illustration of the SNMP model.

Internet	OSI RM
	Application
SNMP	Presentation
	Session
Transport	Transport
Internet	Network
Link	Data Link
Physical	Physical

Figure 5.26 Conceptual layering of SNMP.

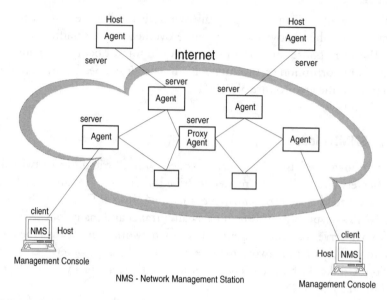

Figure 5.27 SNMP client/server model.

The information maintained by agents is structured in eleven different categories, each identified by a unique group number (described in detail later). For example, group 1 (the system group) contains such items as a device description and the elapsed time since the last restart, group 2 (the *IP* group) contains such items as the number of octets received by an interface and inbound packets discarded due to error by an interface, and group 5 (the *ICMP* group) contains such items as destination unreachable messages received and redirect messages sent. Since all agents (supplied by different vendors) are required to maintain these tables, they must be structured in a standardized manner [31].

Now that we know the information exists, the matter at hand is examining how the NMS obtains the information in order to construct reports, determining whether there are parameters that should be changed, and changing them as required. The protocol between the NMS and agents is the SNMP, which in its raw form (as a LAN monitor would see it) amounts to the exchange of messages (octet strings) that represent commands from NMS and responses from agents. The commands and responses are constructed using the formal language *abstract syntax notation one* (ASN.1). Encoding the message for transmission as an octet string and decoding of the octet string to its original form of ASN.1 is performed in accordance with the *basic encoding rules* (BER) [32].

There are only five different types of SNMP messages, which are described in Sections 5.12.4 and 5.12.5. There are three message types initiated by the

NMS to obtain information or to set variables. The agent uses a common response type message for these three messages to the NMS. In addition, the agent may generate unsolicited messages to alert the NMS when the condition of a trap is met. The NMS may enable/disable (with a set variable type message) an agent from initiating either all failure trap messages (i.e., globally) or each failure trap message individually. See Table 5.25 for a list of trap messages.

Proxy agents extend the network management function to devices that are not capable of communicating with the NMS. For example, modems, multiplexers, and devices that support a different network management framework.

5.12.3   SNMP Transport

SNMP was designed to use the connectionless UDP transport layer, although other transport mechanisms may be used (e.g., TCP and OSI CLTS). All SNMP client (NMS) and server (agent) messages are received on UDP port number 161, except trap messages, which are received on UDP port number 162. The IP address and UDP port number make up the transport address. If a different transport is used, the transport address must be changed accordingly. The data portion of a UDP packet is called the SNMP message, which consists of an octet string with a maximum size of 484 octets. That is, there is no classical header as with UDP and IP, although the octet string defining the version number and the community string could be considered the header portion of the SNMP message. (See Section 5.12.4.) The maximum size of 484 octets corresponds to an accepted maximum for IP datagram of 512 octets, which is required to be handled by all elements [(512)-(20,IP HD)-(8,UDP HD)=(484)].

5.12.4   SNMP Message Format

There are three parts to the SNMP message.: the version number, a community string, and the data portion containing the command and response verbs, called *protocol data units* (PDUs). The version number is used to identify the SNMP level, which is currently equal to one. (Version 2 of SNMP is being tested—see Section 5.12.12.2.) Both the client and servers may discard an SNMP message if the version number is incorrect or incompatible with the version number being used. The enhancement process for the MIB has a phase out mechanism for objects, which is done by marking the object as *deprecated* [33]. This means that the object is supported now but will likely not be supported in the next version number. (For example, the "atTable" object is marked deprecated in MIB-2.) Because of restrictions placed on the enhancement process, it is easier to support multiple versions of the MIB.

The community string is used as a password mechanism to limit access to the network management function [34]. It permits general access by multiple clients but can limit modification of variables to a single client with responsibility for the particular agent.

The PDU portion of the SNMP message contains the commands and responses, also called verbs, which are identified in Table 5.23. The PDU contains a request-id, error-status, error-index, object identifier, and instance, which is the contents (or value) of an object identifier.

The request-id is used by the originator to correlate requests and responses, since multiple requests may be outstanding. It also enables the originator to identify duplicates or missing responses. Although the ratio of requests and responses is one-to-one (ideally), the dialogue is asynchronous. That is, the originator may send many requests without waiting for a response. The responses may be duplicated, lost, or out of order. The error-status and error-index are each set equal to zero in a request and may contain values other than zero in the response. Possible values of the error-status are the following:

- No error occurred;
- A field was too big;
- No such name;
- Bad value in field;
- Read only value requested;
- General format error.

The error-index is always zero if the error-status is zero. When the error-status is nonzero, the error-index may provide a pointer to the offending portion of the message.

Table 5.23
SNMP PDU Verbs

SNMP	Value	Meaning
GetRequest	0	Obtain value of specific variable
GetNext Request	1	Obtain value of non-specific variable
GetResponse	2	Response to Get, GetNext and Set verbs
SetRequest	3	Store value in specific variable
Trap	4	Unsolicited response

The last portion of the SNMP message is identification of the subject query and the instance.

The actual format of the SNMP message is an octet string containing all three parts defined above and encoded with the BER [35] of ISO 8825, or CCITT X.210. Examples of using BER to encode different types of SNMP messages are described in Section 5.12.8.

5.12.5 SNMP Commands and Responses

For the NMS to maintain current status and statistical information, it periodically polls each agent with either the GetRequest or GetNextRequest command. The SetRequest may be used to change the value of a variable. The variable may range from a counter to a routing table entry. The GetResponse is used by the agent to acknowledge the GetRequest, GetNextRequest, or SetRequest commands. The acknowledgment may be a positive confirmation or a negative notice if for any reason the requested verb cannot be performed. See Table 5.24 for error codes.

The GetRequest command identifies a specific object name and variable, and the associated GetResponse contains the object name, variable and value. The GetNextRequest may identify the object identifier and variable. The associated GetResponse returns the object name, variable, and value of the next entry of the MIB group. NMS may then use the provided object and variable names in succeeding GetNextRequests to cycle through an entire group. The trap command enables unsolicited responses from the agent, when the condition of the trap is met. The enabling of the trap code by the agent is implementation-specific, but it is recommended that the SNMP object of the MIB that controls the dynamic enabling and disabling of the trap code by the NMS be

Table 5.24
Error Codes

Error States	Value	Meaning
noError	0	No error
tooBig	1	Size of PDU exceeds local limitation
noSuchName	2	Object name not found in MIB
badValue	3	Set value is not an acceptable range
readOnly	4	Cannot set—variable is read only
genErr	5	Not retrievable for other reasons

maintained in nonvolatile memory. The SetRequest command is used to toggle the MIB object (snmpEnableAuthenTraps), which is identified later. The generic-trap fields are identified in Table 5.25.

5.12.6 SNMP Object Identification

SNMP messages from the client reach the appropriate server as a result of the IP address and UDP port number. The naming convention used to select an object at the destination is similar to the DNS in that it involves a hierarchical structure commencing with the root (unnamed). However, objects are named commencing with the root rather than with the object, as in the DNS. The first subname spaces (or directories) under the root are: CCITT (#0), ISO (#1) and a joint directory of ISO and CCITT (#2), as illustrated in Figure 5.28.

Following the tree, the MIB(#1) is reached by addressing: ISO (#1), ORG (#3), DOD (#6), Internet (#1), MGMT (#2) and finally, MIB(#1). This sequence of subidentifiers constitutes the object identifier of the MIB name space. Each of the eleven subtrees under the MIB name space has its own name space defined by the associated ordinal value. Each of the eleven name spaces (tables) is defined in Section 5.12.10 and illustrated in Tables 5.26–5.35 [36].

By convention, an object identifier is a sequence of digits (or names) traversing from the root to the managed object, or fragment of interest (also called leaf). The value of a managed object is called an instance. The instances may be simple (integer, octet string, object identifier) or of a constructed type (list or table). The structure may also be thought of as a set of tables containing elements, each with an element name. In this scenario, the object identifier is a

Table 5.25
Trap PDUs

Generic Trap Type	Value	Meaning
coldStart	0	Reinitiallizing with altered tables
warmStart	1	Reinitializing without altered tables
linkDown	2	Link fail in agent's domain
linkUp	3	Link restore in agent's domain
authenticationFailure	4	Protocol message not authenticated
egpNeighborLoss	5	EGP peer relationship lost
enterpriseSpecific	6	Enterprise-specific event ocurred

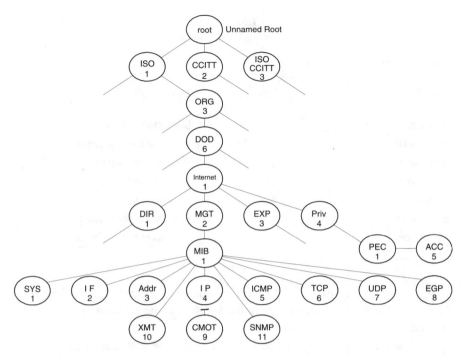

Figure 5.28 Object naming tree.

pointer to the correct table, the object identifier is an element name, and the instance is the value of an element.

For example, consider the *system* name space (SYS), which is MIB-2, Group 1. (See Section 5.12.10.1 and Table 5.26.) Using the convention described above and illustrated in Figure 5.28, SYS is identified by, "1.3.6.1.2.1.1.1," or, in symbolic form it is, "iso.org.dod.internet.mgt.mib.system." The object identifier of a particular entry in SYS is the name space identifier, plus the element name (or ordinal value). To obtain the instance of the object, a subidentifier of zero is appended to the object identifier. The one and only instance for the first entry in MIB-2 Group 1 (sysDescr) is obtained with the following: "1.3.6.1.2.1.1.1.0," or in symbolic form it is, "iso.org.dod.internet.mgt.mib.system.sysDescr.0." The instance of the next object (sysObjectId, ordinal value 2) is, "1.3.6.1.2.1.1.2.0," and so on.

Note that the first three characters of the object name (sys) denotes a value (1). The second part of the object name (e.g., Descr) also denotes a value. For example, the ordinal value of "sysName" is 1.5.

The next MIB-2 group is the Interface (if), ordinal value 2. The if name space is addressed as: "1.3.6.1.2.1.2," and the first element is ifIndex. (See

Table 5.26
Objects in System Group

systemOBJECT IDENTIFIER::={mib-2 1}		
Object Name	**Value**	**Meaning**
sysDescr	1	Description of device
sysObjectID	2	Identity of agent software
sysUpTime	3	Elapsed time since last restarted
sysContact	4	Name of person to contact
sysName	5	Device name
sysLocation	6	Physical location of device
sysServices	7	Services offered by device

Table 5.27.) The instance of "ifIndex" may be obtained by addressing "1.3.6.1.2.1.2.1.0," or stated symbolically as: "iso.org.dod.internet.mgt.mib. interface.ifIndex.0."

When a nested table is contained in a group, the nested table identifier is added to the object identifier. For example, in group 4 (IP) ordinal number 20 identifies a nested table of five entries. To address the second entry of this nested table, the object identifier is, "1.3.4.1.2.1.4.20.2.0." (Only single elements of the structure may be obtained, as described.)

5.12.7 Object Types

Each managed object has a type that defines its structure. MACROs are used to provide a uniform specification of types in the MIB-2. The following lists the object types.

1. INTEGER: The value of the object is represented by a cardinal integer.
2. OCTET STRING: The value of the object is represented by a string of zero or more octets.
3. OBJECT IDENTIFIER: The value of the object is represented by a sequence of digits that define the path from the root to an object.
4. SEQUENCE: The value of the object is the identification of other object types that are used to form a list, or column in a table—for example, INTEGER, OCTET, STRING.

5. SEQUENCE OF *type:* The value of the object is two-dimensional and used to form a table. The first represents the rows and the second represents the columns, with each row having the same number of columns.

6. DisplayString: The value of the object is represented by less than 256 NVT ASCII characters.

7. IpAddress: The value of the object is represented by a four-octet string that is in network octet order.

8. NetworkAddress: The value of the object may represent an address of different families. Presently only the family of IpAddress is defined.

9. Counter: The value of the object cycles from 0 to 2^{32}-1. (Designed for a 32-bit machine.)

10. Gauge: The value of the object oscillates from 0 and 2^{32}-1.

11. TimeTicks: The value of the object represents the time in hundredths of a second since some epoch, up to $(2^{32}-1)/100$ seconds;

12. NULL: A data type for reserving space only.

The object types in all capital letters are basic and described in the BER of ASN.1 as *universal* class. The universal class object types are used in a MACRO to construct the other object types, which are shown in mixed case [37].

An escape mechanism called tagging may be used to wrap a type and essentially bypass the limitations placed on types. This is accomplished by referencing a single existing type. The new type is the same as (isomorphic) the existing type, but is distinct from it by tag and value.

In Figure 5.28, note that the first node (#1) under MIB-2 is SYS. This is the system group and contains objects of information pertinent to the system. See Table 5.28 for a list of all objects contained in the system group.

5.12.8 Basic Encoding Rules (BER) for SNMP Messages

The format of SNMP messages is reasonably simple once the BER are understood. Although reasonably simple, the bit level manipulation is undesirable for the development of upper layer programs. Hence, the necessity for ASN.1, ISO 8824 (same as CCITT 208), which provides a symbolic language for specifying a data structure without concern for bit manipulation. (See Section 5.12.9.)

SNMP uses a subset of the BER specified by ISO, and only that subset is described here. All data is described with BER in three parts. The first part is a tag defining the type data (e.g., octets). The second part is the length of data in the number of octets. The last part is the data being defined, as illustrated in Figure 5.29 (Tag with a single octet of data.)

The high-order two bits (bits +6 and +7) of the tag field are used to describe the class of data. If both bits are each equal to zero, a universal class is specified. These are the types defined above as basic object types in capital letters (e.g., SEQUENCE, INTEGER, OCTET STRING). If bit +7 is equal to one, and bit +6 is equal to zero, a context-specific class is defined. There is also an application-wide class and a private class (bits 01 and 11); these are not used by the SNMP. These different classes may be either simple or constructed type data. Constructed type data is identified by bit +5 of the tag field being equal to one. Bits +0 through +4 define the type of tag.

The length field denotes the total number of octets to follow for this tag field. If the number of octets of data is between 1 and 126 octets, the length field occupies a single octet. Bit +7 of the length field is used to extend the number of octets in the length field. If this bit is equal to one, the next (and possibly the next, etc.) octet is part of the length field. (The value of zero for the length octet is reserved for the NULL type identifier.) The data field is unrestricted with regard to the BER. That is, each octet may have a value of 0 to 255.

Coding illustrations that follow use the convention defined in Figure 5.29.

5.12.8.1 Encoding of Version Number

All SNMP messages commence with a universal tag identifying the total length of the SNMP message. This octet string has the tag field equal to 30 hex and the length field equal to the number of octets in the entire SNMP message. The value 30 hex actually defines: bits +6 and +7 each equal to zero to denote *universal class,* bit +5 equal to one to denote *constructed type,* and bit +4 equal to one to identify the value 16, which denotes the universal SEQUENCE type. The

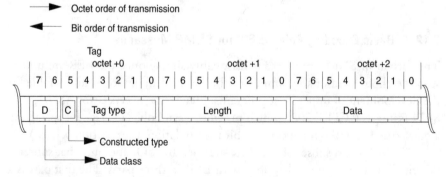

Figure 5.29 Basic format of SNMP data.

opening SNMP tag is encoded as:

> 30,<total length>

The tag is redlined for identification. Following the length octet, (or octets) of the constructed SEQUENCE type defining the entire SNMP message there is another tag field. This tag field is equal to 02 hex (defines a universal INTEGER type). It is followed by a length octet (equal to one) and an octet equal to the version number (02). Note that the tag, length, version number, and all to follow, are part of the original constructed SEQUENCE type. The version of the SNMP message encoded is:

> 02,<length=01>,02

5.12.8.2 Encoding of Community String

The community string commences with an octet equal to 04 hex, which identifies the universal octet string tag type. The tag type is followed by an octet identifying the length of the community string. The BER do not restrict the length of the community string. It seems unlikely that the number would exceed 126 (maximum for a single octet length identifier). The length field is followed by an octet string, typically ASCII printable characters. The community string is encoded as (use "SECRET" for illustration):

> 04,<length=06>,53,45,43,52,45,54

5.12.8.3 Encoding of PDU

Encoding of the PDU is done with the class *context-specific* type tag of constructed type. Hence, the tag field is a single octet with bit +7 equal to one, bit +6 equal to zero (context-specific), and bit +5 equal to one (constructed type). The remainder of the bits in the tag field is used to describe the PDU type. See Table 5.23 for the value used to specify each of the five verbs. Since only the three high-order bits are used to define overhead and there are only five verbs, the value of the tag field for PDUs will always be equal to A hex followed by the value assigned to the verb. That is A0, A1, A2, A3, or A4. The tag field is followed by a length field, which may contain more than a single octet, as described earlier. The length of the PDU is the total number of octets following the length field. The PDU starts with the following octet string:

> A0,<length>

A getRequest, type zero (or A0 hex code) was used for illustration. Following the context-specific tag and length fields, there is a tag field equal to 02 hex (defines a universal INTEGER type) used to specify the request-id. The length octet that follows defines the number of octets in the request-id field. Since the originator of a request is the only party who will use this field, it may be anything desired that enables the originator to correlate commands to responses. A length field defining a single octet that cycles between 0 and 255 works fine. The request-id is coded as:

02,<length=01>,21

A length of one octet and the request-id equal to 33 (21 hex) was used for illustration. The error-status and error-index fields are each universal INTEGER tag types and coded the same as the request-id above. That is, each has a tag field equal to 02 hex, a length octet equal to one, and a data field equal to zero. (Equal to zero for a getRequest, but may be nonzero for getResponse.) The error-status and error-index fields are coded as:

02,<length=01>,00,02,<length=01>,00

The error-index field is followed by a variable binding list (VarBindList), which is a list of one or more variable bindings (VarBinds). Each VarBind contains a variable name and its associated value. Since the value of a GetRequest is unknown in this example, the NULL tag is used to reserve space only. The VarBindList is double wrapped with a constructed SEQUENTIAL tag, which is followed by an octet equal to the length of the entire VarBindList (remainder of the message). The double wrapping tag is encoded as:

30,<length A>

where A is equal to the length of the VarBindList.

The VarBindList of this example consists of a single VarBind, which is an object identifier, and a NULL tag to reserve space for the value. (The NULL tag is replaced in a GetResponse PDU with a specific tag, length octet, and value.) Each VarBind is wrapped with a constructed SEQUENTIAL tag (equal to 30 hex) and a length octet. This is followed by a specific tag type (universal object identifier equal to 6 hex in this example), a length octet, an object identifier (name of variable), and a NULL tag to reserve space. Since the example only contains a single VarBind, the length associated with the constructed SEQUENTIAL tag is two less than the length of the double wrapping constructed SEQUENTIAL tag. That is, $A - 2$.

Assume, for example, that the object identifier is 1.3.6.1.2.1.1.1.0, or in symbolic form, iso.org.dod.internet.mgt.mib.system.sysDescr.0. Then the VarBind, and in this example the remainder of the VarBindList, is encoded as:

30,<A-2>,06,<length=08>,2B,06,01,02,01,01,01,00,05,<length=00>

The number of VarBinds is limited by the size of the SNMP message, which was previously noted as 484 octets.

So, where did the object identifier, "2B,06,01," come from? Should it not be, "01,03,06,01,"? It should be. However, the ISO Standard 8825, BER for ASN.1, combines the first two object identifier components by multiplying the first by 40 and adding the result to the second. Hence, the first encoded object identifier component is equal to the first (01), times 40 (equals 40), plus the second (03) is equal to 43, which is 2B hex [38].

Figure 5.30 combines all the above partial components of a getRequest message. Note the lengths have been inserted but the <>markers remain for ease of identification. Also, the tag types are boxed for ease of identification.

Excepting the 2B or not to be, using BER to construct an SNMP message is reasonably straightforward, but it is time consuming and prone to error. The description in Section 5.10.9.1 produces the same SNMP message using the ASN.1 language.

5.12.9 ASN.1 Language for SNMP Message

A subset of the ASN.1 (ISO 8824)[39] and BER (ISO 8825) was selected to simplify the effort required to produce SNMP messages and an SNMP MIB.

```
30,<26>,02,<01>,02
04,<06>,53,45,43,52,45,54
A0,<19>
02,<01>,21
02,<01>,00,02,<01>,00
30,<0E>,30,<0C>
06,<08>,2B,06,01,02,01,01,01,00,05,<00>
```

Figure 5.30 A getRequest message.

The general definition of an MIB, taken directly from RFC1157, is presented as Figure 5.31.

5.12.9.1 getRequest Example With ASN.1

The example from Section 5.12.8 is presented in Figure 5.32 as ASN.1 language statements. These statements produce the same code as illustrated in Figure 5.30, with the aid of a compiler (see Section 5.12.9.2 for compiler discussion)—that is, an SNMP message with a PDU that obtains the instance of "sysDescr" with a GetRequest command.

Specifying the SNMP message with ASN.1 is considerably easier, and it is not necessary to multiply 40 by 1 and add it to 3 to obtain the first object identifier component.

The receiver of the getRequest obtains the requested information from tables constructed per MIB-2. For example, the particular "sysDescr" data requested is under the MIB-2 group "system." The MIB-2 statements defining the data are of the form described in Figure 5.33.

The SYNTAX of "DisplayString" is derived from the object type of OCTET STRING, with a subtype that restricts the length to 255 characters and to a subset of ASCII printable characters [40].

5.12.9.2 SNMP Development Aids

Various development packages are available to encode and decode ASN.1 SNMP messages and produce a simple machine readable file from the ASN.1 definitions of an MIB. A UNIX-based package called 4BSD/ISODE SNMP may be used to handle both the SNMP messages and the MIB. The *ISO Development Environment* (ISODE) contains compilers that generate table structures in compliance with MIB-2, read ASN.1 modules, and produce octet strings for transmission and routines to produce "C" structure type statements from the octet string at the receiver. This enables the application programmer to deal exclusively in "C" type structures. There are other sources for compilers that work between ASN.1 and BER (for example, FTP software). The downside to the gain in functionality with "C" structures is inefficiency. With access to the Internet, the 4BSD/ISODE SNMP program may be obtained with FTP (binary mode) from "nisc.nyser.net," directory "pub/isode," "isode-6.tar.z." [41]. The current release of ISODE is included in a package of information on a CD from SRI.

5.12.10 MIB-2 Object Identification/Description

Now that the technique of addressing objects in MIB-2 has been defined, we will discuss the detail identification of the objects in each group. As part of the

```
RFC1157-SNMP DEFINITION ::= BEGIN
IMPORTS
    ObjectName, ObjectSyntax, NetworkAddress, IpAddress, Timeticks
    -- top-level message
    Message ::=
        SEQUENCE {
            Version              --version-1 for this RFC
                INTEGER {
                    version-1(0)
                }
            community            --community name
                OCTET STRING,
            data                 --e.g., PDUs if trivial
                ANY              --authentication is being used
        }
    --protocol data units
    PDU ::=
        CHOICE {
                        get-request
                            GetRequest-PDU
                        get-next-request
                            GetNextRequest-PDU
                        get-response
                            GetResponse-PDU
                        set-request
                            SetRequest-PDU
                trap
                        Trap-PDU

        }
    GetRequest-PDU ::=
        [0]
                IMPLICIT PDU
    GetNextRequest-PDU
        [1]
                IMPLICIT PDU
    GetResponse-PDU
        [2]
                IMPLICIT PDU
    SetRequest-PDU
        [3]
                IMPLICIT PDU
    PDU ::=
                SEQUENCE {
                    request-id
                        INTEGER,
                    error-status    --sometimes ignored
                        INTEGER {
                            noError(0),
```

Figure 5.31 SNMP definitions (RFC #1157).

```
                              tooBig(1),
                              noSuchName(2),
                              badValue(3),
                              readOnly(4),
                              genErr(5)
                    }
              error-index        --sometimes ignored
                    INTEGER,
              variable-bindings  --sometimes ignored
                    VarBindList
         }
   Trap-PDU ::=
       [4]
              IMPLICIT SEQUENCE {
              enterprise         --type of object generating
                                           --trap, see sysObject ID
              OBJECT IDENTIFIER,
                    agent-addr --address of object generating
                       NetworkAddress,  --trap
                    generic-trap --generic trap type
                         INTEGER {
                              coldStart(0)
                              warmStart(1)
                              linkDown(2)
                              linkUp(3)
                              authenticationFailure(4)
                              egpNeighborLoss(5)
                              enterpriseSpecific(6)
                         },
                    specific-trap  --specific code, present even
                       INTEGER,    --if not enterpriseSpecific
                    time-stamp     --time elapsed between last
                       TimeTicks,  --network (re)initialization
                    Variable-bindings--"interesting" information
                       VarBindList

                    }
   -- variable binding
   VarBind ::=
         SEQUENCE {
              name
                    ObjectName,
              value
                    ObjectSyntax
         }
   VarBindList ::=
         SEQUENCE OF
              VarBind
   END
```

Figure 5.31 (continued) SNMP definitions (RFC #1157), continued.

```
SNMP Sample Message ::=
      {
              version version-2,
              community "secret",
              data {
                      get-request {
                              request-id 33,
                              error-status noError,
                              error-index 0,
                              variable-bindings {
                                      {
                                              name 1.3.6.1.2.1.1.1.0,
                                              value {
                                                      null
                                              }
                                      }
                              }
                      }
              }
      }
```

Figure 5.32 Example using ASN.1.

```
sysDescr OBJECT-TYPE
          SYNTAX  DisplayString (SIZE (0..255))
          ACCESS  read-only
          STATUS  mandatory
          DESCRIPTION
                  "Name and version of hardware/software.
                  (Only printable ASCII characters.)"
          ::= {system 1}
```

Figure 5.33 Example of "sysDescr" data.

OBJECT-TYPE definition, the STATUS and ACCESS are defined. The status can be mandatory, optional, deprecated, or obsolete. The access options are read-only, read-write, write-only, and not-accessible.

5.12.10.1 MIB-2 Gp 1—System

There are seven objects in the system group, and each is identified in Table 5.26 with the associated ordinal number. The system group is mandatory for all implementations of SNMP. Yet, it is required to enter the "STATUS manda-tory" clause for each OBJECT-TYPE of the system group.

5.12.10.2   Objects in System Group

The objects from the MIB-2 group *system* are intuitive except *sysServices*. The instance is calculated by adding 2 (L - 1) for each layer, L, to form a sum. For example, a node that performs primarily routing functions would have a value of 4 (2(3 - 1)).

5.12.10.3   MIB-2 Gp 2—Interfaces

The MIB-2 group, Interfaces (if), is illustrated in Table 5.27 [42]. There is a column for each interface and a row for each of the objects identified.

The "ifAdminStatus" object provides a means for NMS to request the agent to perform an administrative change. For example, if the instance is changed from down (2) to up (1), the agent would interpret this as a request to initialize the interface. This would cause the "ifOperStatus" instance to change to up (1).

The "ifDescr" object provides a textual string containing information about the interface. The ordinal position of instances has been assigned through 32—for example, ddn-X25(4), ethernet-csmacd(6), iso88023-csmacd(7), fddi(15), frame-relay(32). The "ifSpecific" object provides a reference to MIB-2 definitions specific to the media being used for the interface. For example, if the media is Ethernet, the value is an object identifier of a document defining ob-jects specific to Ethernet. If the information is not available, the object identifier must be {0 0}.

5.12.10.4   MIB-2 Gp 3—Address Translation

The address translation ("at") group in MIB-2 is deprecated. That is, it is in-cluded solely for compatibility with MIB-1 nodes. The information will be moved to the "ip" group in MIB-3. The group contains a single table used to map IP addresses into media-specific addresses. Each row of the table contains three columns, as illustrated in Table 5.28.

As an example, for an Ethernet media, the instances could look like: atIfIndex=1, atPhysAddress=00:00:22:08:2A:C3 and atNetAddress=192.33.4.4.

5.12.10.5   MIB-2 Gp 4—Internet Protocol

The IP group is mandatory for all managed nodes and consists of 20 simple objects and three objects that are identifiers to subtables. The simple objects, either integer or counter types, are illustrated in Table 5.29.

The IP object with ordinal number 20 defines a subtable of five objects containing destination addresses and associated interface number, subnet-

Table 5.27
Objects in Interface Group

interfaces OBJECT IDENTIFIER::={mib-2 2}

Object Name	Value	Meaning
ifIndex	1	Interface number
ifDescr	2	Interface description
ifType	3	Interface type
ifMtu	4	Maximum octets in datagram
ifSpeed	5	Bandwidth in bits per second
ifPhysAddress	6	Lowest layer physical address
ifAdminStatus	7	Interface status desired
ifOperStatus	8	Interface status current
ifLastChange	9	Value of sysUpTime at reset
ifInOctets	10	Total octets received on interface
ifInUcastPkts	11	Number of packets to $n + 1$ Layer
ifInNUcastPkts	12	Non-unicast packets to $n +$ Layer
ifInDiscards	13	Inbound discarded packets/flow control
ifInErrors	14	Inbound packets discarded due error
ifInUnknownProtos	15	Inbound packets with protocol error
ifOutOctets	16	Total octets transmitted on interface
ifOutUcastPkts	17	Transmit requests from layer $n + 1$
ifOutNUcastPkts	18	Number non-unicast transmit requests
ifOutDiscards	19	Outbound packets discarded/flow control
sysServices	20	Services offered by device
sysServices	21	Services offered by device
sysServices	22	Services offered by device

masks, broadcast address, and maximum size for reassembly. The IP object with ordinal number 21 defines a subtable of 13 objects containing destination addresses and routing metrics. The IP object with ordinal number 22 defines a subtable of four objects containing equivalent value of ifIndex, media-dependent physical address, corresponding IpAddress, and a mechanism to invalidate the corresponding ipNetToMediaTable. Each IP subtable is identified at the top of Table 5.30, and each object name associates itself with a one subtable.

Table 5.28
Address Translation Group

Address translation OBJECT IDENTIFIER := {mib-2 3}	
Object Name	**Meaning**
atIfindex	Interface number
atPhysAddress	Media address of mapping
atNetAddress	IP address of mapping

Table 5.29
Simple Objects in IP Group

ip OBJECT IDENTIFIER ::= {mib-2 4}		
Object Name	**Value**	**Meaning**
ipForwarding	1	Acting as gateway (1) or not (2)
ipDefaultTTL	2	Default time-to-live for packets
ipInReceives	3	Datagrams received from interfaces
ipInHdrErrors	4	Datagrams discarded due format error
ipInAddrErrors	5	Datagrams discarded due invalid address
ipForwDatagrams	6	Datagrams forwarded
ipInUnknownProtos	7	Datagrams directed to invalid protocols
ipInDiscards	8	In datagrams discarded for flow control
ipInDelivers	9	Datagrams delivered to $n + 1$ layer
ipOutRequests	10	Datagrams received from $n + 1$ layer
IpOutDiscards	11	Out datagrams discarded for flow control
IpOutNoRoutes	12	Datagrams discarded due no route
IpReasmTimeout	13	Timeout value for reassembly queue
IpReasmReqds	14	Fragments received needing reassembly
IpReasmOKs	15	Datagrams successfully reassembled
ipFragFails	16	Failures detected by reassembly algorithm
ipFragOKs	17	Datagrams successfully fragmented
ipFragFails	18	Fragmentation required but prohibited
ipFragCreates	19	Number fragments created at this entity
ipRoutingDiscards	23	Valid routing entries that were discarded

Table 5.30
Objects in Tables of IP Group

Object Name	Value	Meaning
ipAdEntAddr	1	IP address of this entry
ipAdEntIfIndex	2	Interface number
ipAdEntNetMask	3	Subnet mask for IP address
ipAdEntRcastAddr	4	LSB of broadcast address
ipAdEntReasmMaxSize	5	Largest datagram able to re-assemble
ipRouteDest	1	Destination IP address
ipRouteIfIndex	2	Interface number
ipRouteMetric1	3	Routing metric #1
ipRouteMetric2	4	Routing metric #2
ipRouteMetric3	5	Routing metric #3
ipRouteMetric4	6	Routing metric #4
ipRouteNextHop	7	Gateway IP address for next hop
ipRouteType	8	Type (direct, remote, valid, invalid)
ipRouteProto	9	Mechanism used to determine route
ipRouteAge	10	Age of route in seconds
ipRoutemask	11	Subnet mask for route
ipRouteMetric5	12	Alternate routing metric (-1 if not used)
ipRouteInfo	13	MIB reference for protocol, or {0,0}
ipNetToMediaIndex	1	Interface number
ipNetToMediaPhysAddress	2	Media address of mapping
ipNetToMediaNetAddress	3	IP address of mapping
ipNetToMediaType	4	How mapping was determined

Notes: ipOBJECT IDENTIFIER::={mib-24}; ipAddrEntry OBJECT IDENTIFIER::={ipAddrTable 1}; ipRoute Entry OBJECT IDENTIFIER::={ipRouteTable 1}; ipNetToMediaEntry OBJECT IDENTIFIER::={ipNetToMedia Table 1}

5.12.10.6 MIB-2 Gp 5—ICMP

Implementation of the ICMP group is mandatory for all systems. There are 26 counters in the ICMP group, and each is identified in Table 5.31.

5.12.10.7 MIB-2 Gp 6—TCP

Implementation of the TCP group is mandatory for all systems. There are 14 simple objects (counters) and one subtable identified by object with

Table 5.31
Objects in ICMP Group

Object Name	Value	Meaning
icmpInMsgs	1	Messages received (including errors)
icmpInErrors	2	Messages received with errors
icmpInDestUnreachs	3	Destination unreachable messages received
icmpInTimmeExcds	4	Time exceeded messages received
icmpInParmProbs	5	Parameter problem messages received
icmpInSrcQuenchs	6	Source quench messages received
icmpInRedirects	7	Redirect messages received
icmpInEchos	8	Echo request messages received
icmpInEchoReps	9	Echo reply messages received
icmpInTimestamps	10	Timestamp request messages received
icmpInTimestampReps	11	Timestamp reply messages received
icmpInAddrMasks	12	Address mask request messages received
icmpInAddrMaskReps	13	Address mask request messages received
icmpOutMsgs	14	Messages sent (including errors)
icmpOutErrors	15	Messages not sent due flow control
icmpOutDestUnreachs	16	Destination unreachable messages sent
icmpOutTimeExcds	17	Time exceeded messages sent
icmpOutParmProbs	18	Parameter problem messages sent
icmpOutSrcQuenchs	19	Source quench messages sent
icmpOutRedirects	20	Redirect messages sent
icmpOutEchos	21	Echo request messages sent
icmpOutEchoReps	22	Echo reply messages sent
icmpOutTimestamps	23	Timestamp reply messages sent
icmpOutTimestampReps	24	Timestamp reply messages sent
icmpOutAddrMasks	25	Address mask request messages sent
icmpOutAddrMaskReps	26	Address mask reply messages sent

Note: icmp OBJECT IDENTIFIER::={mib-2 5}.

ordinal value of 13 in the TCP group. The subtable is a connection table and only contains instances during connection state. The connection table is identified at the top of Table 5.32. Although all objects are illustrated in the same table, the object names containing "tcpCon" are contained in the connection table.

Table 5.32
Objects in the TCP Group

Object Name	Value	Meaning
tcpRtoAlgorithm	1	Identifies retransmission algorithm
tcpRtoMin	2	Minimum retransmission timeout
tcpRtoMax	3	Maximum retransmission timeout
tcpMaxConn	4	Maximum TCP connections allowed
tcpActiveOpens	5	State changes (SYN SENT from CLOSED)
tcpPassiveOpens	6	State changes (SYN-RCVD from LISTEN)
tcpAttemptFails	7	State changes (CLOSED from SENT/RCVD)
tcpEstabResets	8	State changes (CLOSED from ESTAB/WAIT)
tcpCurrEstab	9	Connections with ESTAB or CLOSE-WAIT
tcpInSegs	10	Segments received, including InErrors
tcpOutSegs	11	Segments sent, including current
tcpRetransSegs	12	Segments retransmitted
tcpInErrs	14	Segments received in error
tcpOutRsts	15	Segments sent with RST flag
tcpConState	1	State of this TCP connection
tcpConnLocalAddress	2	IP address for this TCP connection
tcpConnLocalPort	3	Port number for this TCP connection
tcpConnRemAddress	4	Remote IP address for TCP connection
tcpConnRemPort	5	Remote port number for TCP connection

Notes: tcp OBJECT IDENTIFIER::={mib-2 6}; tcpConEntry OBJECT IDENTIFIER::= {tcpConnTable 1}.

5.12.10.8   MIB-2 Gp 7—UDP

Implementation of the UDP group is mandatory for all systems. There are four simple objects (counters) and one listener table identified by the object with ordinal value of 5. The listener table contains two objects. All six UDP objects are defined in Table 5.33. The two object names beginning with "udpLocal" are contained in the listener table.

5.12.10.9   MIB-2 Gp 8—EGP

Implementation of the EGP group is mandatory only if the EGP is implemented. There are five scalar objects and one neighbor table with 15 entries.

Table 5.33
Objects in UDP Group

Object Name	Value	Meaning
udpInDatagrams	1	Datagrams dellivered to UDP users
udpNoPorts	2	Datagrams directed to unknownn port
udpInErrors	3	Datagrams not deliverable/format error
udpOutDatagrams	4	Datagrams sent from this entity
udpLocalAddress	1	IP address for this UDP listener
udpLocalPort	2	Local port number of this UDP listener

Notes: udp OBJECT IDENTIFIER::={mib-2 7}; udpEntry OBJECT IDENTIFIER::={udpTable 1}.

Both are defined in Table 5.34. The 15 object names beginning with "egpNeigh" are contained in the neighbor table and identified by the object with ordinal value of 5.

5.12.10.10 MIB-2 Gp 9—CMOT

CMOT defines CMIP over TCP, and MIB-2 Group 9 was designed for this purpose. However, from the comments in MIB-2 regarding Group 9, it may not be implemented soon [43]. An IETF working group has renamed the group to "OSI Internet Management" (OIM) [44].

5.12.10.11 MIB-2 Gp 10—Transmission

Typically, definitions for transmission reside in an experimental portion of the MIB until proven, and then they are given an object identifier under the transmission group. The name assigned is:

type OBJECT Identifier ::= {transmission number}

where *type* is the symbolic value used for the media in the ifType column of the ifTable object, and *number* is the actual integer value corresponding to the symbol.

Consider the Ethernet-like interfaces defined by the ifType object—namely ethernet-csmacd(6), iso88023-csmacd(7), and starLan(11). For these interfaces, the value of the ifSpecific variable has the OBJECT IDENTIFIER, "dot3 OBJECT IDENTIFIER ::= {transmission 7} " [45]. Compare this to the

Table 5.34
Objects in EGP Group

Object Name	Value	Meaning
egpInMsgs	1	Messages received without error
egpInErrors	2	Messages received in error
egpOutMsgs	3	Locally generated EGP messages
egpAs	6	Autonomous system number
egpNeighState	1	State with respect to EGP neighbor
egpNeighAddr	2	IP address of this entry's EGP neighbor
egpNeighAs	3	Autonomous system of this EGP peer
egpNeighInMsgs	4	Messages received with error from peer
egpNeighInErrs	5	Messages received from peer in error
egpNeighOutMsgs	6	Messages generated to this peer
egpNeighOutErrs	7	Messages not sent due to flow control
egpNeighInErrMsgs	8	Error messages received from peer
egpNeighOutErrMsgs	9	Error messages sent to peer
egpNeighStateUps	10	State transition to UP with this peer
egpNeighStateDowns	11	State transitions from UP to any other
egpNeighIntervalHello	12	Interval between Hello commands (T1)
egpNeighIntervalPoll	13	Interval between poll retransmissions (T3)
egpNeighMode	14	Polling mode, active (1), passive (2)
egpNeighEventTrigger	15	Variable to trigger Start/Stop events

Notes: egp OBJECT IDENTIFIER::={mib-2 7}; egpNeighEntry OBJECT IDENTIFIER::={egpNeigh-Table 1}.

OBJECT IDENTIFIER for DS1 media type, v.i.z., "ds1 OBJECT IDENTI-FIER ::= {experimental 2} " [46].

5.12.10.12 MIB-2 Gp 11—SNMP

Implementation of the SNMP group is mandatory only if SNMP is implemented. There are 28 simple objects (27 counters and one integer) in the SNMP group, and each is identified in Table 5.35. (Two ordinal values are unused.)

Table 5.35
Objects in SNMP Group

Object Name	Value	Meaning
snmpInPkts	1	Messages received from $n+1$ layer
snmpOutPkts	2	Messages passed to $n+1$ layer
snmpInBadVersions	3	Received messages with bad version
snmpInBadCommunityNames	4	Received messages with bad community
snmpInBadCommunityUsers	5	Message verbs not allowed by community
snmpInASNParseErrs	6	ASN.1 or BER errors found by SNMP
—Not used—	7	—Not used—
snmpInTooBigs	8	PDUs with error-status of "tooBig"
snmpNoSuchNames	9	PDUs with error-status of "noSuchName"
snmpInBadValues	10	PDUs with error-status of "badValue"
snmpInReadOnlys	11	Valid PDUs with error-status of "readOnly"
snmpInGenErrs	12	Valid PDUs with error-status of "genErr"
snmpInTotalReqVars	13	Retrieved objects with Get and Get-next
snmpInTotalsSetVars	14	Altered objects with Set-Request PDUs
snmpInGetRequests	15	Get-Request processed by SNMP
snmpInGetNexts	16	Get-Next Requests processed by SNMP
snmpSetRequests	17	Set-Requests processed by SNMP
snmpInGetRespoonses	18	Get-Responses processed by SNMP
snmpInTraps	19	Traps processed by SNMP
snmpOutTooBigs	20	PDUs generated with error-status "tooBig"
snmpOutNoSuchNames	21	PDUs sent with error-status "noSuchName"
snmpOutBadValues	22	PDUs sent with error-status "badValue"
—Not Used—	23	—Not Used—
snmpOutGenErrs	24	PDUs sent with error-status "genErr"
snmpOutGetRequests	25	Get-Requests sent by SNMP
snmpOutGetNexts	26	Get-Next Requests sent by SNMP
snmpOutSetRequests	27	Set-Requests sent by SNMP
snmpOutGetResponses	28	Get-Responses sent by SNMP
snmpOutTraps	29	Trap PDUs sent by SNMP
snmpEnableAuthenTraps	30	Toggle agent authentication-failure traps

Note: snmp OBJECT IDENTIFIER::={mib-2 11}.

5.12.11 SNMP Diagnostic Tools

Diagnostic tools are used during the implementation of a network, during normal operation of the network and to aid with the network failure analysis. The diagnostic tools range from a PC software program with a standard Ethernet interface card to sophisticated analyzers that are more powerful than many of the network elements being analyzed [47].

A diagnostic tool used during the development and implementation phase is normally a LAN monitor that looks at an interface like a microscope, i.e., at the bit level. It is being used, possibly without interpretation, to assure that all communications on an interface meet the specification.

During normal operation of the network, the system performance is monitored to determine the day-to-day characteristics of the system. This provides a benchmark or reference point for analysis during erratic behavior of the network (e.g., slow file transfer). Normal characteristics begin with such simple activities as sending echo request to all host, from all hosts—that is, everybody pings everybody. A NMS (e.g., X Window workstation) running SNMP, may be used to collect all the standard statistics plus vendor specifics.

5.12.12 SNMP Changes and Enhancements

Since SNMP's introduction, it has been installed worldwide, and, thus, its shortcomings are discussed worldwide. The primary drawbacks of SNMP are its lack of security (see Section 5.4.12) and the amount of bandwidth it uses in larger systems, which is caused by the polling mechanism and the mechanism used to obtain large amounts of data (iterative process of single elements). Various experiments have been made with SNMP riding an OSI stack with CLTS or COTS, SNMP riding the Ethernet directly using type code 814C hex, CMIP riding the LLC directly (called CMOL) [48], and operating the management portion of ISO's network management system over the Internet's TCP—CMIP/CMIS over TCP, or CMOT and dual environments [49, 50].

5.12.12.1 Secure SNMP

The IETF responded to the security problem of SNMP with proposed standards for authentication and encryption, called Secure SNMP [51] Secure SNMP involves a local database for each SNMP entity that contains access privilege information. The information is used to reliably identify sender and receiver limitations on type of SNMP verbs and to determine the security protocol. A received secure SNMP message that cannot be properly authenticated is simply discarded. The security protocol may protect against disclosure, which is accomplished with the cipher block chaining mode of operation as described in the *data encryption standard* (DES). Encryption requires secret cryptographic keys for both entities (manager and agent). The encryption process is similar to the PEM process described in Section 6.7.11.

The authentication process reduces the capacity of SNMP by approximately five percent. The combination of authentication and encryption reduces the capacity by approximately 50 percent. That is the bad news. The good news is that several large users (e.g., carriers) that previously excluded SNMP for network management because of lack of security now find it a workable option.

5.12.12.2 SNMPv2

Although secure SNMP solves the security problems associated with SNMP, several problems remain—capacity, for example. The IETF solicited proposals for solutions to the remaining problems and received responses from IETF members Jeffery D. Case, Keith McCloghrie, Marshall T. Rose, and Steven Waldbusser. Their proposal, called *simple management protocol* (SMP), is for a reasonably major revision of SNMP that will become SNMP version 2 if approved by the IETF. SMP contains nearly the same features of secure SNMP plus many new features to address other problems. The following is a list of new features contained in SMP.

1. It provides device locking by management station while SET requests are being issued to configure a device.

2. It provides a GETBULK request that enables retrieving a range of variables without cycling through a group using the GetNext request, resulting in a more than 1,000 percent improvement in total transfer speed.

3. It provides a new data type for OSI addresses to enable network management of OSI networks.

4. It provides 64-bit counters for certain variables that would, with the present 32-bit variables, wrap with higher speed services of the 1990s.

5. It provides an *exception* response to a request when the response is incomplete. (Currently, the entire request is rejected.)

6. It provides an enhanced set of error codes to enable understanding of why certain requests are rejected.

7. It provides an *inform* command that permits management stations to communicate with one another. From this is formed a hierarchical management structure that may be used to delegate management activities to mid-level management stations.

8. SMP may use transport protocols other than UDP, such as Apple-Talk, IPX, and an OSI protocol stack.

As might be expected, SMP is not backward-compatible with SNMP. There are two proposed methods of handling the incompatibilities. The first involves a bilingual *management station* (MS) that can work with either SNMP or SMP. Bilingual MSs exploit the capabilities of SMP but the MS host suffers. The second proposed solution involves a proxy MS to convert from one format to the other. This permits existing MSs to continue operation without change, but it does not exploit the features of SMP.

The SMP specifications may be obtained via ANON-FTP with the nnsc.nsf.net host from the subdirectory, "draft-rose-smp." The eight documents available are tm-00.txt, tc-00.txt, proto-00.txt, smi-00.txt, intro-00.txt, mib-00.txt, m2m-00.txt, and coex-00.txt.

The name SMP was dropped for SNMP version 2 (SNMPv2), which is now on the standards track. The SMI for SNMPv2 is defined in RFC 1902, MIB-II is defined in RFC 1213, and the SNMPv2 protocol is defined in RFC 1902 (with textual conventions defined in RFC 1903 and conformance statements in RFC 1904).

5.12.13 Future of SNMP

The arch rival of SNMP is the NMS designed by the ISO. The ISO version of network management consists of *common management information service* (CMIS) and *common management information protocol* (CMIP) operating over an OSI stack. This is referred to as *CMIS/CMIP.*

Both SNMP and CMIS/CMIP were started at approximately the same time, but SNMP has been on the market for many years while CMIS/CMIP remained on the drawing board until recently. SNMP is slow, adds 10% (or more) overhead network traffic, and is not simple (as touted in its name). Although CMIS/CMIP is more complex, it provides many powerful PDU verbs, has a better authentication mechanism, and is an ISO standard—for whatever that is worth. (Fortunately, publishing a standard does not make it so; otherwise, there would be huge RS449 connectors hanging everywhere.)

There are over 50 vendors offering software (some with hardware) to manage an internet using SNMP—from A (ACC) to Z (Zenith Electronics Corporation). In addition, there are foreign vendors supplying SNMP products—e.g., Spider, U.K. The products include bridges, routers, gateways, workstations, PCs, and mainframes, and the number of vendors supplying SNMP is nearly doubling each year [52].

With the ISO stack (including CMIP) essentially dead, SNMP has enjoyed a boom in the marketplace and replacement during the 90's would be difficult to impossible. However, SNMPv2 never progressed past the draft status and was sent back to the drawing board. SNMPv3 draft specifications were being produced through the end of 1997. (See draft-ietf-snmpv3-usm-v2-00 dated 12/97. The inability to get past the draft status for enhancements to SNMP for more than five years (1993–1998) could open the door to other competing network management protocols.

5.13 Network Time Protocol (NTP)

When older machines boot, they often require operator action to set the date and time. Newer machines with battery backup typically retain the date and time even when turned off. Without a method of synchronization, even hosts with a reasonably accurate time, including backup, distort calculations of such things as response time and transient time. Since these timestamps are used for diagnostics and performance analysis (e.g., ICMP), it is important to have the time between hosts synchronized.

The newest *network time protocol* (NTP) servers maintain time to within nanoseconds. The goal of the original TIME server was to synchronize host times to within the second. Depending on the configuration, a reasonable expectation with the current NTP is a few milliseconds [53].

Small internets may have a single server, while the Internet has a group of NTP servers that are in a hierarchically distributed configuration.

A client sends an NTP message to one or more servers and processes the replies as received. The server interchanges addresses and ports, overwrites certain fields in the message, recalculates the checksum, and returns the message immediately. Information included in the NTP message allows the client to determine the server time with respect to local time and adjust the logical clock accordingly. In addition, the message includes information to calculate the expected timekeeping accuracy and reliability, thus selecting the best from possibly several servers.

Since the time is given in elapsed seconds from a fixed date and time in *universal time coordinated* (UTC), previously called *Greenwich mean time* (GMT), hosts in different time zones will remain in synchrony. The fixed date and time (epoch) for UNIX systems is January 1, 1970. For the TCP/IP Internet, it is January 1, 1900. The field will overflow in 2036 (136 year cycle). The high order bit was set in 1968.

The client/server model described in Section 1.3 may be followed, although a symmetric mode is available that can be used for peer-to-peer communication between hosts. In the conventional client/server mode, the sequence is similar to the echo request. The client uses destination port number 123 in a UDP message to the server. The server, in addition to flipping addresses and port numbers, provides the following information in the response message. (In the peer mode, both the source and destination port numbers are equal to 123.)

5.13.1 Network Time Message Format

The NTP message format is illustrated in Figure 5.34. The following is an explanation of the fields in Figure 5.34.

- *Leap indicator* (LI): This is a two-bit field indicating leap correction at the end of the last day of the current month, which would be June or December. When set to require advancing by one second (loss of a second), at 23:59:59 the clock is advanced to 23:59:60. This microscopic adjustment is used to correct for the small changes in the mean rotation period of the Earth. The possible values are illustrated in Table 5.36. Since 1972, the value "01" (minus one second) has been used 14 times. The value "10" (plus one second) has never been used.

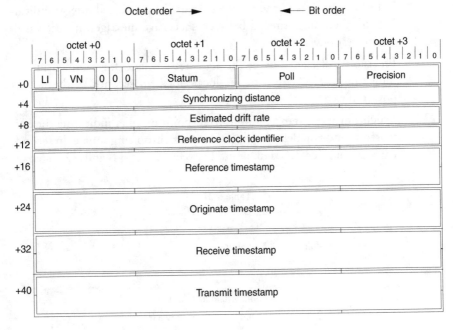

Figure 5.34 Network Time Message Format

Table 5.36
Leap Indicator

Value	Meaning
00	No warning
01	-1 second (following minute has 59 seconds)
10	+1 second (following minute has 61 seconds)
11	Alarm condition (clock not synchronized)

- Version number (VN): Three-bit code indicating the version number, which is currently equal to 1. That is, bit +3 of octet +0 is equal to 1, and bits +4 and +5 are each equal to zero.

- Reserved: Reserved, three-bit field equal to zero. That is, bits +0, +1, and +2 of octet +0 are each equal to zero.

- Stratum: All eight bits of the +1 octet are used to identify the stratum level of the local clock. The possible values are identified in Table 5.37. There are five primary servers in the United States, each with a WWVB radio clock and one or more Internet interfaces.

- Poll: All eight bits of the +2 octet are used for a signed integer indicating the maximum interval between successive messages, in seconds to the nearest power of two.

- Precision: All eight bits of the +3 octet are used for a signed integer indicating the precision of the local clock, in seconds to the nearest power of two.

- Synchronizing distance: 32-bit, fixed-point number indicating the estimated round trip delay to the primary synchronizing source, in seconds with an implied fraction point between octets +1 and +2. That is,

Table 5.37
Stratum Level of Local Clock

Value	Meaning
0	Unspecified
1	Primary reference (e.g.
2...n	Secondary reference (via NTP)

octets +0 and +1 represent the integer, and octets +2 and +3 represent the fraction.

- Estimated drift rate: 32-bit fixed-point fraction number indicating the estimated drift rate of the local clock.
- Reference clock identifier: 32-bit field identifying the particular reference clock as illustrated in Table 5.38. In the case of type 0 (unspecified) or type 1 (primary reference), this is a left-justified, zero-filled ASCII string.
- Timestamps: There are three timestamps. Each is a 64-bit, unsigned fixed-point number, in seconds relative to 0000 UT on 1/1/1900. The integer part is in the first 32 bits and the fraction part in the last 32 bits. The format allows convenient multiple-precision arithmetic and conversion to time protocol representation (seconds) but does complicate the conversion to ICMP timestamp message representation (milliseconds). The precision of this representation is about 0.2 nanoseconds, or 200 picoseconds.
- Reference timestamp: Local time at which the local clock was last set or corrected.
- Originate timestamp (t1): Local time at which the request departed the client host for the service host.
- Receive timestamp (t2): Local time at which the request arrived at the service host.
- Transmit timestamp (t3): Local time at which the reply departed the service host for the client host.
- Symbols t1, t2, and t3 assigned to each of the timestamps: If the symbol t4 is assigned to the local time, the response is received by the client from the server, the round trip delay is equal to $[(t4-t1)-(t3-t2)]$ and the clock offset is equal to $[[(t2-t1)-(t4-t3)]]/2$.

Table 5.38
Reference Clock Identifier

Stratum Level	Code	Meaning
0	DCN	Determined by DCN routing
1	WWVB	WWVB radio clock (60 KHz)
1	GOES	GOES satellite clock (468 MHz)
1	WWV	WWV radio clock (5/10/15 MHz)

5.14   Simple Network Time Protocol (SNTP)

The SNTP is a weakened version of the NTP that offers adequate time synchronization for Internet users that do not require the ultimate accuracy of the full NTP implementation. SNTP uses the same interface and format as NTP but with timestamp accuracy closer to the original time protocol [54]. The sheer size and complexity of NTP makes it less desirable for some applications, such as most applications in a PC, with less memory and need for the ultimate accuracy in time [55]. So, if your accuracy expectations are on the order of a second, SNTP is the answer. (Under ideal conditions with a dedicated design, SNTP can synchronize time to an order of micro-seconds.)

Although the message format is the same as that of NTP, clarification of the response is needed. For example, only the transmit timestamp has explicit meaning in the SNTP response. The field contains the server time of day in the same format as the time protocol. The time protocol (original) is much simpler than NTP or SNTP, but only provides a single 32-bit timestamp, which repre-sents the first 32 bits of the present NTP 64-bit transmit timestamp. The frac-tion part of the field may be valid, but the accuracy achieved with SNTP does not justify its use. Several of the fields in the SNTP message (same as the NTP message format) are not used and set to zero.

Note that another similar sounding time protocol is the daytime proto-col [56]. The daytime protocol listens on either UDP or TCP port number 13 and when a connection is made, it sends a message containing the cur-rent date and time as an ASCII character string, discards the received mes-sage, and closes the connection. The format of the ASCII character string is: "month, day, year, time zone"—for example, "Tuesday, August 11, 1992 07:16:33PST."

5.15   Echo Protocol

The echo protocol was designed for debugging and as a measurement tool. The echo server uses well-known UDP port number 7 to listen for client echo requests. The client allocates an unused UDP port number for the source port number and sends a message via UDP to the echo server. The echo server receives the request, flips the source and destination addresses, flips the UDP port IDs, and returns the message to the client. The server works from a queue of requests, so multiple requests from the Internet may arrive nearly simultaneously [57].

5.16 Remote Authentication Dial-In User Service (RADIUS)

Remote authentication dial-in user service (RADIUS) is an upper layer program developed by Livingston Enterprises to centralize the functions of user authentication and accounting [58]. A single enterprise-wide database may be used to validate users, supply configuration parameters, and collect statistics. This eliminates the individual password files and statistical information usually maintained at various points of access. The network access may be accomplished with IP, IPX, AppleTalk, etc. Although designed primarily for dial-in users, it may be used for Telnet sessions or any user validation such as a Netware or Microsoft NT.

There are three categories of RADIUS components. The first is *users*. Users are illustrated in Figure 5.35 as laptops that require validation to access an enterprise network. There are dial-up users (typically PPP), remote users connected via a router, directly connected users (validating logon only), and Internet users. All users are clients to the NAS. The NAS is, in turn, a RADIUS client to the RADIUS server. The RADIUS client supports and maintains user connections during the authentication process and then routes the authenticated users directly onto the network. The RADIUS server accepts a request from a RADIUS client and returns either an accept or reject packet to the RADIUS client. It may also include session configuration information, such as an IP address and provide accounting services.

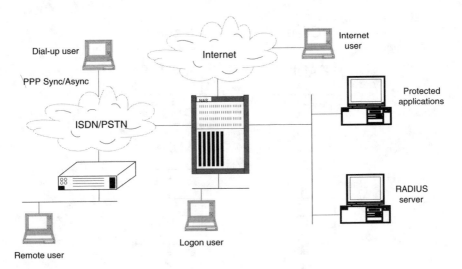

Figure 5.35 RADIUS elements.

When multiple RADIUS servers are configured, each may perform a client role by forwarding requests to another RADIUS server that is responsible for a particular user. This is referred to as proxy RADIUS. A single RADIUS server that handles both authentication and accounting is illustrated in Figure 5.35. Each function may have a separate server.

Initiating the RADIUS process is implementation-specific and may be dictated by the type of authentication. For directly connected users, the logon process of the NOS recognizes the configured RADIUS client. For remote connections, such as PPP, the configured RADIUS server is detected during LCP PAP/CHAP negotiation. (Depending on the implementation, RADIUS may be circumvented if a valid password is configured for LCP.)

5.16.1 RADIUS Architecture

RADIUS is an upper layer program and uses the UDP transport layer with port number 1812 for authentication and port number 1813 for accounting. RADIUS authentication is illustrated in Figure 5.36. UDP was selected in preference to TCP because many of the features of TCP would be duplicated with RADIUS, the high overhead of TCP and a faster response time using UDP.

5.16.2 RADIUS Operation

Detecting the presence of a RADIUS server and initiating an authentication request varies depending on whether the user is directly or remotely con-

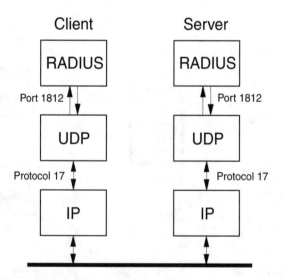

Figure 5.36 RADIUS architecture.

nected. The NOS detects the presence of RADIUS during the logon of directly connected users and presents a customized login prompt. A RADIUS server is detected during the LCP PAP/CHAP negotiation phase of remote users. Initiating RADIUS may be based solely on having a RADIUS server configured. It may also require additional conditions, such as the supplied user name, and password cannot be associated with a configured dial port. In this case, RADIUS is optional depending on the LCP configuration.

Once the determination is made to use RADIUS, the RADIUS client creates an access-request packet that contains the user name, password, and the port ID being accessed. Transactions between the RADIUS client and server are authenticated through the use of a shared secret, which is never sent over the network. The password is encrypted using a method based on the RSA *message digest algorithm* (MD5) [59]. This is the payload portion of the UDP datagram, which is directed to the RADIUS server via the network.

An access-request from a client that the RADIUS server does not have a shared secret for decrypting the password is silently discarded. Otherwise, the RADIUS server finds the database entry associated with the user and validates as required for the particular user. This always includes the password and may include other information, such as the port number. If any condition is not met, an access-reject packet is sent to the RADIUS client, which may display a textual message for the user that identifies the reason for rejection.

The user is not home free yet. The RADIUS server may choose to solicit additional information from the user in the form of an access-challenge packet, which is sent to the RADIUS client. The challenge packet contains an unpredictable, random number. The challenge for the user is to return the number properly encrypted. The RADIUS client prompts the user for the additional information and formats the user response in an access-request packet containing a new request ID. The user-password attribute has been replaced by an encrypted response, which is the access-challenge random number properly encrypted. Unauthorized users will lack the facilities (smart card, software, and shared secret) to calculate the proper response. The RADIUS server may respond with an access-accept, an access-reject, or another access-challenge.

When the access-challenge cycle is completed, the RADIUS server responds with an access-accept packet that authorizes the user network access, but sometimes with restrictions (i.e., stay out of the chat rooms because the user is on probation). Other parameters provided in the access-accept packet include an IP address, subnet mask, MTU, compression, and packet filter identifiers.

5.16.3 RADIUS Packet Format

Each RADIUS packet is contained in the payload portion of a UDP datagram and commences with a single octet code field that defines the type of RADIUS packet. The fixed 20-octet header format of the RADIUS packet is illustrated in Figure 5.37. Each field of the RADIUS packet is described as follows.

- Code: The code is an eight-bit field that identifies the type of RADIUS packet. The various RADIUS packet types are identified in Table 5.39.

- Length: The two-octet length field indicates the overall length commencing with the code field and includes the attribute field (described in Section 5.16.4). If an attribute is outside the length field, it is ignored. However, if the length field is longer than the number of attributes, the packet is silently discarded. The minimum length is 20 octets (the fixed header part), and the maximum length field is 4,096 octets.

- Identifier: The eight-bit identifier field is used in correlating responses received to the requests sent.

- Authenticator: The 16-octet authenticator field in the access-request packet contains a random, unpredictable number, which is used in conjunction with the shared secret between the NAS and RADIUS server and the user supplied password. These three numbers are put through a one-way MD5 algorithm to produce the real validation number (hash). This number is stored in an attribute described later.

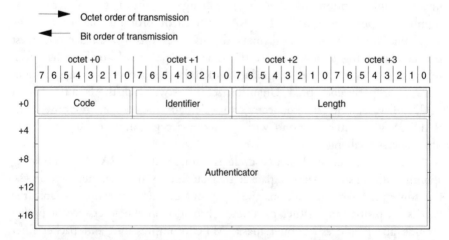

Figure 5.37 RADIUS packet format.

Table 5.39
RADIUS Packet Types

Code	Packet Type
1	Access-Request
2	Access-Accept
3	Access-Reject
4	Accounting-Request
5	Accounting-Response
11	Access-Challenge
12	Status-Server (experimental)
13	Status-Client (experiemental)

The same authenticator field in the access-accept, access-reject, and access-challenge packets contains a hash value derived by using the MD5 algorithm with the concatenated values of the Code, ID, Length, RequestAuth, Attributes, and the shared secret.

5.16.4 Attributes and Format

RADIUS attributes immediately follow the fixed 20-octet header and consist of an octet string of tagged data. The length of each attribute field is defined by the second octet of the attribute. The overall length of the attribute field is limited by the length field in the fixed header. The format of an attribute is illustrated in Figure 5.38.

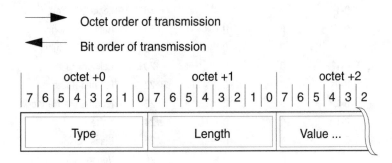

Figure 5.38 Attribute field format.

The single-octet type field identifies the function of the attribute. A RADIUS client or server may ignore an attribute with an unknown type. The attributes in use are identified in Table 5.40, which also identifies the RADIUS packet type using the attribute. Access-request packets must contain attributes 1 and either attribute 2 or (exclusive) attribute 3.

Other RADIUS attributes for authentication and accounting are defined to support the provisions of mandatory tunneling in dialup networks. [60] These attribute types begin with number 64, which is used to denote the tunnel

Table 5.40
RADIUS Attributes and Packet Types

Request	Accept	Reject	Challenge	#	Attribute Description
1	0	0	0	1	User-Name
0-1	0	0	0	2	User-Password
0-1	0	0	0	3	CHAP-Password
0-1	0	0	0	4	NAS-IP-Address
0-1	0	0	0	5	NAS-Port
0-1	0-1	0	0	6	Service-Type
0-1	0-1	0	0	7	Framed-Protocol
0-1	0-1	0	0	8	Framed-IP-Address
0-1	0-1	0	0	9	Framed-IP-Netmask
0	0-1	0	0	10	Framed-Routing
0	0+	0	0	11	Filter-Id
0	0-1	0	0	12	Framed-MTU
0+	0+	0	0	13	Framed-Compression
0+	0+	0	0	14	Login-IP-Host
0	0-1	0	0	15	Login-Service
0	0-1	0	0	16	Login-TCP-Port
0	0+	0+	0+	18	Reply-Message
0-1	0-1	0	0	19	Callback-Number
0	0-1	0	0	20	Callback-Id
0	0+	0	0	22	Framed-Route
0	0-1	0	0	23	Framed-IPX-Network
0-1	0-1	0	0-1	24	State
0	0+	0	0	25	Class
0+	0+	0	0+	26	Vendor-Specific

Table 5.40 (continued)
RADIUS Attributes and Packet Types

Request	Accept	Reject	Challenge	#	Attribute Description
0	0-1	0	0-1	27	Session-Timeout
0	0-1	0	0-1	28	Idle-Timeout
0	0-1	0	0	29	Termination-Action
0-1	0	0	0	30	Called-Station-Id
0-1	0	0	0	31	Calling-Station-Id
0-1	0	0	0	32	NAS-Identifier
0+	0+	0+	0+	33	Proxy-State
0-1	0-1	0	0	34	Login-LAT-Service
0-1	0-1	0	0	35	Login-LAT-Node
0-1	0-1	0	0	36	Login-LAT-Group
0	0-1	0	0	37	Framed-AppleTalk-Link
0	0+	0	0	38	Framed-AppleTalk-Network
0	0-1	0	0	39	Framed-AppleTalk-Zone
0-1	0	0	0	60	CHAP-Challenge
0-1	0	0	0	61	NAS-Port-Type
0-1	0-1	0	0	62	Port-Limit
0-1	0-1	0	0	63	Login-LAT-Port

Notes: 0: Not allowed; 0+: zero or more instances may be present; 0-1: zero or one instance may be present; 1: exactly one instance must be present.

type. When tunnel types are used, the field identified in Figure 5.38 as value, becomes a tag octet that is used to group attributes in the same packet that belong to the same tunnel. The tag octet is followed by the value octet, which identifies the type of tunnel as identified in Table 5.41.

5.16.5 RADIUS Accounting

The accounting function, like authentication, is best handled by a centralized database of user statistics. RADIUS does this using the same mechanism as described for authentication, except the port number is 1813 instead of 1812, and 11 new attributes are defined for accounting. The new attributes are identified in Table 5.42.

Table 5.41
Tunnel Types

Attribute number 65	Used to identify the tunnel medium type and there are 4 defined; (1) IP, (2) x.25, (3) ATM and (4) Frame Relay.
Attribute number 66	Used to identify the NAS end of the tunnel.
Attribute number 67	Used to identify the server end of the tunnel.
Attribute number 68	Used to identify the identifier assigned to the call.
Attribute number 69	Contains the tunnel password (hidden).

Table 5.42
RADIUS Accounting Attributes

Number	Attribute Description
40	Acct-Status-Type
41	Acct-Delay-Time
42	Acct-Input-Octets
43	Acct-Output-Octets
44	Acct-Session-Id
45	Acct-Authentic
46	Acct-Session-Time
47	Acct-Input-Packets
48	Acct-Output-Packets
49	Acct-Terminate-Cause
50	Acct-Multi-Session-Id
51	Acct-Link-Count

The same 40-octet fixed header is used with RADIUS accounting packets, followed by attributes. The first accounting transaction involves a RADIUS packet type code 4, accounting-request from the NAS to the RADIUS server to start service. The response from the RADIUS server to the NAS contains a RADIUS packet type code 5 (accounting-response). The accounting server may be the same as the RADIUS authentication server, or different. Starting and stopping accounting service is accomplished with different options of the same attribute 40 (acct-status-type). The four options to attribute 40 include start, stop, accounting-on, and accounting-off. At session conclusion the NAS (RADIUS client) generates a stop packet for the RADIUS server describing such statistics of the session as elapsed time, input and output octets, and input

Table 5.43

Cause Codes for Session Termination

Value	Cause Code
1	User request
2	Lost carrier
3	Lost service
4	Idle timeout
5	Session timeout
6	Admin reset
7	Admin rReboot
8	Port error
9	NAS error
10	NAS request
11	NAS reboot
12	Port un-needed
13	Port pre-empted
14	Port suspended
15	Service unavailable
16	Callback
17	User error
18	Host request

and output packets. After the accounting service is stopped, the attribute 49 (acct-terminate-cause) indicates how the session was terminated. The field value of the attribute identifies the cause (or error type) for session termination, as illustrated in Table 5.43. The RADIUS server uses this information to update the accounting database.

Endnotes

[1] The ICMP specification is contained in RFC 792.

[2] See RFC 896, Congestion Control in IP/TCP Internetworks, and RFC #1254 on survey of congestion control.

[3] See RFC 1191, Path MTU Discovery.

[4] See RFC 1112, Host extensions for IP multicasting.

[5] The specification for DVMRP is contained in RFC 1075.

[6] The specification for ARP is contained in RFC 826, and RFC 1009 contains implementation specifics.

[7] The specification for Proxy ARP is contained in RFC 826. RFC 1027 describes how to implement transparent gateways using ARP.

[8] Subnet addressing is described in Section 3.3.3, and the specification is contained in RFC 950.

[9] The specification for RARP is contained in RFC 903.

[10] The specification for IARP is contained in RFC 1293.

[11] The specification for BOOTP is contained in RFC 951 with updates in RFCs 1395, 1497, and 1542. The specification for DHCP is contained in RFC 2131, and a description of the interoperation is contained in RFC 1534. The superset of options is described in RFC 2132.

[12] See RFC919 for discussion on broadcasting datagrams and RFC 1812 for discussion on limited and directed broadcasting.

[13] Routers are the most common diskless machine today. Although the BOOTP specification indicates that hosts use BOOTP, a router acts like a host during the loading process. That is, they use BOOTP.

[14] The default values for many of these options are defined in RFC 1122 and RFC 1127, Host Requirements.

[15] For more detail on the Options (Vendor-specific) field, see RFC 2132.

[16] RFC 2132 indicates this number was made up, which made it magic.

[17] The UNIX implementation of RIP is called *routed* and pronounced "route-d". (UNIX programs end with "d" for daemon). The Berkeley 4BSD system contained *routed*.

[18] See RFC 1058, Routing Information Protocol.

[19] The specification for OSPF is contained in RFC 2178. There are several related RFCs identified in Section 5.8.4.

[20] Reference RFC 1584, Multicast Extensions to OSPF.

[21] Reference RFC 1597, OSPF NSSA Option.

[22] The specification for EGP is contained in RFC 904, which updates RFC 827 and RFC 888.

[23] See RFC 1166, Internet Numbers, for a list of autonomous systems.

[24] The current specification for BGP (Version 4) is contained in RFC 2178, which obsoletes earlier versions.

[25] RFC 1034, Domain Names Concepts and Facilities, provides an introduction to DNS. RFC 1035 is the specification for DNS. Both are updated by RFC 1348 with a new resource record to handle NSAP addresses.

[26] See RFC 1033, Local System Administrator.

[27] See "Introduction to Administration of an Internetbase Local Network," 46 pages, July 24, 1988, by Charles L. Hedrick, Rutgers University Computer Science Facilities Group, Piscataway, NJ. It is a guide to setup and administer a network using the TCP/IP protocol suite.

[28] The UNIX program providing DNS name server services is called *named.*

[29] See RFC 1237 (NSAP Guidelines) and RFC 1348 (DNS NSAP RR).

[30] The original network management effort for the Internet was called SGMP and described in RFC 1028 (November, 1987). The present SNMP (RFC 1157) was derived from RFC 1028 and reached the status of "Standard" in May 1990.

[31] The types of information maintained by agents is defined in the Management Information Base (MIB-2), specification in RFC 1213 and augmented with "Concise MIB Definitions," RFC 1212. The rules for defining the MIB-2 are contained in the *structure of management information* (SMI). The specification for the SMI is contained in RFC 1155.

[32] The SMI calls for using a subset of ISO 8824, specification for ASN.1, to define objects in the MIB-2. Further, RFC 1155 specifies that ISO 8825, Basic Encoding Rules (BER), will be used for encoding and decoding ASN.1 messages.

[33] The Internet Standard 17 defines MIBII (RFC 1213), with the status *recommended.* MIB-I is also defined with the status *recommended,* but without a standard number. There is a lot of activity in the MIB with many new enhancements being defined (and many obsoleted).

[34] This is considered a weak protection since network monitors easily reveal the code. Secure and SNMP-2 use much stronger tools.

[35] Not to be confused with bit error rate, which is also BER.

[36] MIB-II has three more groups than MIB-I. Namely, CMOT, transmission and snmp. The number of objects in several of the existing groups has been increased. SMI refinements were also made in MIB-II. MIB-II is a superset of MIB-I and each object in common has the same name.

[37] See Concise MIB Definitions, RFC 1212, for description of the OBJECT-TYPE macro.

[38] This could have resulted from just, severe criticism of the efficiency of ASN.1 and BER. (There are commercial products available that operate many times more efficiently.) The response could have been to pick two straws from the pile and combine them in a less than straightforward manner.

[39] The adapted subset is precisely defined in RFC 1155, Structure of Management Information (SMI).

[40] See MIB-II, RFC 1213 and Concise MIB Definitions, RFC 1212.

[41] *The Simple Book,* by Marshall T. Rose provides an address/telephone number for queries regarding ISODE.

[42] An Extension Generic Interface Table is defined in RFC 1229 and updated in RFC 1239 but is still considered experimental.

[43] The definition of Group 9 in MIB-2 has only the comment, "...historical (some say hysterical)."

[44] RFC 1214 is a specification of an MIB designed for use with CMIP, either over pure OSI or with the CMIP over TCP. The maturity level has been changed to *historic* to indicate it is obsolete.

[45] The OBJECT IDENTIFIER "transmission 7" is described in detail in RFC 1284, Definitions of Managed Objects for the Ethernet-like Interface Types.

[46] The OBJECT IDENTIFIER for DS1 is defined in RFC 1232, Definitions of Managed Objects for the DS1 Interface Type. RFC 1232 is updated by RFC 1239.

[47] RFC 1147, FYI on a Network Management Tool Catalog, provides a tutorial on network management and a catalog of many tools available to assist network managers in debugging and maintaining the Internet. The catalog tells what a tool does, how it works, and how it can be obtained. (Some are free.)

[48] The specification for SNMP over Ethernet is contained in RFC 1089.

[49] See RFC 1195, OSI ISIS for IP and Dual Environments. RFC 1195 is updated by RFC 1349.

[50] See RFC 1189, CMOT and RFC 1214, OSI Internet Management: MIB-2.

[51] See RFC 1351, SNMP Administrative Model and RFC 1352, SNMP Security Protocols.

[52] The explosion is characterized by the number of SMI network management private enterprise codes recently assigned. (This is the prefix 1.3.6.1.4.1.) In less than two years, nearly 500 were assigned.

[53] The original Internet time server (TIME) specification is contained in RFC 868, which is Internet Standard 26, with the status of *elective*. The current network time protocol (NTP) is Standard number 12, with "Recommended" status. Its specification is contained in RFC 1119, with updates in RFC 1305.

[54] The Time protocol (original) is much simpler than NTP or SNTP, but only provides a single 32-bit timestamp, which represents the first 32 bits of the present NTP 64-bit transmit timestamp. Time protocol listens on port number 37 and simply responds to a connection with the 32-bit timestamp representing the number of seconds since 1/1/1970, and closes the connection, either UDP or TCP.

[55] The specification for SNTP is contained in RFC 1361.

[56] The specification for the daytime protocol is contained in RFC 867.

[57] The echo protocol is Internet Standard #20 with a status of recommended. The specification is contained in RFC 862.

[58] The specification for RADIUS (authentication) is contained in RFC 2138 and the companion specification for accounting is contained in RFC 2139.

[59] RFC 1321, The MD5 Packet-Digest Algorithm.

[60] See draft RFC "dft_radius-tn.txt."

6

Upper Layer User Service Protocols

This chapter contains descriptions of the application layer programs that provide a direct service to users. Some of the programs use a lower layer program that may be used by other programs. For example, NFS uses RPC, which is independent and may be used by other programs. TELNET provides a remote login capability. FTP uses TCP to provide a secure file transfer facility, while TFTP uses UDP to provide the same, but less secure, file transfer facility. NFS provides remote file management capability, and the SMTP provides an electronic mail service. The X window system provides an independent graphics interface between a user and multiple applications.

6.1 TELNET

TELNET is a terminal login service that allows a user to make a TCP connection with an application on a remote host. The TELNET service (as seen by the user) is simple, but the protocol is much more complex. TELNET may also be used to connect terminal-to-terminal and application-to-application. The classic client/server model is followed for the TELNET service. That is, a TELNET client in the host serving the user opens a TCP connection with the TELNET server in the remote host on well known port 23. The client accepts data from the user and sends it to the remote server using TCP. The TELNET client also accepts response data from the remote server and displays it on the user's terminal. Since the server accepts requests from multiple clients, it has master and slave portions, as described in Section 1.3. The master opens the TCP connection with the application, and the slave performs the data transfer. The connection may be made with either a domain name or an IP address [1].

The client resides in the host to which the physical terminal is normally attached, and the server resides in the host that normally provides the application program. As an alternate point of view, applicable even in terminal-to-terminal or application-to-application connection, the client resides in the host that initiated the communication.

6.1.1   Network Virtual Terminal (NVT)

The characteristics of a locally attached terminal are known to the application, but the characteristics of a remote terminal will likely be unknown to the application. Hence, when TELNET is used, the client converts the terminal characteristics of the user to those of a universal terminal type, called pseudoterminal. The server (slave) converts the pseudoterminal characteristics to make them appear as though generated by a local terminal. The pseudoterminal is also referred to as a *virtual terminal*, and the feature just described is called *network virtual terminal* (NVT). Figure 6.1 illustrates the mechanism.

The application formats data just as it would for the locally attached terminal and sends data to the server, which converts it to the NVT format and sends to the client. The client formats the data for the user terminal and sends it.

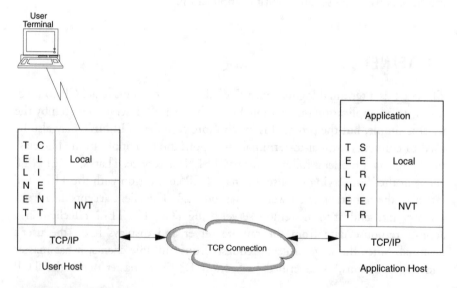

Figure 6.1   TELNET/NVT architecture.

The NVT represents the lowest common denominator of existing terminal features: it is a bidirectional character device with a printer and a keyboard; it operates in scroll-mode; it has unlimited line/page length; and it uses seven-bit ASCII characters encoded in eight-bit octets. Since ASCII control characters vary between systems, TELNET has a precise definition of NVT control characters, which is illustrated in Table 6.1. (The code number is the decimal value of the control character.)

6.1.2 Dynamic Mode Negotiation

Either the user or the application at any time during the connection may negotiate enhanced characteristics beyond those offered by the NVT. This is accomplished by embedded commands in the data stream. TELNET command codes are one or more octets in length and are preceded by an *interpret as command* (IAC) character, which is an octet with each bit set equal to one (FF hex). Should an octet equal to FF hex occur in real data, it must be preceded by an IAC character to prevent it from being mistaken as an IAC character. The receiving side discards the added IAC character, and data integrity is preserved. The IAC character may be followed by a single command, or by a command and negotiable options. Table 6.2 illustrates the TELNET command codes.

Although the underlying service to TELNET (TCP/IP) is full duplex, it can be placed in *half duplex* (HDX) mode to accommodate users with physically HDX terminals that have *lockable keyboards* (e.g., IBM 2741). When one side of the TELNET connection cannot proceed without input from the other side, that side must transmit the TELNET go ahead (GA) command, which is

Table 6.1
NVT Printer Codes

Mnemonic Name	Code Number	Meaning
NUL	0	No Operation
LF	10	Move to next print line
CR	13	Move to beginning of print line
BEL	7	Produce audible signal
BS	8	Move print position one space left
HT	9	Position to horizontal tab stop
VT	10	Position to vertical tab stop
FF	12	Position to top of page

Table 6.2
TELNET Command Codes

Mnemonic Name	Code Number	Meaning
SE	240	End of subnegotiation
NOP	241	No operation
DM	242	Data mark
BRK	243	Break
IP	244	Interrupt process
AO	245	Abort output
AYT	246	Are you there
EC	247	Erase character
EL	248	Erase line
GA	249	Go ahead
SB	250	Begin subnegotiation
WILL	251	Sender request enabling option
WONT	252	Sender rejects enabling option
DO	253	Sender request other side enable option
DONT	254	Sender rejects other side enabling option
IAC	255	Interpret as command

suppressed by systems that do not require the HDX mode. For example, applications do not normally require such action. (See Option 3 in Table 6.3; the suppress go ahead option *must* always be accepted.) The TELNET GA is encoded in the data stream as "IAC,GA" (FF,F9 Hex).

TELNET also provides an out-of-band signaling capability with embedded command codes. When desired to have commands bypass the normal data stream, the IAC code and a DM code delimit the special data. Commands in this sequence are given to TCP to be handled as urgent data (IAC, BRK, and DM, for example).

Each negotiable option has an ID, which immediately follows the command for option negotiation, that is, IAC, <command>, <option code>. The negotiation process involves commands and responses, because commands may be accepted or rejected. See Table 6.3 for a list of characteristics that may be negotiated.

There are only four commands for option negotiation, but because the commands and responses may easily be confused, they require clarification.

Table 6.3
TELNET Option Codes

Option ID	RFC Number	Meaning
0	856	Binary transmission
1	857	Echo
2	DDN HB	Reconnection
3	858	Suppress go ahead
4	DDN HB	Message size negotiation
5	859	Status
6	860	Timing mark
7	726	Remote controlled transmission and echo
8	DDN HB	Output line width
9	DDN HB	Output page size
10	652	Output carriage return disposition
11	653	Output horizontal tab stops
12	654	Output horizontal tab disposition
13	655	Output form feed disposition
14	656	Output vertical tab stops
15	657	Output vertical tab disposition
16	658	Output line feed disposition
17	698	Extended ASCII
18	727	Logout
19	732	Byte macro
20	735	Data entry terminal
21	736	SUPDUP
22	749	SUPDUP output
23	779	Send location
24	930	Terminal type
25	885	End of record
26	927	TACACS user identification
27	933	Output marking
28	947	Terminal locator number
255	861	Extended options list

There are two major divisions of negotiation: negotiating what I want for my side and what I want for your side. (Each side may have different characteristics.)

When I am negotiating for my side, the verb *will* is used for enabling (response *do* for accept and *don't* for reject), and the verb *won't* is used for disabling (response *don't* for accept—it cannot be rejected).

When I am negotiating for your side, the verb *do* is used for enabling (response *will* for accept and *won't* for reject), and the verb *don't* is used for disabling (response *won't* is affirmative becuase the basic NVT must be supported).

Using the terminal type option (option 24), the mode may be changed from scrolling to page. This permits screen users to inhibit further output until the present page is released. Also, a forms mode may be negotiated, which requires only that the fields of forms be transmitted by users and that output to users is added to a selected template (form) on the screen.

6.1.3   IBM 3270 Terminal Support (TN3270)

Some implementations provide a set of TELNET options for IBM 3270 class terminal support. This is called TN3270 Client and TN3270 Server. The negotiation for TN3270 essentially implements a bypass of NVT, and the binary data is handled transparently. Even in transparent mode, the data is scanned for the NVT IAC command character. Data equal to the IAC character (FF hex) has an embedded IAC to signal the receiver that it is data [2].

6.1.4   TELNET Variations

An implementation of TELNET under UNIX called *rlogin* provides greater flexibility and permits direct access to the remote application's command interpreter. The UNIX server daemon is called *telnetd,* and the command to use telnetd is "rsh." The following is an example using telnetd with the command "rsh."

    rsh <domain name><command>

This command executes the <command> the host identified by <domain name> just as though the user were directly connected to <domain name>. With this kind of uncontrolled access to the command interpreter, rlogin has serious security problems.

6.2 File Transfer Protocol (FTP)

The ability to transfer files between users and applications satisfies several needs. Some examples are diskless workstations that rely on a centralized file server for disk storage, workstations with limited storage that rely on a file server for archive, and a group of workstations (or applications) that uses a file server for a common database. The common database is an online sharing by members of a group while the other examples perform a transfer of files. In both cases, the details of FTP become complex to handle access protection, authorization, file ownership, and data formats.

FTP is the second-most-used application on the Internet. It may be implemented on small PCs or large mainframes, and it uses the transport layer TCP to provide connection set-up, reliable data transfer, and data pacing. The specification for FTP is contained in RFC 959, which may be retrieved from the InterNIC Directory and database services server using FTP. For users with a Windows FTP client, it is a point-and-click, intuitive process to obtain RFCs. For DOS-based users, the initial connection to the FTP client software is accomplished with a DOS *FTP* command. Depending on the client software, it may be possible to identify the host in the FTP command. If not, it will be necessary to enter an "Open" command and specify the destination. For example, the possible destinations are "rs.internic.net" for the U.S. east coast, "ftp.isi.edu" for the U.S. west coast, "ftp.is.co.za" for Africa, "nic.nordu.net" for Europe, and "munnari.oz.au" for the Pacific Rim [3]. You will be prompted for a user name and password, which is "anonymous" (user name), and your e-mail address (password). The FTP client will make a TCP connection with the server. Normally, a server would require identification in the form of "user@host," but the server accepts the user name *anonymous* for general public access. The RFCs (specifications) are under the directory *RFC,* so it is necessary to change to that directory with the command, "cd rfc." See Figure 6.2 for an illustration of a client/server model with a user accessing its FTP client software, the Internet, and the FTP server.

The user command to transfer a file containing text for an RFC is "get rfcnnn.txt", and to obtain the specification for FTP the user command is, "get rfc959.txt".

Once FTP is invoked, the FTP client responds to the user as in an interactive system. To obtain a list of all commands, enter the command, "ftp help." The response to the general help command is a list of all commands. Help for a particular command may be obtained with the command, "help <command>." Each command entered is followed by a response beginning with a three-digit number. The first digit of the number provides a status of the entered command. That is: (1) denotes positive preliminary reply, (2) denotes positive

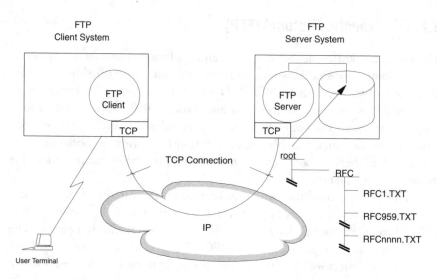

Figure 6.2 FTP client/server model.

completion reply, (3) denotes positive intermediate reply, (4) denotes transient negative completion reply, and (5) denotes permanent negative completion reply. During the login process the first digit of the banner requesting login is "3"—positive intermediate reply. The first digit of a file transfer completion notice is "2." The following example, illustrated in Figure 6.3, uses FTP to obtain the specification for FTP. The user (local) commands entered are illustrated in bold font.

Other FTP commands may be determined by entering a help command. Figure 6.4 is an example of using the FTP help command to obtain a list of all the remote FTP commands. Figure 6.4 also shows a help command for a particular command "dir."

If interested, there are many topics that may be browsed by changing to a different directory than RFC. For example, "cd protocols" for TCP-IP & OSI Documents, or "cd std" for a list of standard numbers with associated RFCs.

The FTP architecture allows concurrent access by multiple clients. Hence, a master/slave server is used, just as described in the model client/server. See Figure 6.5 for an illustration of the control and data connections.

The master awaits new connections and creates a slave to handle each connection. The slave accepts and handles the control connection from the client, and it initiates a separate process to handle the data transfer connection. As illustrated in Figure 6.2, the client control connects to the server control using one TCP connection and the associated data transfer portion uses a separate TCP connection. The control connection remains open for the duration of the

```
>ftp rs.internic.net
Connected to rs.internic.net
220 *****Welcome to InterNIC Registration Host *****
*****Login with username "anonymous"
*****You may change directories to the following:
ddn-news          -DDN Management Bulletins
domain            -Root Domain Zone Files
ien               -Internet Engineering Notes
iesg              -IETF Sterring Group
ietf              -Internet Engineering Task Force
internet-drafts   -Internet Drafts
netinfo           -NIC Information Files
netprog           -Guest Software (ex.whois.c)
protocols         -TCP-IP & OSI Documents
rfc               -RFC Repository
scc               -DDN Security Bulletins
std               -Internet Protocol Standards
220 And more!
User (rs.internic.net(none)): anonymous
331 Guest login ok. Send your complete e-mail address
password: XXXXX [echo turned off]
230 Guest login ok. access restrictions apply.
ftp> cd rfc
250 CWD command successful
ftp> get rfc959.txt
200 PORT command successful
150 Opening data connection for rfc959.txt (147316 bytes).
226 Transfer complete.
local rfc959.txt remote rfc959.txt
14716 bytes received in 30 seconds (4.7 Kbytes/s)
ftp> close
221 Goodbye
ftp> quit [or bye]
```

Figure 6.3 Example using InterNIC to obtain RFCs.

connection, but the data transfer connection may be closed and another opened for successive file transfers. To protect against a failed server, the client sets an inactivity timer. If it expires, both connections are closed. The default value of the timer is five minutes.

FTP transmits TELNET character strings over the control connection to convey commands and responses. The commands begin with a command code followed by an argument field that is separated by one or more spaces. The argument field consists of a variable length character string ending with the character sequence <CRLF>. Every command must generate at least one reply, although there may be more than one. An FTP reply consists of a three digit

ftp>| help |
Commands may be abbreviated. Commands are:

!	close	lcd	open	rmdir
$	cr	ls	prompt	runique
account	delete	macdef	proxy	send
append	debug	mdelete	sendport	status
ascii	dir	mdir	put	struck
bell	disconnect	mget	pwd	sunique
binary	form	mkdir	quit	tenex
bufsize	get	mls	quote	trace
bye	glob	mode	recv	type
case	hash	mput	remotehelp	user
cd	help	nmap	rename	verbose
cdup	image	ntrans	reset	?

ftp>| help dir |
dir   list contents of remote directory

Figure 6.4 Example using FTP help.

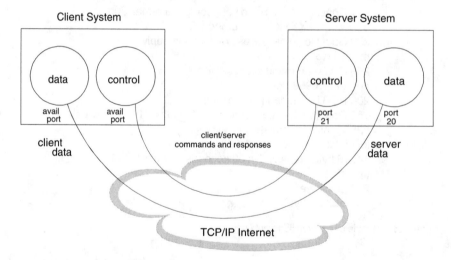

Figure 6.5 FTP control and data connections.

number (transmitted as three alphanumeric characters) followed by text. The number is used to determine the next state to enter. The three-digit code, followed by space <SP>, followed by one line of text, and terminated by the TELNET end-of-line code constitutes an FTP response. The server takes no action until the CRLF is received. Many of the available FTP commands are not used. The most used FTP commands and the ones designated as RE-QUIRED are identified below (in capital letters).

1. Solicit user name (USER);
2. Solicit password (PASS);
3. User's account number (ACCT);
4. Change working directory (CWD);
5. Change working directory to parent directory (CDUP);
6. Logout (QUIT);
7. Data representation: ASCII, EBCDIC, IMAGE, byte size (TYPE);
8. Structure as file, record, page (STRU);
9. Transfer as stream, block, compressed (MODE);
10. Port used for data connection (PORT);
11. Request server to listen on port (PASV);
12. Retrieve from server (RETR);
13. Store at server (STOR);
14. Append to specified server file (APPE);
15. Rename a file (RNFR and RNTO);
16. Abort the transfer of data (ABOR){must accept};
17. Delete a file (DELE);
18. Remove a directory (RMD);
19. Make a directory (MKD);
20. Print working directory (PWD);
21. Send name list from server to user (NLST);
22. Send directory list or file information (LIST);
23. Determine type of operating system (SYST);
24. Solicit status of present command (STAT);
25. Obtain help from server (HELP);
26. No action (NOOP).

Figure 6.6 illustrates the primary sequence of commands and responses between the client and server to obtain the same FTP specification previously illustrated for the user commands only. It assumes the user command.

"ftp ns.internic.net" has been entered and responded to with, "Connected to rs.internic.net, , 220 Welecome to the rs.internet.net Registry host. (Responding with "anonymous" for the user name yields a response code of 231 and with the correct password 230, Guest Login OK, Restrictions apply.") User commands are illustrated in lower case, and client commands are in upper case.

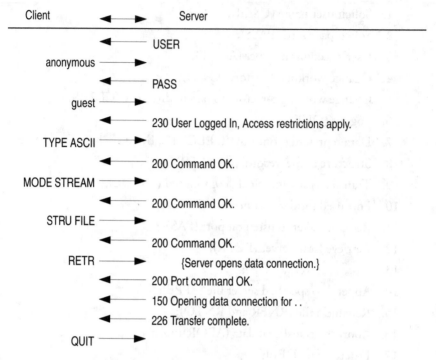

Figure 6.6 Example of FTP client/server commands.

All commands and responses prior to the retrieve command (RETR) use the TCP command connection. The actual data transfer uses the TCP data connection. RETR by the client was preceded by a *get* command from the user. Upon completion of the data transfer, the data connection is closed. If the user enters a new *get* command, a new data connection is opened. The QUIT command from the client was preceded by a *close* command from the user, which resulted in the response message, "221 Goodbye." from the client to the user (not shown). The user enters the command "quit" (or "bye") to disconnect from the server.

6.3 Trivial File Transfer Protocol (TFTP)

The service provided by TFTP is a no-frills file transfer facility. Unlike the FTP, TFTP does not logon to the server host using TCP. It uses the connectionless UDP for the transport layer and must specify the source and destination port numbers for the UDP header to provide a multiplexing service. That is, the convenient mechanism of interfacing TCP with a TCB to open, maintain, and close the connection is not available. Since UDP is an unreliable service, TFTP must also be responsible for transmission reliability, which it accomplishes with a timeout and retransmit mechanism.

If a more reliable, sophisticated file transfer program is available, why should one use TFTP? FTP is complex and requires a lot of memory as compared to TFTP. TFTP is small and can easily be encoded on ROM to provide an initial memory image (bootstrap) when a diskless workstation is powered on. Once the initial bootstrap program is loaded from ROM, the remainder of the bootstrap may be obtained using FTP/TCP from any server in the Internet. (This is also discussed in Section 5.6.1.)

6.3.1 Architecture of TFTP

TFTP is an applications layer program that interfaces directly with the transport layer UDP; the architecture is illustrated in Figure 6.7 [4].

In the interface to UDP, the client specifies the well-known destination port number 69 and a unique source port number. The TFTP server listens on port number 69 and makes responses to the unique source port specified in the received UDP datagram. By using different source port numbers, the client can

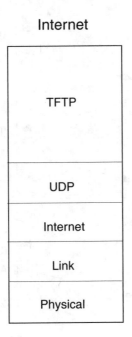

Internet

Figure 6.7 Conceptual layering of TFTP.

handle multiple users, each communicating with the same TFTP server. The TFTP service follows the basic client/server model described in Section 1.3.

6.3.2 Mechanics of TFTP

With a total of only five commands, the rules of TFTP are simple. An example of a user performing a write to the TFTP server is illustrated in Figure 6.8, with an explanation of each transmission.

The commands are 1) read, 2) write, 3) send, 4) acknowledge, and 5) error; they are explained later. The TFTP process begins with the user sending either a read request (RRQ) or a write request (WRQ). The RRQ is used to obtain a file from the server and the WRQ is used to send a file to the server. (The example above uses WRQ, code 2.) The communication is synchronous in that every command sent must be acknowledged before the next command

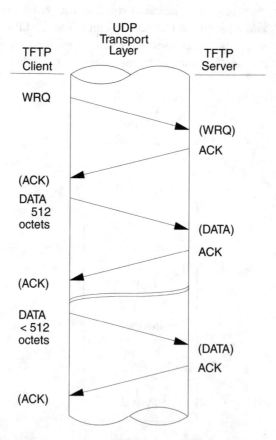

Figure 6.8 TFTP process.

can be sent. The WRQ request is acknowledged with an ACK response from the server that contains a block number equal to zero. This is first used to signal the user that the connection is initialized. In data blocks it is used for message integrity. Data blocks, each 512 octets, follow the RRQ or WRQ, and each is acknowledged by the server with an ACK. This process of sending and acknowledging continues until a block with less than 512 octets is received by the server, which is the signal that this is the last block of the message. The server acknowledges the end of transmission indication by sending an ACK response, which terminates the transfer. If the sender times out waiting for an ACK, it retransmits the last unacknowledged data block. If the receiver times out waiting for a data block, it retransmits the last ACK block.

6.3.3 Operation Codes

Each of the five operation codes is identified with the associated code number in Table 6.4.

6.3.4 Format of TFTP Messages

The format of TFTP messages varies by operation code. Each format has in common that it starts with a two-octet field that contains the operation code.

6.3.4.1 Read (RRQ) and Write (WRQ)

The TFTP header for either read or write contains the appropriate operation code (1 or 2), followed with the file name and mode identification that are separated by an octet equal to zero, and terminated with an octet equal to zero. The format is illustrated in Figure 6.9.

Table 6.4
TFTP Operation Codes

Mnemonic Name	OPS Code	Meaning
RRQ	1	Read data from server
WRQ	2	Write data to server
DATA	3	Data
ACK	4	Acknowledgement
ERR	5	Error

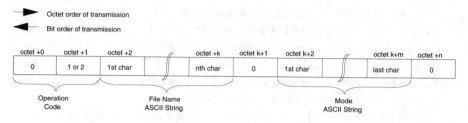

Figure 6.9 Read/write operation codes.

The operation code is a two-octet field with possible values of 01 (hex) or 02 (hex). The limited number of operation codes effectively begins each RRQ or WRQ request with an octet equal to zero.

The filename field is a string of ASCII characters that specify the name of the file to read or write and is terminated with an octet equal to zero.

The mode field is a string of ASCII characters that specify the mode of the message to be read or written. TFTP supports two modes: "octet" and "netascii."

The "netascii" (as defined in the specification) mode is the standard, 7-bit ASCII, but with each character occupying eight bits. "netascii" data is used to send messages (character strings).

The "octet" mode is eight bits of raw data and is used to transfer files.

The receiver, for either mode, is responsible for interpreting the data and storing it in the appropriate format dictated by the receiving host operating system. (For example, the word length in some machines is not equal to an even multiple of eight bits (number of octets), which means that every word will contain filler (pad).

6.3.4.2 Data

The TFTP header for a data block contains a two-octet field for the operation code, a 16-bit block number, and up to 512 octets of data. The subtle message (and connection) terminator is a block with less than 512 characters. The block number is initialized to the value zero with the ACK to the read or write request. The first data block has a block number equal to one and is sequentially incremented for each block of the message. This enables the receiver to identify duplicate blocks and missing blocks. The format of the data block with header is illustrated in Figure 6.10.

6.3.4.3 Acknowledgment

The TFTP ACK block is essentially the same as a data block, but without the data and the operation code equal to 04 instead of 03. See Figure 6.10. The

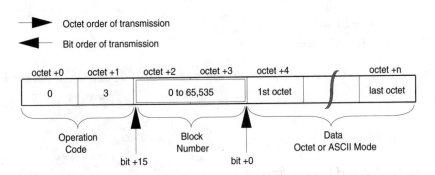

Figure 6.10 TFTP data block format.

block number in the ACK is equal to the block number of the message received. The server uses an ACK to acknowledge data blocks received, and the client uses a data block to acknowledge an ACK, except for duplicate ACKs and the ACK terminating the connection.

The exception for termination is obvious, but for duplicate ACKs it is subtle and requires special mention. By design, receipt of a duplicate ACK can only occur when the first ACK was delayed and resolicited with the original (duplicate) data block. To break the cycle of duplicate data transmissions (and ACKs), the current data block is never retransmitted as a result of receiving a duplicate ACK [5].

6.3.4.4 Errors

A TFTP error message is identified by an operation code equal to five. The ERROR block is a negative message acknowledgment, which may cause the block to be retransmitted, or the connection to be broken. Many error conditions, such as timeout, simply result in dropping the connection. The error message has four fields. The operation code field is equal to 05 hex and is immediately followed by a two-octet error code field. See Table 6.5 for a list of error codes. The error code field is followed by an ASCII string for human consumption that explains the error code. The ASCII string (and error message) is terminated with an octet equal to zero. The error message is illustrated in Figure 6.11. Table 6.5 lists the error codes.

The specification for TFTP does not discuss how to provide protection from illegal access, which leaves the topic for local specific implementation. The file attributes (read only, read and write, or restricted access) normally provide the only protection from illegal access.

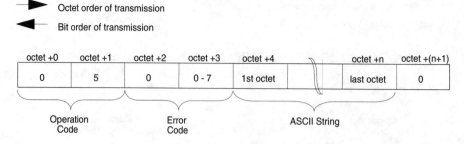

Figure 6.11 Format of TFTP error message.

Table 6.5
TFTP Error Codes

Error Code	Error Message (ASCII String)
00	Not defined (see message if present)
01	File not found
02	Access violation
03	Disk full
04	Invalid operation code
05	Unknown port number
06	File already exists
07	No such user

6.4 Remote Procedure Call (RPC)

Remote procedure call (RPC), as the name suggests, provides the programmer with the capability of having procedures (tasks, subroutines, and subprocedures) executed remotely. The programmer declares to the compiler that a particular procedure is remote. The compiler inserts the proper code for a RPC to the client. On the server side, the procedure is declared to the compiler that it is the remote procedure. When the RPC is made, the client obtains control and forms a message to the server that contains the arguments for the remote procedure. The server unpacks the arguments, performs the procedure, packs the results of the procedure in a message, and returns it to the client. The client unpacks the message and returns control to the calling program [6]. Having the compiler insert the RPC and the client/server provide results to the calling program shields the programmer from the complexities of the network. In fact, the process may be less complicated for the programmer than some local procedures that use stacks for arguments and place the results of the local procedure in a different hardware stack (e.g., extended memory).

RPC is a session layer protocol that may use either UDP or TCP transport protocols. The selection between UDP or TCP is typically determined by the length of the messages between the client and the server. When RPC uses the transport layer UDP, the calling application program must be aware in order to perform its own timeout, retransmission and duplicate detection as the RPC does not provide these services. Some versions of UDP arbitrarily truncate messages larger than 4,096 octets, which may necessitate the use of TCP transport. When TCP transport is used, RPC inhibits any retransmissions since TCP assures a reliable data transfer.

The session layer header for RPC contains the transaction ID, the identifier for the direction of transfer, the protocol version number, the program number, the procedure number, and the authorization (both information and verification). From this information, the server is able to select and execute the proper procedure and return the results to the client.

An appendix to the specification of RPC describes another protocol necessary for using RPC—namely the *port mapper protocol* (PMP). As mentioned earlier, several port numbers are reserved for specific server functions. (A list of all reserved ports is noted in Section D.2 of Appendix D.) The allocated bits for reserved port numbers are limited to the low order 10 bits, which limits the number of reserved ports to 1,024. (This number was recently expanded from 512 to 1,024.) To permit expansion beyond the 1,024 limit, the PMP allocates a dynamic port number for a specific server function. This permits a reasonable expansion capability for the *fixed* server port numbers in the dynamic range (210 through 2^{16}-1), which makes dynamic binding of remote programs possible. PMP itself uses a fixed port number (111 for both TCP and UDP) to permit registration of remote programs with dynamic port numbers. When a program first becomes available on a machine, it registers itself with the PMP on the same machine. The response is accepted or rejected if unable to establish the mapping or one already exists for the program. When a program becomes unavailable, it unregisters itself with the PMP on the same machine. The PMP also provides the capability of getting a port number that is already in use, listing all programs (with version, protocol, and port values), and calling another remote procedure on the same machine without knowing the remote procedure's port number.

6.5 External Data Representation (XDR)

External data representation (XDR) is a sublayer function to the NFS protocol, which is described in Section 6.6. It was developed as an independent function that can be used by other upper layer protocols. XDR, which is a presentation layer protocol in the OSI reference model, provides uniformity in the manner in which data is represented. For example, both Intel 80x86 and Motorola 680x0 machines (e.g., IBM 370 and VAX) have a 32-bit word. However, byte one of the Motorola machine represents the high-order byte, while byte one of the Intel 80x86 machine represents the low-order byte. This characteristic is sometimes referred to as *little-endian* when the lsb is stored at the lowest memory address, and *big-endian* when the msb is stored at the lowest memory address. Another characteristic that causes incompatibility between different machines is the alignment of elements. For example, the 68010 processor cannot work with unaligned addresses. A potential problem also exists with the representation of bits within a byte and the representation of floating point numbers within a word. Fortunately, manufacturers have complied with ISO 646 in transmitting the lsb of an octet first. Note that this may be displayed (illustrated) either from left to right or from right to left, but the lsb is the first transmitted. Manufacturers have also adhered to IEEE 754 in storing floating point numbers in a word. For example, a 32-bit word has one sign bit, 8 bits for the exponent, and 23 bits for the mantissa.

When an application layer wishes to communicate with the application layer of a different machine, it first converts the data elements into XDR format and sequentially stores each in a data buffer. The buffer is sent to the Internet for delivery to the server—typically by RPC, TCP/UDP, and IP. The conversion from XDR format is the responsibility of the receiving machine.

Although these incompatible machines remain incompatible, their data can be exchanged by using XDR. The downside to using XDR is the unnecessary conversion of data when identical machines are involved that do not have XDR as their internal format [7].

6.6 Network File System (NFS) Protocol

The NFS protocol basically permits an application to use the storage of a remote location the same as local storage. The NFS product was introduced by Sun Microsystems in 1985 and takes its name from the NFS protocol. This was a logical product for Sun Microsystems since its main product was diskless workstations. Because it was developed to accommodate different operating systems, NFS is open. That is, it is not biased to a particular manufacturer. It is available for MS-DOS, VM, MVS, VAX, etc. At the beginning of 1990, more than 250 computer manufacturers had licensed versions 4.0 of NFS from Sun Microsystems. Other implementations of the NFS protocol are available. For example, FTP Software offers an implementation of NFS for DOS-based systems that is called Interdrive. An implementation for DOS based systems with a serial line interface is offered by Information Modes.

NFS, when configured properly, is nearly transparent to the user. When installed in a DOS-based PC, the user employs NFS to access files on another host by simply changing the directory to the NFS drive and issuing normal read/write commands. For example, if a PC running DOS and TCP/IP network software has two floppies and a hard disk configured as drives A, B, and C, and drive D is configured as the storage of a different machine, entering the command "CD D:" results in the storage of the remote machine being mounted for use just as though it were local. If the hard disk "C" is old and slow, the transfer rate for the remote storage ("D") may even be faster—typically several hundred kilobytes per second on a 10M Ethernet LAN, or approximately 80% of the directly connected transfer rate. Even on a 115-Kbps serial line, the transfer rate is amazingly fast and it is not easy to identify that the transfer is slower than a locally connected drive.

The NFS product was designed to permit other usage of the XDR (presentation layer) and RPC (session layer). That is, other application programs may use either XDR or RPC. (These protocols were described in Sections 6.4 and 6.5.)

There are several competing remote file type protocols but only NFS has been classified as a "proposed standard" protocol [8]. Other competitors include Remote File System (RFS) by AT&T and a medley created by IBM and Microsoft (PC-NET, MS-NET, NETBIOS, and LAN Manager). The OSF is preparing a *distributed file service* (DFS), which may be a contender for an official Internet standard. DFS solves many of NFS's problems (including data integrity and security).

6.6.1 NFS Architecture

NFS is an application layer protocol. If comparing it to the OSI reference model, there is a one-to-one correspondence. See Figure 6.12 for an illustration of conceptual layering of NFS [9].

6.6.2 NFS Client/Server Model

NFS follows the classical client/server model described in Section 1.3. The NFS client is complex relative to the typical client function because it must be

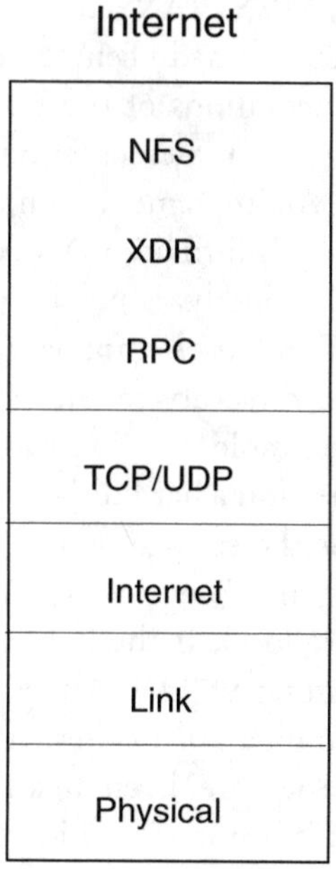

Internet

NFS
XDR
RPC
TCP/UDP
Internet
Link
Physical

Figure 6.12 Conceptual layering of NFS.

integrated into the operating system. When an application program requests a file operation, the operating system must distinguish between local storage and remote storage (NFS client software). See Figure 6.13 for an illustration of NFS client system functions.

The NFS client is transparent to the application programs, which require no alteration or special preparation for using NFS. When an application program issues a file command for a drive that is not local, the operating system passes the command to the NFS client. The NFS client establishes communication with the NFS server using a protocol called *mount*. The protocol XDR is used to normalize the file content, and the protocol RPC is used to transfer data between the NFS client and NFS server.

Multitasking systems, such as UNIX, may exercise NFS in both directions. That is, each system may have both a client and server function that are simultaneously active. Single-tasking systems, such as basic MS-DOS in a PC, only permit either the client or server function to be active, but not both simultaneously. Quarterdeck's new multi-tasking DOS environment permits PCs to exercise NFS in both directions.

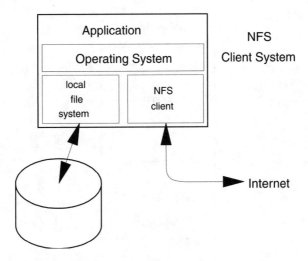

Figure 6.13 NFS client.

6.7 Simple Mail Transfer Protocol (SMTP)

The function of a mail program is to facilitate the exchange of electronic messages between users on a network. There are many mail programs in use that vary from complex and difficult to reasonably straight forward and simple to use. The Internet uses *simple mail transfer protocol* (SMTP) to provide a mail service. Some networks provide a gateway to the Internet's SMTP for exchanging mail [10]. Hence, the user interface is dictated by their home network. The user interfaces are different, and switching from one to another may be difficult. The user interface for some require entering the destination first and then the message. Others require entering the message first and then the destination (e.g., CompuServe). Imagine typing a long message and then finding that the destination address is invalid! The goal of SMTP is to provide a simple, straightforward message service.

6.7.1 Architecture of SMTP

SMTP is an applications layer program that interfaces directly with the transport layer TCP [11]. The destination, well-known port number 25, is used in the TCP header, which causes TCP to direct incoming segments to SMTP for processing. The SMTP architecture is illustrated in Figure 6.14.

Internet

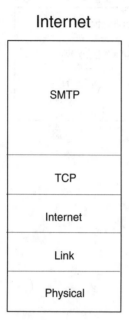

Figure 6.14 SMTP architecture.

6.7.2 General Mechanics of SMTP

The SMTP process commences with a user, assisted by an agent submitting a message to an SMTP client for delivery to another user. The destination user may be directly connected to the same SMTP system or to another SMTP system, which is connected via the Internet. That is, the two end users may or may not be in the same autonomous system. See Figure 6.15 for an illustration of the mechanics. In this illustration, user B may either send to user A or user C.

6.7.2.1 User Message Preparation

The message, called memo, is typically sent interactively by a user with its agent (the portion of the client that interfaces the user). The agent is, or should be, a user-friendly program that accepts minimum required elements of the message in two parts. The header part contains answers to prompts for "To," "From," and "Date" (required) and other optional fields such as Sender, Message-ID, Reply-To, cc, Comment, and In-Reply-To. The message part is simply text. Depending on the friendliness of the agent, some of the prompts for header items may be implied or symbolic. The agent may reside in a DOS-based PC and only require an address (e.g., Bob) and text from the user.

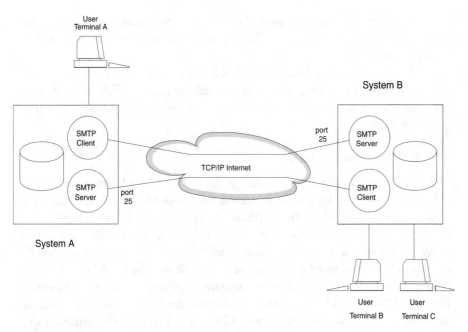

Figure 6.15 SMTP client/server model.

6.7.2.2. User Agent Functions

In addition to assisting the user with message input (prompts), the agent builds a formatted message, US ASCII characters only, with the information supplied by the user. The formatted message consists of a variable number of header lines. The header is terminated by a blank line (CRLF)(CRLF) and followed by the message as composed by the user. The message is of the following form:

Date :{date/time given to SMTP client} (CRLF)

From :Richard<RichardHarms@This-host.This-net> (CRLF)
To : John Meyer<Group@Sys.net> (CRLF)
Subject : {Subject provided by user} (CRLF)
Sender : Dicksec@Other-Host (CRLF)
Reply-To : Hughlan.Warren@Tool-Sup (CRLF)
cc : Larry Dorf <Banks@Big.mon>
Comment : {What ever comment user has in mind} (CRLF)
In-Reply-To : <news@Register.com> (CRLF)
Message-ID : <2345.123.lm-The@Other-Host>](CRLF)(CRLF)
{User text, converted to US ASCII if required.}

The "Date :," "From :," and "To :" lines in the header are required. All others are optional. Note, the header is terminated with a blank line "(CRLF)(CRLF)."

The user agent then builds a standardized list of required destinations (derived from user inputs and converted to machine readable format) and sends both the standardized list and the formatted message (tagged together) to a queue for the SMTP client.

6.7.2.3 SMTP Client Functions

The SMTP client reads the next message from the queue of user/agent inputs. (The queue may be large, since it is used for the collection of messages from all user agents serviced by the particular SMTP client.) If an address in the standardized list is for this SMTP system, it is handed off directly to the SMTP server. Otherwise, a TCP connection is established with each remote SMTP server machine warranted by the addresses in the standardized list. (If all addresses are for the same SMTP server domain, only one connection is opened.) The SMTP client sends each address in the standardized list, one at a time, to the SMTP server associated with the address. The SMTP server sends an OK reply for each address received and the SMTP client marks the address in the standardized list as sent. When all addresses in the standardized list are sent, the SMTP client sends a single copy of the message to each SMTP server with a TCP connection. The SMTP server sends a positive acknowledgment for the message, and the SMTP client is no longer responsible for the message.

6.7.2.4 SMTP Server Functions

The SMTP server constructs a header for each address received and places it in the queue of the appropriate mailbox, or an output queue of another SMTP server if the message is being forwarded. The header contains a pointer to the user's text and typically five lines of header constructed by the SMTP server. When the user opens the mail box and reads the message, it will appear as follows.

> Return-Path: {Should be same as From: by client} (CRLF)
> Posted-Date: {date/time actioned by server} (CRLF)
> Received-Date: {date/time received from origin} (CRLF)
> Received: from {address of relay or origin} (CRLF)
> Date: {date/time placed in user's mailbox} (CRLF)
> From: Richard <RichardHarms@This-host.This-net>(CRLF)
> To: John Meyer<Group@Sys.net<(CRLF)

[Optional fields supplied by SMTP client](CRLF)(CRLF)
{Text as supplied by user}

Note: Lines in bold font are constructed by SMTP server.

6.7.3 SMTP Memo Addressing

An Internet SMTP client expects the destination address of a message to be of the form, "local part@domain.name." The local part is the mailbox address at the destination machine. The domain name is of the destination SMTP server.

The destination domain name may not be serviced by the receiving SMTP server. This is handled by adding relay addresses in front of the final address. The final address is separated from the relay addresses with a colon (:). The format of the address with relays is of the form:

"@relay name 1@,,,@relay name n:local part@domain name." The receiving SMTP server removes its domain name and, if not the last relay, forwards the message to the next address in the chain. Private gateways vary in their treatment of relay addresses. The following examples illustrate the variations in address treatment by some relays.

- Internet user to Internet user: username@domain name. For example,

  gsmith@nisc.sri.com

- Internet user to BITNET user: user%site.BITNET@BITNET-GATE-WAY. For example,

  gsmith%emoryul.BITNET@cunyvan.cuny.edu
  gsmith%emoryul@CORNELLC.CIT.CORNELL.EDU

- Internet user to UUCP user: user%host.UUCP@uunet.uu.net. For example,

  user%domain@uunet.uu.net

- Internet user to SprintMail user (case-sensitive):

  /G=Mary/S=Anderson/O=co.abc/ADMD=SprintMail/C=US/
  @SPRINT.COM

or

  /PN=Mary,Anderson/O=co.abc/ADMD=SprintMail/C=US/
  @SPRINT.COM

- Internet user to CompuServe user:

Replace the comma in the CompuServe userid with a period and add the compuserve.com domain name. For example,

71360.1142@compuserve.com

- CompuServe user to Internet user: Insert "INTERNET:" before an Internet address.

INTERNET:user@host

- Internet user to MCIMail user: accountname@mcimail.com; mciid@ mcimail.com; and fullusername@mcimail.com. For example,

10500002@mcimail.com

6.7.4   SMTP Connection

The SMTP client reads the next message in queue for delivery and obtains a destination address from the formatted list of destination domain address(es) following the message body. Note that this list is generated by the SMTP client from the names supplied by the user. The SMTP client opens a TCP connection with the configured SMTP server using destination port number 25. (This is the port number that the SMTP server listens on.) The connection is synchronous in that the SMTP client must receive a reply for each command sent before the next command can be sent. The connection is broken as a result of the SMTP client sending a QUIT command. See Section 6.7.5.

6.7.5   SMTP Command Sequence

The sequence of SMTP commands is paramount, and at best, the mail will not be transferred properly if the recommended sequence is not followed—for example, if in the data transfer state, all commands are treated as data until the sequence "(CRLF).(CRLF)" is sent. The problem of the connection being picked up by a different SMTP client may be caused by not terminating properly (QUIT). The recommended sequence is illustrated in Figure 6.16 and described in Section 6.7.6.

6.7.6   SMTP Protocol Example

Table 6.6 contains an example of the SMTP protocol. The first column is a reference number for description. The sender of the communication is identified as client or server. (This is redundant since the reply is always from the server and always begins with a three-digit number.) The numbers of the following description correspond to the example in Table 6.6.

1. The SMTP server listens on port number 25 and when the TCP connection with the SMTP client is made, it sends a greetings message that indicates readiness and identifies itself with the official name of

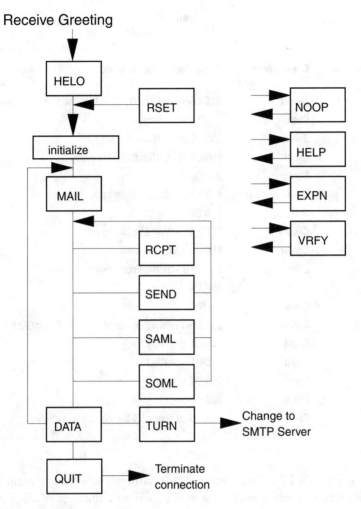

Figure 6.16 SMTP client command sequence.

the service host. The number 220 in the greeting message is the SMTP code for "the system is ready." After this initial poll (or kickoff), the SMTP client is master of the synchronous protocol.

2. The SMTP client sends a hello (HELO) command with its domain host name for validation. It is the same as a person saying, "Hello. My name is Pete. Will you accept my mail?"

3. The SMTP server sends a code 250 reply (OK) with its domain name. (It could send any one of six error codes, which are identified in Table 6.8.)

Table 6.6
SMTP Protocol Example

Item Number	Client/Server	Reply Code Communication
1	Server	220 {Server Name B} Mail Transfer Service Ready
2	Client	HELO {Host Name A}
3	Server	250 {Service Name B}
4	Client	MAIL FROM:<Smith@Test.sys>
5	Server	250 OK
6	Client	RCPT TO:<Jones@Mfg.tst>
7	Server	250 OK
8	Client	RCPT TO:<Williams@Mfg.tst>
9	Server	550 No such user here
10	Client	RCPT TO:<Bob@Engr.dev>
11	Server	250 OK
12	Client	DATA
13	Server	354 Start mail input; end with <CRLF>.<CRLF>
14	Client	{ASCII character text}
15	Client	<CRLF>.<CRLF>
16	Server	250 OK
17	Client	QUIT
18	Server	221 {Host Name B} Service closing channel

4. The SMTP client sends a MAIL command, which tells the SMTP server to initialize for new mail by resetting the state tables. The command also gives the SMTP server the <reverse path> (Smith@Test.sys) that can be used for addressing error messages resulting from this mail command.

5. The SMTP server sends an OK reply (code 250).

6. The SMTP client sends a RCPT command, which sets up the SMTP server with one recipient, which is "Jones," at mail station "Mfg.tst."

7. The SMTP server sends an OK reply (code 250).

8. The SMTP client sends another RCPT command, except with a different <forward path>. (This destination name is invalid.)

9. The SMTP server sends a reply code 550, which means "No such user here."

10. The SMTP client sends another RCPT command with the forward path of <Bob@Engr.dev> .

11. The SMTP server accepts the forward path <Bob@Engr.dev> and returns an OK reply (code 250).

12. The SMTP client has completed the setup of all forward paths associated with this message and sends a DATA command, which will alert the SMTP server of data to come.

13. The SMTP server sends the reply, "Start mail input; end with <CRLF>.<CRLF>. The go-ahead reply code is 354.

14. The SMTP client sends the entire memo (string of ASCII characters supplied by the user).

15. The SMTP client sends the data terminate command, "<CRLF>.<CRLF>"—or 0D0A2E0D0A hex.

16. The SMTP server replies with an OK (code 250).

17. The SMTP client sends the QUIT command to terminate the SMTP server connection. The SMTP client could continue with another MAIL command, or turn the line around and assume the role of SMTP server.

18. The SMTP server places a tag in each mailbox of Smith, Jones, and Bob. Each may read the message during his or her next session. The SMTP server sends a reply 221 that contains its service name and the message "Service closing channel."

6.7.7 SMTP Commands and Replies

An SMTP command (sent by SMTP client) and reply (sent by SMTP server) consist of a string of seven-bit ASCII characters with the high order bit (parity) of each cleared to zero. The commands and replies are not case-sensitive, except for the mailbox user name (local name), forward-path names, and reverse-path names. All commands are four characters in length and terminated with an end-of-line (EOL), or by a space (SP) character (20 hex) if an argument is present. (EOL) is composed of two ASCII characters, carriage return (CR) and line feed (LF) and typically illustrated as (CRLF). Most present keyboards create the (CRLF) from the single ENTER key stroke (in hex it is 0D0A). When a return path or forward path is part of an argument, it is enclosed in angle brackets (<path>), which are transmitted as part of the command. For clarity, any other string or domain name is illustrated as {string} or {name} and the "{" and "}" characters are not transmitted as part of the command or reply. Sections 6.7.7.1–6.7.7.14 describe the commands.

6.7.7.1 Hello (HELO) Command

The hello command, the first command sent by the SMTP agent in a connection, is used to initialize the connection and identify the SMTP client. The argument field contains the host name of the SMTP client. The command is of the form: "HELO(SP){domain host name} (CRLF)."

6.7.7.2 Mail (MAIL) Command

The mail command is the first command in the process after connection establishment and the argument identifies the sender. The command is of the form: "MAIL(SP)<reverse path>(CRLF)." The <reverse path> may be the originator only, or a list of relay names with the first name being the most recent.

6.7.7.3 Recipient (RCPT) Command

The recipient (RCPT) command is used to identify an individual recipient of the mail. The argument consists of a destination mailbox address and is optionally preceded by a list of forward paths. If forward paths are present, the SMTP server will remove its own name from the list and place the mail (yet to be received) in queue for the next name in the list. The RCPT command is repeated for each intended recipient. The RCPT command is of the form: "RCPT(SP)TO:<forward path>(CRLF)." For the command to be successful, the message must be placed in queue for the destination mail box.

6.7.7.4 Data (DATA) Command

The DATA command is of the form: "DATA(CRLF)." This command informs the SMTP server that the phase of sending forward paths is complete and that the next communication will be data. The normal reply from the SMTP server is a 354 code with the text "Start mail input; end with <CRLF>.<CRLF>." (See reply codes by command illustrated in Table 6.7 for error actions.)

Upon receipt of the 354 code reply, the SMTP client sends the entire message (associated with the forward paths supplied in previous commands) as a string of octets. Each octet contains a seven-bit ASCII character with the parity bit set to zero. The mail data is terminated with (CRLF).(CRLF), (0D0A2E0D0A hex). To prevent a user from inadvertently using this sequence in text, the SMTP client checks the first character of every line. If a period (.) is found, another period is added to the string of octets immediately following the user's period. This assures that the sequence 0D0A2E0D0A hex can not appear in the transmission of the text. The SMTP server also checks the beginning of every line and if the sequence "(CRLF)..", (0D0A2E2E hex) is found, it removes the second period (.).

Table 6.7
Reply Codes

Reply Code	Meaning
211	System status
214	Human information about how to use SMTP
220	<domain> Service ready
221	<domain> Service closing channel
250	Requested mail action okay, completed
251	User not local, forwarded to forward path
354	Start mail input, end with <CRLF>.<CRLF>
421	<domain> Service not available
450	Requested action aborted/mailbox unavailable
451	Requested action aborted/error in processing
452	Requested action aborted/insufficient storage
500	Syntax error, command unrecognized
501	Syntax error in parameters or arguments
502	Command not implemented
503	Bad sequence number
504	Command parameter not implemented
550	Requested action not taken/mailbox unavailable
551	Requested action aborted/error in processing
552	User not local/please try <forward-path>
553	Action not taken/mailbox name not allowed
554	Transaction failed

Upon termination of the data command, the SMTP server processes the message by placing a tag (containing a pointer to the message) in appropriate mailbox(es) and forward queue(s) if required. The SMTP server then sends an OK reply (code 250) to the SMTP client. (See reply codes by command and error actions in Table 6.8.)

6.7.7.5 Send (SEND) Command

The SEND command is used to identify an individual terminal to receive the mail. The command is the same as the RCPT command except that the message

Table 6.8
Replies for Each Command

Command Code	Standard Reply Code	Abnormal Reply Codes
CONN	220	F:421
HELO	250	E:500, 501, 504, 421
MAIL	250	F:552, 451, 452/E:500, 501, 421
RCPT	250	F:550-553, 450-452/E:500-503, 421
DATA	354 > data 250	F:552, 554, 451, 452
RSET	250	E:500, 501, 504, 421
SEND	250	F:552, 451, 452/E:500-502, 421
SOML	250	F:552, 451, 452/E:500-502, 421
SAML	250	F:552, 451, 452/E:500-502, 421
VRFY	250, 251	F:550, 551, 553/E:500-502, 504, 421
EXPN	250	F:550/E:500-502, 504, 421
HELP	211, 214	E:500-502, 504, 421
NOOP	250	E:500, 421
QUIT	221	E:500
TURN	221	F:502/E500, 503

Legend: F (Fail) and E (Error)

is being sent to a terminal instead of a mailbox. The RCPT command is of the form: "SEND(SP)TO:<reverse path>(CRLF)." The command is successful if the message is delivered to the specified terminal.

6.7.7.6 Send OR Mail (SOML) Command

The SOML command is used to identify either a mailbox address or an individual terminal to receive the mail. The SOML command is the same as the SEND command except that the message is being sent to either a terminal or a mail box. If the addressed terminal is active, the message is sent to the terminal. Otherwise, it is sent to the mailbox. The SOML command is of the form: "SOML(SP)TO:<reverse path>(CRLF)." The command is successful if the message is delivered to either the specified terminal or the mailbox.

6.7.7.7 Send AND Mail (SAML) Command

The SAML command is used to identify both a mailbox address and an individual terminal to receive the mail. The SAML command is the same as the SOML command except that the message is being sent to both a terminal and a mail box. If the addressed terminal is active, the message is sent to the terminal and mailbox. Otherwise, it is sent to the mailbox. The SAML command is of the form: "SEND(SP)TO:<reverse path>(CRLF)." The command is successful if the message is delivered to the mailbox.

6.7.7.8 Reset (RSET) Command

The RESET command causes the SMTP connection to be initialized and the new state is as though the 250 reply code had been received in response to a HELO command at the beginning of the connection. The RSET command is of the form: "RSET(CRLF)."

6.7.7.9 Verify (VRFY) Command

The VRFY command acts as a directory assistance command. It requests the SMTP server to verify the supplied argument represents a user. If the argument is a user, the SMTP server provides (in the reply) the fully specified mailbox address. The VRFY command is of the form: "VRFY(SP){string} (CRLF)."

6.7.7.10 Expand (EXPN) Command

The EXPN command is another directory assistance command like the VRFY command, except that the argument is a mailing list instead of a user. If the argument can be identified by the SMTP server, the list of fully specified mailbox addresses is returned. The EXPN command is of the form: "EXPN(SP){string} (CRLF)."

6.7.7.11 Help (HELP) Command

The HELP command is an operator assistance command that may have an argument. If an argument is present, the SMTP server provides a description of the specified command in the argument (if it is recognized). If an argument is not present in the command, the reply from the SMTP server is a list of all commands. The HELP command is of the form: "HELP[(SP){string}] (CRLF)."

6.7.7.12 No Operation (NOOP) Command

The NOOP command does not affect any parameters or the state of the SMTP connection. The normal response to the NOOP command is the code 250, OK. For systems implementing an inactivity timer, the NOOP command resets the timer. The NOOP command is of the form: "NOOP(CRLF)."

6.7.7.13 Quit (QUIT) Command

The QUIT command advises the SMTP server that the SMTP client is finished and that it should close the channel. The SMTP client should not close the channel until receipt of the positive reply, code 251. This prevents the possibility of leaving the channel open in the event of a garbled QUIT command, which is responded to with a code 500. The QUIT command is of the form: "QUIT(CRLF)."

6.7.7.14 Turn (TURN) Command

The TURN command presents the SMTP server with the option of changing roles—that is, becoming the SMTP client. If the SMTP server does not wish to change roles, it sends a refusal reply (code 502). If the SMTP server wishes to change roles, it sends an OK (code 250) reply and waits for the service ready greeting (code 220) from its new SMTP server. It must then send the HELO command. The TURN command is of the form: "TURN(CRLF)."

6.7.8 Reply Codes

Replies are always sent by the SMTP server in response to an SMTP client command. The reply consists of a three-digit code, a space code (SP), and a textual description of the code. Table 6.7 identifies all SMTP server reply codes with a brief description of each. The reply is of the form: "nnn(SP){message text}."

6.7.9 Possible Replies for Each Command

Table 6.8 identifies all the possible codes that can be sent for each command code. There are three classes of reply codes identified: 1) standard, 2) fail, and 3) error. A standard code indicates everything is okay. For example, the domain address you supplied was located—code 250. An error code indicates an error, but the session may continue. For example, the domain address you supplied was not located—code 500. A fail code indicates the inability to continue. For example, requested action aborted due insufficient storage—code 452.

6.7.10 Multipurpose Internet Mail Extensions (MIME)

SMTP, as originally defined, provides an electronic mail system for text messages, with text limited to US ASCII characters and short lines [12, 13]. MIME is a recent enhancement to SMTP that provides a mechanism to extend the type of text in a message to include audio, image, video, different character sets that

accommodate other languages (e.g., Asian and European languages), and mixtures of these [14]. MIME also makes it possible to properly interface the Internet with other mail systems that handle nontext messages (e.g., X.400).

To remain compatible with older SMTP systems, MIME defines a header for the SMTP message that identifies the message as MIME type for MIME compliant hosts and gateways. Noncompliant hosts and gateways ignore the MIME header when present. If the header is not present, the default is conventional SMTP (preMIME, or standard RFC #822). The MIME header is placed in the top level of the normal SMTP header and commences with (or flagged by) the following character sequence.

MIMEVersion: 1.0

Since it is unlikely that an older SMTP mail system, including a gateway, would use this sequence of characters, only MIME compliant gateways and hosts will recognize the header. This makes it backward compatible with older SMTP systems [15].

When the "MIME-Version: 1.0" header is present, it is followed by other format specifiers such as the following:

- Content-type header: Overall type of header;
- Subtype identifiers: Specify the nature of the data in the body;
- Content-transfer-encoding header: Used to identify the type of transformation required for decoding the message;
- Parameters: Specified in an attribute/value notation that varies by type;

These are the tools used by MIME to provide the enhancements. Since they are tools, it follows that the tools could be used in a different manner than presented to transport a nontext message.

Sections 6.7.10.1 through 6.7.10.3 provide a description with examples is of the MIME tools. This is followed a complex example that illustrates various combinations of the MIME tools and how they may be used. (Reference Section 6.7.10.4.)

6.7.10.1 Content-Type Headers

The content-type header is used to specify the general type and format of data. Subtypes of each general content-type are used to further define the specific format and data. There are seven content-type headers initially defined, and some of the subtypes have parameters to enable further refinement. The content-type headers and the associated subtypes presently defined are the following:

- Text (textual information): The primary subtype, "plain", indicates plain (unformatted) text. The subtype "richtext" is a common set of typical word processor formats, which is defined later. Parameters may be used to indicate a character set—for example, Contenttype: text/plain; charset=usascii.

- Multipart: data consisting of multiple parts of independent data types. The four defined subtypes are the following:
  - Mixed: for multiple different parts to be viewed serially;
  - Alternative: For representing the same data in multiple formats (e.g., one in plain text and one in richtext);
  - Parallel: For parts intended to be viewed simultaneously—otherwise, the same as mixed;
  - Digest: For multiparts with each part equal to type "message/ RFC822." (Same as mixed except the default content-type value is changed from "text/plain" to "message/RFC822"—for example, possibly multiple parts of same subject.)

- Message: An encapsulated message that may contain its own different content-type header field. There are three subtype fields as follows.
  - rfc822: an ASCII text message;
  - Partial: segment of segmented message;
  - External-body: external data source.

- Image: image data. Image requires a display device (e.g., graphical display, a printer, or FAX) to view the information. Initial subtypes are defined for two widely used image formats, "jpeg" and "gif."

- Audio: audio data, with initial subtype "basic." Audio requires an audio output device (e.g., speaker or telephone).

- Video: video data. Video requires the capability to display moving images, typically including specialized hardware and software. The initial subtype is "mpeg."

- Application: Other kinds of data, typically either uninterpreted binary data. The subtypes are the following.
  - Octet stream: used for uninterpreted binary data, in which case the information may be written to a file;
  - ODA: transporting ODA documents in bodies.

- PostScript: transporting Postscript documents in bodies. (Note that security is a consideration when executing PostScript commands.)

6.7.10.2 Content-Transfer-Encoding

The content-transfer-encoding field is used to indicate the type of transformation that has been used in order to represent the body in an acceptable manner for transport. The content-transfer-encoding field's value is a single token specifying the type of encoding, as enumerated below in BNF [16]. (The values are case-insensitive.)

Content-transfer-encoding := "BASE64" / "QUOTEDPRINTABLE" / "8BIT" / "7BIT" / "BINARY" / xtoken

These are exclusive or values and indicate that there are six possible choices for the content-transfer-encoding as follows.

1. The value of "7BIT" is the default type used when the content-transfer-encoding is not specified and is the same as preMIME SMTP. That is, no encoding has been performed.

2. The value of "8BIT" is used to indicate that there may be nonASCII characters in the text. That is, the high order bit of octets may be set.

3. The value of "BINARY" means that there may be nonASCII characters in the text and the length of lines may be longer than standard SMTP.

4. The value of "BASE64" indicates an encoding that is the same as the one used in privacy enhanced mail applications [17]. See description in Section 6.7.11. Basically, the process involves grouping each three octets of the message into 24-bit groups, then dividing each 24-bit group into four, six-bit groups, each used as an index into a table of 64 printable, ASCII characters. Decoding involves grouping each four-octet group back into the original three-octet form.

5. The value of "QUOTED-PRINTABLE" is used to denote an encoding of largely ASCII data (or pure US ASCII) for the purpose of ensuring the integrity of data that must pass through a character-translating, and/or line-wrapping gateway. This largely involves replacing the characters that are potentials for corruption with hexadecimal numbers that are flagged by preceding each with an equal (=) character. If an actual character in the text is an equal character (=), it is replaced with its appropriate hexadecimal value 3D. That is,

receiving the characters "=3D" would be decoded as an equal character (=). To assure that the line length maximum of 76 characters is not exceeded, an equal character followed by a return (3D0D hex) is interpreted as a "soft" return, which satisfies a menacing gateway and is discarded by the receiver [18].

6. The value of "x-token" is used to define a new content-transfer-encoding type. To flag the nonstandard type, it is necessary that the first two characters be "X-"—for example, "X-newname." The use of this option is discouraged (except for development between cooperating user agents), as it would likely cause interoperability problems.

Content-transfer-encoding is not permitted with content-types "multipart" or "message." Although "eight-bit" or "binary" mail is not legal on the Internet, in the event that such an eight-bit or binary-capable transport becomes available, both eight-bit and binary bodies should be identified using this mechanism.

6.7.10.3 Subtypes and Parameters

The subtype specifies a specific format for the specific content-type. Each subtype is affiliated with a specific content-type and would not be meaningful with others. The subtypes are described as follows.

- Text/plain subtype: The default content-type and subtype for Internet mail is: "text/plain; charset=usascii";

- Text/richtext subtype: All characters are US ASCII, unless otherwise specified with "charset", and represent themselves, with the exception of the "<" character (ASCII 60) and ">" (ASCII 62), which are used to mark formatting commands. Commands within the "<>" may be negated by preceding them with a forward slash ("/", ASCII 47). The initially defined formatting commands, which may not be supported by all, are the following:

    - Bold: Commence with bold font;
    - Italic: Commence with italic font;
    - Fixed: Commence with fixed-width font;
    - Smaller: Commence with a smaller font;
    - Bigger: Commence with a larger font;
    - Underline: Commence with underlined text;
    - Center: Center text;
    - FlushLeft: Place text left justified;

- FlushRight: Place text right justified;
- Indent: Place text indented from left margin;
- IndentRight: Place text indented from right margin;
- Outdent: Outdent text from the left margin;
- OutdentRight: Outdent text from the right margin;
- SamePage: Keep text intact on same page;
- Subscript: Commence with subscript;
- Superscript: Commence with superscript;
- Heading: Commence with a page heading;
- Footing: Commence with a page footing;
- ISO-8859X: Commence with appropriate "charset";
- US-ASCII: Commence with US-ASCII "charset";
- Excerpt: Commence exerpt from another source;
- Paragraph: Commence a new paragraph;
- Signature: Commence signature line;
- Comment: Commence comment to sender only;
- No-op: No operation;
- lt: To be replaced by return character (<)
- nl: Causes a line break;
- np: Causes a page break.

Example: <bold><italic><the-text</italic></bold>

Rich text differentiates between *hard* and *soft* line breaks. The US ASCII line break (CRLF) is a soft break and the command "nl" is a hard break, as in the following body fragment:

```
<bold>Now</bold> is the time for
<italic>all</italic> good men
<smaller>(and women)</smaller> to
come
to the aid of their
<nl>
beloved <nl><nl>country. <comment> Stupid
quote! </comment>—the end
```

would appear as the following:

Now is the time for *all* good men (and women) to come to the aid of their beloved

country.—the end

The multipart/mixed subtype requires an encapsulation boundary parameter to separate the message bodies. The encapsulation boundary is defined as a line consisting entirely of two hyphen characters ("", decimal code 45) followed by the boundary parameter value from the content-type header field. For example:

ContentType: multipart/mixed;
boundary=SAMPLE BOUNDARY/Sample Boundary

This indicates that the entity consists of several parts, each itself with a structure that is syntactically identical to an RFC 822 message, except that the header area might be completely empty and that the parts are each preceded by the following line.

—SAMPLE BOUNDARY/Sample Boundary

The last body part is followed by the same boundary except that it is ended with two hyphen characters. e.g.,

—SAMPLE BOUNDARY/Sample Boundary—

The EOL immediately preceding the boundary is part of the boundary. To provide separation between body parts requires an EOL to terminate a body part and a second EOL to begin the boundary. Figure 6.17 provides an example of a multipart message that has two plain text, body parts. The first is explicitly typed and the second is implicitly typed.

The multipart/digest subtype is syntactically identical to multipart/mixed, but the semantics are different. The default content-type value for a body part is changed from "text/plain" to "message/rfc822" to allow a more readable digest format. Figure 6.18 provides an example of the digest subtype.

The multipart/parallel subtype is syntactically identical to the multipart/mixed, but the semantics are different. The purpose of the parallel subtype is to permit simultaneous presentation of multiple messages. For users that lack the required hardware and software to utilize this feature, the parts will be presented serially.

The message/rfc822 (primary) subtype indicates that the body contains an encapsulated message, with the syntax of an RFC822 message.

The message/partial subtype provides the capability for a large message to be delivered as several smaller messages, which is similar to fragmentation and reassembly. There are three parameters used with Message/Partial subtype. The

From: Nathaniel Borenstein <nsb@bellcore.com>

To: Ned Freed <ned@innosoft.com>

Subject: Sample message

MIME-Version: 1.0

Content-type: multipart/mixed;

                boundary=simple boundary

This is the preamble. It is to be ignored,

though it is a handy place for mail composers to

include an explanatory note to non-MIME compliant

readers.

--simple boundary

This is implicitly typed plain ASCII text.

It does NOT end with a linebreak.

--simple boundary

Content-type: text/plain; charset=us-ascii

This is explicitly typed plain ASCII text.

It DOES end with a linebreak.

--simple boundary--

This is the epilogue. It is also to be ignored.

Figure 6.17 Multipart/mixed subtype example.

first parameter is an ID number used to reassemble the parts. The second parameter is the sequential number of this fragment (beginning with one). The third parameter is the total number of fragments in the message. It is optional on intermediate fragments and mandatory on the last fragment. The parameters are illustrated in Figure 6.19.

The encapsulation of the message in fragments is transparent to the user. Hence, a large audio message would be fragmented and reassembled in a manner that would appear to the recipient as a simple audio message without pauses.

The message/externalbody subtype is used to send data that is not part of the message body. The parameters describe a mechanism for accessing the external data. When a message body or body part is of type "message/external-body", it consists of a header, two consecutive CRLFs, and the message header for the encapsulated message. If another pair of consecutive CRLFs appears, this ends the message header for the encapsulated message. However, since the

From: Moderator-Address

MIME-Version: 1.0

Subject: Internet Digest, volume 42

Content-Type: multipart/digest;

boundary=---- next message ----

------ next message ----

From: someone-else

Subject: First person's opinion

...body goes here ...

------ next message ----

(And, so on with other opinions.)

------ next message ------

Figure 6.18 Multipart/digest subtype example.

encapsulated message's body is external, it does not appear in the area that follows. See the example in Figure 6.20 and "THIS IS NOT REALLY THE BODY."

The area at the end, which might be called the "phantom body," is ignored for most external-body messages.

The parameter "access-type" is mandatory, and others may also be mandatory depending on the value of access-types. Present assigned values include: "FTP," "ANONFTP," "TFTP," "AFS," "LOCALFILE," and "MAILSERVER". The parameters "EXPIRATION," SIZE, and PERMISSION are op-

Content-Type: Message/Partial;

number=1; total=3;

id="oc=jpbe0M2Yt4s@thumper.bellcore.com";

Content-Type: Message/Partial;

number=2; total=3;

id="oc=jpbe0M2Yt4s@thumper.bellcore.com";

Content-Type: Message/Partial;

number=3; total=3;

id="oc=jpbe0M2Yt4s@thumper.bellcore.com";

Figure 6.19 Message/partial subtype example.

```
Content-type: message/external-body; access-
type=local-file;
        name=/u/nsb/Me.gif
Content-type:  image/gif
THIS IS NOT REALLY THE BODY!
```

Figure 6.20 Message/external-body subtype example.

tional for all access-types. Expiration is a "date-time" in RFC 822 syntax, size is in the number of octets, and permission may be either "Read" or "Readwrite."

The "ftp" and "tftp" accesstypes indicate that the message body is accessible as a file using the FTP or TFTP protocols, respectively. For these access-types, the following additional parameters are mandatory:

- NAME: The name of the file that contains the actual body data;
- SITE: A machine from which the file may be obtained, using the given protocol.

In addition, the following optional parameters may also appear when the access-type is FTP or ANONFTP:

- DIRECTORY: A directory from which the data named by NAME should be retrieved;
- MODE: A transfer mode for retrieving the information, e.g., "image."

The "mailserver" access-type indicates that the actual body is available from a mail server. The mandatory parameter for this access-type is "SERVER," which specifies the e-mail address that the actual body data can be obtained.

The Application/OctetStream subtype is the primary use of application type and indicates that the body contains binary data. The set of possible parameters include: NAME (file name), TYPE (general), CONVERSIONS (conversion history), and PADDING (number of bits required for octet alignment). An example of a UNIX-specific usage follows.

```
ContentType: application/octetstream;
name=foo.tar.Z; type=tar;
conversions="xencrypt,xcompress"
```

The Application/PostScript subtype indicates a PostScript program, and as such may contain a potential code that could cause havoc. The execution of general purpose PostScript interpreters entails serious security risks. It is usually safe to send PostScript to a printer, where the potential for harm is limited.

The Application/ODA subtype is used to indicate that a body contains information encoded according to the Office Document Architecture (ODA) standards, using the ODIF representation format [19]. The content-type line specifies an attribute/value pair that indicates the document application profile (DAP), using the key word "profile." The following is an example of the ODA subtype.

Content-type: application/oda; profile=Q112

6.7.10.4 A Complex Multipart Example

Figure 6.21 is an outline of a complex multipart message [20]. The message has five parts to be displayed serially: two introductory plain text parts, an embedded multipart message (with two subparts displayed in parallel), a rich text part, and a closing encapsulated text message in a nonASCII character set. The embedded multipart message has two parts to be displayed in parallel, an audio part, and an image part. The total message may be viewed as five parts, with part number three containing two subparts.

Note the technique of nesting within a multipart message. The third multipart (follows third "boundary one") contains another multipart content-type, but this one is parallel with a different boundary specification ("boundary two"). All submessage parts within the third message part are to be delivered in parallel. The first is audio data and the second is image data; both must be decoded before output. The second nested multipart terminates the parallel-multipart. This is flagged by the two hyphen characters following the second "boundary two."

6.7.11 Privacy-Enhanced Mail (PEM)

PEM is an enhancement to the conventional SMTP mail service that provides the following features.

1. Disclosure protection;
2. Authentication of sender;
3. Message integrity.

```
MIME-Version: 1.0
From: Nathaniel Borenstein <nsb@bellcore.com>
Subject: A multipart example
Content-Type: multipart/mixed;
        boundary=boundary one

[This is the preamble area of a multipart message.]
--boundary one

...Some text appears here...
[Note that the preceding blank line means
no header fields were given and this is text.]

--boundary one
Content-type: text/plain; charset=US-ASCII
[Illustrates explicit vs. implicit]
typing of body parts.]

--boundary one

Content-Type: multipart/parallel;
        boundary=boundary two

--boundary two

Content-Type: audio/basic
Content-Transfer-Encoding: base64
... base64-encoded 8000 Hz single-channel
        u-law-format audio data goes here....

--boundary two

Content-Type: image/gif
Content-Transfer-Encoding: Base64
[... base64-encoded image data goes here....]

--boundary two--
--boundary one

Content-type: text/richtext

This is <bold><italic>richtext.</italic></bold>

--boundary one

Content-Type: message/rfc822
From: (name in US-ASCII)
Subject: (subject in US-ASCII)
Content-Type: Text/plain; charset=ISO-8859-1
Content-Transfer-Encoding: Quoted-printable

[... Additional text in ISO-8859-1 goes here ...]

--boundary one--
```

Figure 6.21 Complex multipart MIME example.

Disclosure protection assures that only the addressed recipient will be able to read the message. Authentication of the sender protects against masquerading or bogus senders and prohibits the originator from denying having sent a message. Message integrity assures that the message has not been either accidentally corrupted or intentionally changed.

The PEM services are provided by a PEM-compliant agent and require no changes to the SMTP, TCP, or IP. Users of a PEM compliant agent may, on a message-by-message basis, send and receive messages with the above enhancements. [21, 22].

The enhancement is backward compatible with SMTP. However, should a nonPEM compliant client receive a PEM encoded message, the user would see garbage. A message sender must be cognizant of whether the intended recipient has implemented PEM. PEM compliant clients and servers on dissimilar hosts are able to exchange encrypted messages.

The PEM cryptographic algorithm uses an eight-octet (64 bits) *data encryption key* (DEK), of which eight bits are used to set the parity of each octet of the DEK odd [23]. The remaining 56 bits provide 7 x 1016 (70 quadrillion) possible keys. Since the keys are randomly selected and only one key will correctly decrypt the text, it is unlikely that text could be deciphered by trial and error even with reasonable computer power. It is, however, possible to generate a new DEK for each message. Since the DEK is only used once, it is not feasible for the sender to advise the receiver of the DEK value except in the message header. So, to preclude anyone from reading the DEK except the intended receiver, it is enciphered with an *interchange key* (IK) known only to the receiver.

There are two types of IK operation, symmetric and asymmetric. Symmetric operation requires that both the sender and receiver possess the identical IK, which is used to encrypt the DEK and for computation of the *message integrity check* (MIC). Asymmetric operation requires two parts to the IK. The receiver's public component is used to encrypt the DEK, and the sender's private part is used for computation of the MIC.

Since the algorithm to encrypt (encipher) and decrypt (decipher) data for PEM is a published standard, the real security provided by PEM is gained (or lost) from the technique of assignment and handling of IKs. IKs may be distributed by centralized management servers or by direct distribution between users.

6.7.11.1 General Mechanics of PEM

A user with a PEM-compliant agent may elect on a by message basis to apply encryption to a message that is intended for a recipient who also has a PEM compliant agent. The normal address information plus other data required for encryption is supplied by the user to the agent. The agent uses the data to generate output containing decryption keys and text that is suitable for secured transmission by the SMTP client. The receiving agent, with a special key, is able to decipher the key header fields of the message, which enables the decryption of the message text. The agent is a standalone function that may be in a PC, which could offer a higher degree of security than an agent implementation on a multi-user system.

A new DEK is generated for the encryption process of every PEM message. An IK is then used to encrypt the DEK, which is inserted in the header of the message. As described earlier, there are different techniques of assigning and using IKs (symmetric and asymmetric). There are different algorithms used to encrypt DEKs. When an asymmetric system is used, the mode/algorithm specified is "RSA" to denote the RSA algorithm [24]. When a symmetric system is used, the mode/algorithm specified is DES-ECB, which is the data encryption standard electronic code-book mode. Both systems are described in the PEM specifications, with a high recommendation for the asymmetric systemd.

The text transformation of an outbound PEM-compliant message is characterized by the following.

Transmit_Form = Encode(Encipher(Canonicalize(Local_Form)))

The PEM procedure involves the client collecting the user message in its local form and restructuring it to the interSMTP form (referred to as canonical form). Both the localform and canonical form are called plain text. The client passes the text portion to the standalone PEM function. The text is encrypted using a DEK, and the result is called cipher. The DEK is encrypted using an IK, symmetric, or asymmetric systems. The encrypted DEK and various other fields (depending on whether it is a symmetric or asymmetric system) are inserted at the beginning of text (PEM header) and terminated with a blank line. The enciphered text is encoded in a manner that assures compatibility with SMTP.

The inverse transformations are performed, in reverse order, to process an inbound PEM. This is characterized by the following.

Local_Form = DeCanonicalize(Decipher(Decode(Transmit_Form)))

The transmit form of the message is decoded from its SMTP compatible format to the deciphered format. The IK (common if symmetric system, or the private component if asymmetric system) is used to decipher the DEK. The DEK is then used to decipher the text.

6.7.11.2 PEM Header Fields

Authentication and integrity services are provided by means of several header fields generated by PEM and placed at the beginning of the text.

An MIC is computed on the message text in its canonical form. In asymmetric systems, the MIC is contained in the "X-MIC-Info:" field, which is encrypted under the recipient's public component of the IK. Hence, only one MIC per message is needed instead of one per recipient as with a symmetric system. A single IK is shared between originator and recipient with symmetric key management and the MIC is contained in the "X-Key-Info:" field [25].

The "X-Proc-Type:" field is the first inserted encryption header. The two subfields designate the version of PEM being used and an indicator of full or partial encryption.

An "XSenderID:" field is included in the header to provide the first component of the IK used for message processing. The second IK component is carried in a following "X-Recipient-ID" field, one for each individually named recipient.

Each "XRecipientID:" field is followed by an "XKeyInfo:" field. In symmetric systems, the "X-Key-Info:" field transfers a DEK and an MIC that are encrypted under the IK identified by the preceding "X-Recipient-ID:" and the "X-Sender-ID:" fields. In asymmetric systems, the "X-Key-Info:" field carries the DEK, which is encrypted under the recipient's public component.

6.7.11.3 Enciphering and Encoding

Octets each equal to FF hex are used, as required, for padding to force 8-octet alignment of the text. Since the value FF hex is outside the range of seven-bit ASCII, a count field is not required for the padding. Encryption of the text is performed with the encryption key defined in the encryption header. The resulting text from the encryption process represents binary data and is no longer compliant with SMTP. Hence, it is then encoded in a manner that assures SMTP compliance. The encoding process is summarized below.

- The entire enciphered text is operated on as a continuous bit stream commencing with the +0 octet through the +n octet. The total number of bits in the text is equal to $(n + 1) \times 8$. With k equal to this, the text forms the vector:

$$X^{k-1} + X^{k-2} + X^{k-3} +, , + X^{k-k}$$

- Each successive group of six bits, starting with $Xk-1$, is used as an index into the array of ASCII characters illustrated in Table 6.9.

Table 6.9
Printable Encoding Characters

Value	Code	Value	Code	Value	Code	Value	Code
0	A	16	Q	32	g	48	w
1	B	17	R	33	h	49	x
2	C	18	S	34	I	50	y
3	D	19	T	35	j	51	z
4	E	20	U	36	k	52	0
5	F	21	V	37	l	53	1
6	G	22	W	38	m	54	2
7	H	23	X	39	n	55	3
8	I	24	Y	40	o	56	4
9	J	25	Z	41	p	57	5
10	K	26	a	42	q	58	6
11	L	27	b	43	r	59	7
12	M	28	c	44	s	60	8
13	N	29	d	45	t	61	9
14	O	30	e	46	u	62	+
15	P	31	f	47	v	63	/

- The entire eight bits from the array is inserted into the new encoded text for each group of six bits from the enciphered source text. The result is a printable encoding of characters. Every three octets of the enciphered text results in four octets of encoded text. From the definition of source input being octets +0 through $+n$, there are $n + 1$ octets in the enciphered text. If there are j octets of encoded text, j is equal to: $8(n + 1)/6$.

- The process continues until either there are less than 24 bits in the enciphered text to form four encoded characters or the last group of six bits ends on an octet boundary ($n + 1/3$ is an even number). If there are less than 24 bits, zeros are padded to complete the 24 bits. The zeros are converted to either one or two pad characters (= character) and inserted in the encoded text as required to force ending the enciphered text on an octet boundary. If the remainder is a single octet, two equal characters (==) are entered in the encrypted text. If the remainder are two octets, a single equal character (=) is entered in the encrypted text.

6.7.11.4 PEM Output Format

Encapsulation of PEMs is done by enclosing a layer of headers within the present SMTP text, which forms a new header that is separated from the text by a blank line. The new header contains various fields (See Section 6.7.11.2.) required for encryption and authentication processing. Since the SMTP header is not modified, the feature is transparent to SMTP and amounts to a higher layer protocol. The general format of an encoded, encrypted message is illustrated in Figure 6.22.

6.8 Hypertext Transfer Protocol (HTTP)

FTP and Telnet traffic on the Internet has declined since the early 1990s, while HTTP traffic has increased proportional to the overall traffic (in other words, it

Enclosing Header Portion
    (Contains header fields per SMTP.)

Blank Line
    (Separates enclosing header from encapsulated message.)

Encapsulated Message

    Pre-Encapsulation Boundary (Pre-EB)
        -----PRIVACY-ENHANCED MESSAGE BOUNDARY-----

    Encapsulated Header Portion
        (Contains encryption control fields inserted in
        plaintext. Examples include "X-DEK-Info:",
        "X-Sender-ID:", and "X-Key-Info:".)

    Blank Line
        (Separates Encapsulated Header from encoded text.)

    Encapsulated Text Portion
        (Contains encoded, encrypted message data.)

    Post-Encapsulation Boundary (Post-EB)
        -----PRIVACY-ENHANCED MESSAGE BOUNDARY-----

Figure 6.22 PEM output format.

has exploded) [26]. This is because most new Internet users are busy surfing the Internet, which typically involves a Web browser (client) soliciting information from a HTTP server. This section provides only an introduction to HTTP for the purpose of traffic identification on the Internet [27].

HTTP was developed by researchers at the European Laboratory for Particle Physics (CERN) in Switzerland as a tool for exchanging multimedia information. The first version of HTTP (HTTP/0.9) was used to transfer document files between researchers over the Internet [28]. The document files contained tagged words that were used as pointers to other information, or lists of other information. The *Hypertext Markup Language* (HTML) is used to construct the payload message transported by HTTP, which accounts for the name *hypertext,* which implies text that conveys more than the dictionary meaning of individual textual words. That is, it is not linear and there is more than meets the eye. HTTP is a simple protocol that works fine without HTML for transporting nonhypertext data. Without HTML browsers, however, HTTP would be restricted to text line displays. Furthermore, popular Web browsers such as Netscape and Explorer would not have been developed.

6.8.1 HTTP Architecture

HTTP is an upper layer application that uses the TCP wellknown port 80 to create a connection between a client and server. A client using the early version of HTTP simply sent request messages to a HTTP server, which sent a response message to the client and closed the connection. The architecture of HTTP is illustrated in Figure 6.23.

The connection between the client and server is illustrated without intermediaries, which was typical with the original version of HTTP.

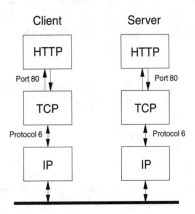

Figure 6.23 HTTP client/server architecture.

6.8.2 HTTP Message Formats

The request message is an ASCII (ISO-8859-1) character string containing a verb (command describing what to do) and an address (where to do it at). The address format is defined by a uniform resource location (URL), which also includes the protocol to be used [29]. The ASCII string of a general HTTP command from the client to a server appears as:

   GET http://server.name/path/file.type

The verb "GET" is a solicitation for service from the client to the server. The remainder constitutes the URL. "HTTP://" defines the protocol to be used (could also be FTP, SMTP, NNTP, Gopher, and WAIS protocols). The next field is the name of the HTTP server containing the requested data, which is followed by the path name and file name separated by a forward slash (/). When a forward slash follows the server name without a path and file name, the default file is requested. This is typically the home page of the HTTP server. The response message contained the requested data, and the server closed the connection. Although new fields have been added to revised versions of HTTP, this basic format is still accepted for backward compatibility.

HTTP revision 1 (HTTP/1.0) began the formalization of HTTP by defining header parts and body parts similar to SMTP mail (RFC 822). The first apparent change was to add two new verbs (HEAD and POST) and called them methods, which is adopted here. The next change was to formally identify the version of HTTP with a major and minor number.

To differentiate between the old and new, the original request is called a *simple request* and the new version is called a *full request*. The full request may have three header parts and a body part, separated by a blank line. That is, the last header part is ended with a double (CR)(LF) sequence, which is the same as SMTP. The full request format is:

   Method(SP)http://server.name/path/file.type(SP)HTTP/1.0(CR)(LF)
      General-header(CR)(LF)
      Request-header(CR)(LF)
      Entity-header(CR)(LF)(CR)(LF)
      Body

The methods are described as follows. (There are other methods defined, but they are not implemented uniformly across all servers. For example, PUT, DELETE, LINK, and UNLINK.)

- GET: The GET method retrieves the data identified by the URL address. If the content of the URL address is a process, the data from that process is returned.

- HEAD: The HEAD method is identical to the GET method except the response contains only the header (general, response and entity). This provides information about the source of information and is used for testing the validity of hypertext links. From the response message, a user may determine information about the web page, such as the last date modified.

- POST: The POST method allows the end user to place data on the server. This may be used in a bulletin board operation, submittal of a form, registration, or to enhance an existing database.

The same differentiation is made between the original response (simple response) and the new version (full response). The full response format is:

HTTP-1.0(SP)Status-Code(SP)Reason-Phase(CR)(LF)
   General-header(CR)(LF)
   Request-header(CR)(LF)
   Entity-header(CR)(LF)(CR)(LF)
   Body

6.8.3 Error Codes

The status code is a three-digit integer describing the result of the operation. The reason phrase (when present) is a textual, ASCII string interpreting the status code. Table 6.10 identifies the assigned status codes.

6.8.4 Header Fields

There may be zero or more header fields and each consists of a name (not case sensitive) followed by a colon (:), a space (SP), and the field value. Table 6.11 identifies the headers by name and whether they are used in a request message, response message, or the body.

6.8.5 Examples

An easy method of viewing the HTTP protocol in operation is Telnet to a Web server with well-known port 80, then enter ASCII characters for the desired activity. For example, use Telnet to connect to an advertised URL and simply

Table 6.10
HTTP Response Error Codes

Status Code	Reason Phrase
1xx	Reserved
200	O.K., succeeded
201	O.K., resource created
202	Accepted but processing incomplete
204	Accepted but no response message
301	Relink to a new assigned URL
302	URL temporarily redirected
304	No new data after date specified
400	Apparent client error
401	User authentication required
403	Forbidden, not available
404	Not found
500	Internal error
501	Not implemented
502	Invalid response obtained
503	Temporarily not available

type, GET /(ENTER). Since the server should be backward-compatible to a simple-request (HTTP/0.9), you should get the HTML body for that Web page. Unless the server has implemented HTTP/1.1 (rare), the TCP connection will be dropped. Now repeat the same exercise using HTTP/1.0. That is, enter, GET /(SP)/HTTP/1.0(ENTER). If the server is HTTP/1.0-compliant, you should receive the same message except with a header. Now repeat the same sequence, except specify HTTP/1.1. That is, enter, GET/(SP)/HTTP/1.1 (ENTER). You will likely receive the identical response message as with HTTP/1.0 and the TCP connection is dropped. This means the server has not implemented HTTP/1.1 and is ignoring the ".1" revision number, as illustrated below. (If compliant with HTTP/1.1, the connection is not dropped.)

Request
    telnet www.acc.com 80
    GET / /HTTP1.1

Table 6.11
HTTP Headers

Header name	Request	Response	Body
Allow			X
Authorization	X		
Content-encoding			X
Content-length			X
Content-type			X
Date	X	X	
Expires			X
From	X		
If-modified-since	X		
Last-modified			X
Location		X	
Pragma	X	X	
Referer	X		
Server		X	
User-agent	X		
WWW-authentication		X	

Response
HTTP/1.0 200 OK
MIME-Version: 1.0
Server: WebSTAR/2.1 ID/42693
Message-ID: <b0abc370.4659@www.acc.com>
Date: Sat, 06 Dec 1997 07:38:33 GMT
Last-Modified: Wed, 03 Dec 1997 16:43:13 GMT
Content-type: text/html
Content-length: 10667

<HTML>
<DEAD>

<TITLE>ACC Insight</TITLE>
[Text omitted]

Notice that the header is HTTP/1.0, but HTTP/1.1 was requested. Now repeat the same sequence except use the method HEAD. That is, enter, HEAD /(SP)/HTTP/1.0. The response message is:

HTTP/1.0 200 OK
MIME-Version: 1.0
Server: WebSTAR/2.1 ID/42693
Message-ID: <b0adc370.4803@www.acc.com>
Date: Sat, 06 Dec 1997 07:55:07 GMT
Last-Modified: Wed, 03 Dec 1997 16:43:13 GMT
Content-type: text/html
Content-length: 10667

Notice that the header is identical to the Get method, but there is no body part.

To view a response error code, enter an invalid method. For example, enter GAT / /HTTP/1.0, which yields the following.

HTTP/1.0 501 Not Implemented
Date: Sat, 06 Dec 1997 15:04:03 GMT
Server: CCO/1.0.5.6
Content-type: text/html

 <HEAD><TITLE>Method not implemented</TITLE><HEAD>
<BODY><H1>Method not implemented
</H1>gat to / not supported.<P></BODY>

The error code is 501 and the response phrase is "not implemented."

Typical users of HTTP do so with the aid of their friendly Web client (e.g., Netscape). A practical example of this is to obtain an RFC, a draft RFC, or a FYI RFC. Bring up your Web client and enter the address, "rs.internic.net". Now select the directory and database services option, followed by the Internet documentation option. This will provide a list of options, including IETF RFC's, IETF Standard RFCs, IETF FYIs, and IETF Internet Drafts. Compare this technique of obtaining RFCs with the FTP method described in Section 6.2.

6.8.6   HTTP Future

Many improvements were made in the new HTTP/1.1 [30]. The specification format is in an augmented Bacus-Naur Form (BNF) similar to MIME in

SMTP, but it is not MIME-compliant [31]. HTTP/1.1 has a MIME-like format containing metainformation describing the request and response messages in more precise terms. It also defines how intermediaries handle request and response messages between clients and servers and make the connection persistent—that is, left open for continued use. Intermediaries include a proxy (forwarding agent), a gateway (reformats as required by server), and a tunnel (e.g., transparent bypass for firewalls).

Although there are a lot of related draft changes and enhancements, HTTP is here for the long haul. It will not have any major changes because of backward compatability. Near turf wars are present about minor details that are akin to whether the circle used to dot the letter "i" is filled or not.

6.9 X Window System

The X Window System is a hardware and software independent graphics interface between a user and a host. The X Window System extends the features of TELNET to a complete graphics interface with applications in local and multiple remote hosts. The X Window System was developed by the Massachusetts Institute of Technology (MIT) in 1984 and has since become an industry standard.

The graphical interface, windowing, multitasking, and switching may be used in a standalone machine, or networked. In the networked mode, concern for this discourse, the X Window System follows the client/server model described in Section 1.3.

The client (called X Client) of an X Window System is typically a host containing applications such as spreadsheet (e.g., Lotus), database (e.g., Dbase), or graphics (e.g., AutoCAD) programs that communicate their video commands to the server by means of a TCP connection [32]. Without the X Window System, these applications would send commands directly to the video RAM. The client sends request messages to the server, which may merit a reply message, an error message or no response. Although a request from a client may merit a specific reply, the communication is typically asynchronous and the server initiates (without solicitation) messages called *events*. A client may specifically request messages for particular types of events associated with a window. These are well-defined, formatted messages that with their prescribed sequence, define the *X Protocol*.

The server (called *X Server*) is a graphics application that controls a display screen as directed by messages received from clients. Each client application has a window on the display screen of the server, and the server concurrently modifies the appropriate window as required by the received client request messages.

When a machine only has a server with the sole function of providing graphical display service to clients, it is called an *X Window Terminal,* or simply an *X Terminal.* A host may have both a client and a server, and in distributed networks, a client application may have a connection with any server—not just its local server. A client application using its local server may route messages directly to the local server and bypass a transport mechanism, which amounts to a standalone application of X Window.

Typically resident (can also be remote) in the server is a *window manager* that controls the size, position, and stacking order of the windows displayed. Characteristics of the window decoration is a function of the window manager, which is common to all the client applications connected to the server. There are several window managers available for an X Window System, which accounts for the minor variations in the display [33].

Access to the server is initially restricted to the users on the local host, which may modify an access control list thereby opening access to remote clients. Figure 6.24 illustrates the basic elements of the X Window client/server system.

Figure 6.24 depicts a simple client and server connected by the Internet. Most intranets have the client and server attached to the same Ethernet segment. The TCP connection between the client and server is made on multiple TCP ports commencing with number 6,000. Hence, there may be *n* displays for a given set of *n* applications, each with a TCP connection. That is, the server listens on port 6,000+ *n* for application *n* and associates the requests with window *n*. Each TCP connection is initiated with the client sending an initial octet of data to identify the byte order that will follow. If the octet is equal to 102 (ASCII B), the transmissions will be sent most significant octet first. If the octet is equal to 154 (ASCII l), the transmissions will be sent least significant

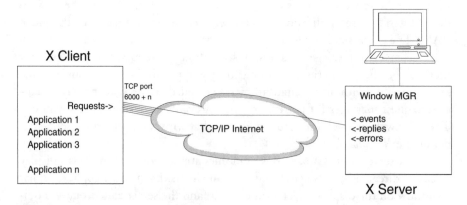

Figure 6.24 X Window client/server model.

octet first. Octet order for image data is defined by the server, and the client must translate (if required). The octet order is defined by the client for all other parts of the X protocol. Further information is exchanged to verify that the client and server are both using the same version of the X Protocol (the current version number is eleven).

The client/server model for X Window is event-driven. The client is essentially asleep until receiving an event message from the server. For example, if a user has a word processor application and a graphics drawing application open on the client but is doing nothing, the client is not checking status or waiting for a specific type event. (The communication is asynchronous.) It is asleep waiting for whatever might arrive. Should the user enter a command to cause the word processor to go from editing a document to print, the sequence of these events would cause the server to change the window reflecting the print options.

Many X Window systems involve special purpose machines with fine resolution displays that function solely as X Terminals. (For example, NCDware X Terminals.) Without the client, they are useless. Recently, with faster DOS-based PCs that may have access to several megabytes of memory and memory managers to use it, the PC is being used not only as an X Terminal, but as a client as well. This permits DOS applications to be exercised by UNIX machines, and vice versa. Examples of such implementations are DESQview/X and Vista-eXceed.

Event messages are 32 octets in length and contain an eight-bit type code. The high order bit of the type code is equal to one if the message is in response to an event request message. Otherwise, the message is an unsolicited event. The event message also contains a 16-bit sequence number that identifies the last request issued by the client and processed by the server.

Request messages are variable in length and contained in a single block. Each request contains an eight-bit "major" opcode and a 16-bit length field, that is a count of four-octet groups (including the header) in the entire block. Every request is assigned a sequence number, starting with one that is used in replies, errors, and events.

Reply messages are sent from the server to the client in response to a request message. (Most requests do not generate replies. Device events may require a synchronization mechanism.) A reply message contains a 32-bit length field that is a count of the four-octet groups contained in the reply message, including the header. The reply message has a fixed length portion of 32 octets followed by zero or more octets of message. The reply also contains the least significant 16 bits of the corresponding request's sequence number.

Error messages are 32 octets in length and contain an eight-bit error code that identifies the reason for error. It also contains the least significant 16 bits of the sequence number of the request message causing the error.

Endnotes

[1] The TELNET protocol specification is contained in RFC 854. The TELNET option specification is contained in RFC 855. See Table 6.3 for a list of negotiable options.

[2] The TN3270 option specification is contained in RFC 1041.

[3] The InterNIC system provides a domain name registration and database of Internet information service. The service is contracted to a private company, Network Solutions Inc. (NSI), through the end of March 1998. A competitive service provided by Emergent Corporation is planned to commence in February 1998. The contract with NSI may not be renewed.

[4] The specification for TFTP (Revision 2) is contained in RFC 1350.

[5] The original specification for TFTP, RFC 783, did not inhibit sending the current duplicate data block and resulted in duplicates for all succeeding blocks. The condition was identified in RFC 1123 and named "Sorcerer's Apprentice Syndrome."

[6] The specification for RPC (RFC 1057, developed by Sun Microsystems in 1988) does not restrict the implementation to a synchronous mode of operation. RPC calls may be asynchronous, which permits the client to do other useful work while waiting for one or more replies from the server.

[7] The specification for XDR is contained in RFC 1014, by Sun Microsystems.

[8] The official specification for NFS is contained in RFC 1094, which also contains (as an appendix) a specification for the mount protocol.

[9] Although there are distinct layers between NFS and the transport layer, the IAB defines the intermediate layers as sublayers to NFS. That is, there are only application, transport, Internet and link layers as defined by the IAB. (Reference RFC 1122.)

[10] See RFC 1327 for the specification of mapping between X.400(1988)/ISO 10021 and SMTP.

[11] The specification for SMTP is written in two parts. The first part (RFC 821) describes the protocol between the client and server, and the second part (RFC 822) defines the format of the user interface. These SMTP descriptions are based on the specifications noted herein. Key parameters may change in new RFCs and should always be checked. Newer RFCs not evaluated for these descriptions include RFCs 2045, 2046, 2047, 2048, and 2049.

[12] US ASCII is seven-bit coded characters, each contained in an octet. Reference ANSI X3.41986.

[13] RFC 821 restricts mail messages to 7bit USASCII data with 1,000 character lines.

[14] The MIME specification is contained in RFC 1521.

[15] The co-author of the MIME Specification, Ned Freed, stated there was only one instance since MIME installation where a host (noncompliant with RFC 822) kicked out a Content-Type header. The fix was easy.

[16] BNF stands for BackusNaur Form meta-language. The naming rules, repetition indication, and local alternatives were changed by Ken L. Harrenstien, of SRI International, for ease of understanding. Hence, it is referred to as "Modified BNF." The format is specified in RFC 822.

[17] An older version of PEM provided a clear text mechanism by delimiting text with asterisk (*) characters.

[18] The mechanisms of "BASE64" and "QUOTED-PRINTABLE" are both used in another feature to provide the capability of sending nonASCII characters in the header. See RFC 1522, Representation of NonASCII Text in Internet Message Headers.

[19] See ISO 8613, Information Processing: Text and Office System; Office Document Architecture (ODA) and Interchange Format (ODIF), Part 18, 1989.

[20] Examples in Section 6.7.10 were taken directly from the MIME Specification contained in RFC 1341.

[21] The specifications for PEM (RFCs 1113, 1114, and 1115) use X.400 nomenclature and refer to both clients and servers as user agents (UAs). It also uses the term message transfer system (MTS), with which the UA interacts. The portion of the UA that interfaces with a user is called the local UA functions.

[22] The PEM specifications permit the exclusion of portions of the text delimited with asterisk characters (*). The specifications are being rewritten by the IETF PEM Working Group to omit this capability and other changes in the security keys.

[23] The cryptosystem specified for use with asymmetric systems was invented by R. L. Rivest, A. Shamir, and L. Adleman and is widely known as *RSA*.

[24] FIPS PUBS 46-1, Data Encryption Standard, 2 January 1988, defines the encipherment algorithm used for message text encryption. IBM (with patent rights to the technology), has granted nonexclusive, royalty-free licenses under the patents to make, use, and sell apparatus complying with FIPS 46-1. Also see FIPS PUB 81, DES Modes of Operation, 2 December 1980 , which defines the modes of operation.

[25] See RFCs 1114 and 1115 for description of asymmetric encryption algorithm and key management.

[26] Until March 1995 the percentage of the total was decreasing for FTP and increasing for HTTP. The total volume of traffic for HTTP exceeded the traffic for FTP in June 1995. See /nsfnet/statistics/1995/nsf-9503.highlights at ftp://ftp.merit.edu/statistics.

[27] Hypertext Markup Language (HTML) is used to encode HTTP messages, and every technical book store is well stocked with books on HTML.

[28] File locking was originally used to resolve contention problems caused by multiple users. This will likely resurface to realize the full potential of HTTP.

[29] Technically, it contains a rniform resource identifier (URI), which is a superset of addresses. Only the URL is used.

[30] See specifications, RFC 1945, Hypertext Transfer Protocol—HTTP/1.0 (Informational Protocol), RFC 2068, Hypertext Transfer Protocol—HTTP/1.1 (Proposed Standard) and RFC 2109, HTTP State Management Mechanism (Proposed Standard)—A limited session.

[31] HTTP/1.1 contains fields not described in MIME and does not use some fields defined for MIME. Although HTTP messages include the MIME level number supported, it requires a proxy/gateway function to ensure compliance when exporting HTTP/1.1 messages to MIME compliant environments. The number of draft proposed changes to HTTP indicates that it is not yet a stable protocol.

[32] The X Window System specification states a transport (generic). Other transports may be used (e.g., IPX with Novell networks). Examples of products supporting TCP/IP are PC/TCP, PC/NFS, WIN/TCP, LAN WorkPlace, 3+Open TCP, HP ARPA Services, TCP BNS/PC and Beame & Whiteside NFS.

[33] Some of the popular window managers are OSF/Motif, OPEN LOOK, DWM, and TWM.

7

Future of TCP/IP Protocol Suite

To identify the TCP/IP protocol changes during the past five years, simply refer to the last 1,000 RFCs and see which ones made it through the standards process. How many of them could have been predicted five years ago? The draft RFCs are not a sure indicator. SNMPv2 appeared to be a given and went back to the drawing board for security improvement.

The most activity on the Internet in the past five years has been with the WWW and ISPs. Although there has been a lot of draft RFC activity with HTTP, the revision level has only recently moved from 1.0 to 1.1.

A change pattern to the Internet involves resources and applications. When a new technology is invented that increases either processing power or transmission bandwidth, an application is developed to use it. When a new application surfaces that is not feasible because of the lack of resources, they are cranked up. That is, increased bandwidth, processing power, and storage capacity. Web browser activity certainly increased the need for resources, which was responded to with ATM and gigabit Ethernet. The limited capabilities of ISPs and corporate intranets was responded to with protocols like PPP, L2TP, and RADIUS. Security needs can dictate the viability of an application even when the resources are available.

In the mid-1990s, many ISPs were started with a simple remote access server that had a connection to the Internet and a separate analog modem for each active dial-up user—for example, a Telebit NB40 with 32 independent analog modems. The modem footprint is getting smaller (twice this number of modems on a single card), it is getting faster (56 Kbps when in conjunction with a digital link), and it is being used in a more effective manner. For example, ISPs or corporate intranets may use an ACC *remote access concentrator*

(RAC) to dynamically allocate analog or digital ports from a modem pool with a *virtual port* (VP) concept. When the RAC is configured for carriers, a *virtual private network* (VPN) concept is used to provide wholesale Internet access to ISPs and intranets or to simply provide a transparent connection between dial-up users and their home network. Multiple RACs are managed by a *virtual port service manager* (VPSM) that is server-based and provides a proxy service to an ISP or home network RADIUS server.

Central to this entire scenario of VPs and VPNs is PPP. PPP-originated calls (ISDN or PSTN) are aggregated on a digital link (T1/E1, PRI, or ATM) to the appropriate ISP, corporate network, or the Internet. With L2TP this may be done with any protocol stack that can bind with PPP (e.g., IPX or AppleTalk). With only a slight variation, the dial-up user may be cellular or a combination of cable and satellite, which increases the user access rate by multiples of the present. If you could buy stock in a protocol, PPP would be a good choice. The extensible design assures its longevity. In this analogy, HTTP is a blue-chip stock.

Government funding can accelerate development of resources and applications. (Development of the original Internet was funded by DARPA.) An initiative called the *Next Generation Internet* (NGI) is a three-year program that started in 1996 with $300M divided among several government agencies (with the lead role going to DARPA). This will involve a test network with 100 sites that are linked at a speed 1,000 times greater than today (proponents hope) and the creation of revolutionary applications (i.e., telemedicine and distance learning).

Which applications will grab the most new resources? For existing applications, Web browsers using HTTP is my first pick. New Internet users are announced in increments of 10,000, and most new users are not using FTP to exchange files with members of the scientific community. They are surfing the Web looking for news, sports, and entertainment.

New development is under way by the IETF to provide multimedia services, such as Internet telephony with *voice over IP* (VoIP). A multimedia application provides (RSVP) with a destination(s) and *quality of service* (QOS) requirement, which are then coordinated with an RSVP agent in each interim router. The setup for VoIP is required in both directions. RSVP is a transport layer protocol and uses the flow label option of IPv6 to queue traffic based on its delay-sensitivity nature. Interim routers may cache for short periods an IP source address with Flow Label to expedite handling. The result is a virtual circuit-switched network that works for LANs as well as WANs.

Another effort under way by the IETF to provide voice transmission over the Internet involves the *real time protocol* (RTP), which uses the transport layer

UDP for end-to-end delivery services. A companion control protocol (RTCP) using a different port number manages the quality of service required for delay-sensitive data.

The TCP/IP protocol suite could not be derailed if made illegal. It was essentially made illegal in the early 1990s by a government mandate to replace it with an OSI stack. (This was called The Government OSI Profile, or GOSIP for short.) It survived. What makes it bomb proof? There is no single answer, but leading the list is that it is truly an open system. It is open in that it is viewed daily by the best minds from around the world—not a few from one company, one standards group, or one country. The TCP/IP protocol suite is also open to change. The original IPv4 (Classic IP) is in the process (slow) of being replaced by IPv6. A new transport layer protocol (RSVP) is available to handle delay-sensitive data because TCP is not suited. Although the TCP/IP protocol suite lives on, the title is often shortened to *IP stack*.

Appendix A: Networking Evolution

Prior to the late 1960s, only message switching and front-ending of mainframes (primarily circuit switches) were available. This was reasonable since every good communications engineer at the time knew that dynamic bandwidth allocation, which involved statistical multiplexing, was not a workable option because of hardware and software cost. Of course, as the price of hardware fell and line costs increased, the picture changed [1].

Industry recognized the need for networking and provided proprietary architectures to accommodate dynamic bandwidth allocation, with the dual goals of selling products and influencing the standards bodies both directly and indirectly. IBM's SNA is today a leader in the field and has strongly influenced the development of standards [2]. DEC, with DECNET, has been active in all standards development, including Internet standards.

DARPA funded a program to develop packet switching using mobile radio, satellite, and cable from 1975 to 1982. During the same time period, Xerox developed packet switching with a coaxial cable. These dissimilar networks (ARPANET, packet radio networks, Atlantic packet satellite, domestic wideband, and sea-going MATNET and Ethernet) were interconnected by DARPA using gateways and a common set of networking protocols. These protocols are now called the TCP/IP protocol suite.

In 1968, the *Consultative Committee on International Telegraph and Telephone* (CCITT), envisioning the need for statistical multiplexing of data, formed a study group to develop standards for packet switching. Although its initial effort (1972 Green Books) was little more than a multivolume set of notes on circuit switching, the revised effort (1976 X.25 Orange Books) was accepted around the world. In 1977, the major *public data networks* (PDNs) were interconnected with another CCITT Recommendation, X.75. Today, X.25 PDNs provide access to users in nearly every country.

The ISO defined the OSI reference model for its network architecture in June 1979 (Version 2, working draft)—about the same time that the design of DARPA's TCP/IP Internet was nearing completion. Implementations of the lower layers of the reference model were made during the late 1980s.

Figure A.1 graphically illustrates the time frame of the development of packet switching, networking, and other communications activity that took place during this era. This is followed by a brief description of the ISOC, the CCITT and the ISO in Sections A.1.1, A.2, and A.3.

A.1 The Internet

The present Internet architecture and protocols were formulated during the period 1977–1979, although the ideas were born much earlier. Initial specification work started in 1973 and went through four versions, culminating with version 4 in 1979. Vinton G. Cerf and Robert Kahn are generally accepted as the inventors of TCP/IP in 1974 [3].

Robert Kahn was one of the principal architects of ARPANET. In late 1972, he conducted a three-day demonstration in Washington, D.C., (ICCC) of interconnecting 12 hosts with 40 terminals using ARPANET packet switching technology. Dr. Kahn worked with Bolt Beranek and Newman (BBN), which developed the ARPANET and the first TCP/IP Internet and continues to support it. Dr. Kahn is the founder and president of the Corporation for National Research Initiatives, a member of the National Academy of Engineering, a Fellow of IEEE, and a trustee of the Internet Society.

Vinton G. Cerf is an IEEE Fellow (1988) and past chairman of the IAB. He led the design team at Stanford that developed TCP/IP and managed the DARPA Internet project from 1976–1982. Dr. Cerf continues to play a major role in the Internet as president of the Internet Society. He is also vice-president of the Corporation for National Research Initiatives.

Cerf and Kahn received the ACM System Software Award in 1991 and the IEEE Koji Kobayashi Award in 1992 for their TCP/IP work.

The following is a list of significant milestones and events associated with the TCP/IP protocol suite evolution.

1. The *Internet Control & Configuration Board* (ICCB) was formed and funded by DARPA in 1979.

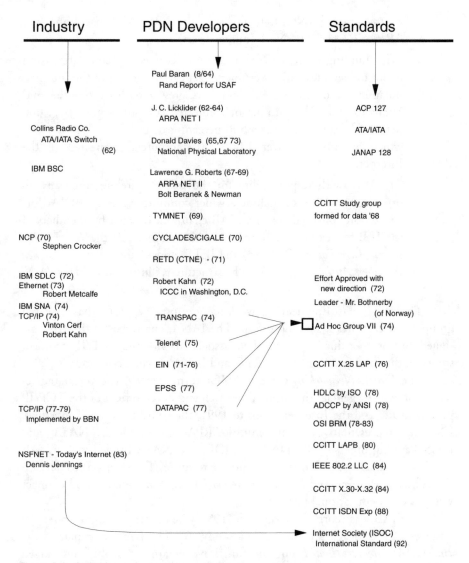

Figure A.1 Switching and networking evolution.

2. Bolt Beranek and Newman (BBN) was funded to implement TCP/IP with UNIX in 1979. Utilities compatible with BSD [4] UNIX were offered. A UNIX-based TCP/IP implementation was made available to all major universities by the BSD.

3. In 1983, the ICCB was reorganized to become the IAB. The *Defense Communications Agency* (DCA) mandated that all computers connected to ARPANET use TCP/IP, which replaced the earlier *network*

control program (NCP). It also split the ARPANET backbone into ARPANET and MILNET (military part).

4. With funding from the NSF to support connecting computers from major universities and government agencies to the rest of the Internet, the number of connected computers grew rapidly. From the mid 1980s to 1990, the number of connected computers grew by 1,000 percent. Estimates are for a 500 percent growth from 1990 to 1993, which amounts to over 1,000,000 computers connected to the Internet.

5. The IAB, with responsibility for the Internet architecture, was re-organized in 1989 to include a wider community. The new IAB organization consisted of: 1) an IAB chairman (elected by members of the IAB for a two-year term) and IAB members, 2) the *Internet Research Task Force* (IRTF), 3) the IETF, and 4) the *Internet Assigned Numbers Authority* (IANA). The structure is illustrated in Figure A.2.

The IETF has Engineering working groups, the IRTF has research groups, and each has a steering group. The IANA, chartered by the IAB, assigned and/or coordinated all numbers associated with the TCP/IP protocol suite and the Internet. The IETF, IRTF, and IANA remain active today.

The *Federal Networking Council* (FNC), formed in 1991, coordinated activities of federal agencies that funded research and development of the TCP/IP protocol suite and the Internet. Prior to 1991, the coordination was informal. FNC has representatives from NSF, DARPA, DOE, DOD, NASA, and HHS—among others. NSF, DARPA, DOE, and NASA funded various IAB and IETF management efforts and continue to do so. The plan was to develop the *National Research and Education Network* (NREN), which would connect virtually every educational institution [5].

The IAB was reorganized again in 1992 when it was brought under the auspices of the ISOC, an international body [6]. The IAB was renamed the *Internet Architecture Board,* but the functions remain reasonably unchanged (with new technologies and problems). The present infrastructure of the ISOC, including the IAB, is described in Section 1.2.1. Many of the RFCs maintained by the IANA are rich with history of the Internet [7]. Table 1.1 provides a list of RFCs describing the ISOC structure and process.

A.1.1 Internet Society (ISOC)

The ISOC, an international professional organization, has under its auspices the IAB. The IAB, since 1983, and the ICCB prior to that, is responsible for the evolving architecture and protocols of the Internet. The ISOC is governed by a board of trustees made up of 14 trustees (including five officers) and an executive director. The officers include the president, two vice-presidents, a secretary, and a treasurer. The board of trustees has an advisory council and a secretariat (with an executive director). See Figure A.2.

The individual members of the ISOC elect trustees for three-year terms.

The *Internet Society News* is the formal publication of record for announcements on matters related to Internet standards. The *ISOC News* is distributed by electronic mail to all ISOC members. At least twice a year, the ISOC publishes a status of the Internet standards process.

Volunteers manage the infrastructure of the ISOC, including members of the IAB and its task forces. Although several agencies continue to support key aspects of the TCP/IP protocol development, the majority of personal activity

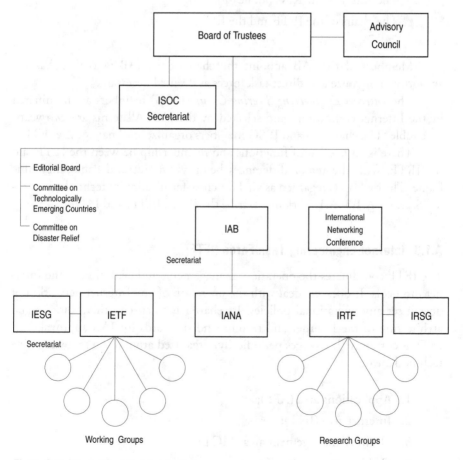

Figure A.2 Internet Society infrastructure.

(e.g., organizing meetings, attending meetings, and writing RFCs) is done on a voluntary basis.

A.1.2 Internet Architecture Board (IAB)

The IAB is the coordinating committee for Internet design, engineering, and management. The IAB has a maximum of 15 members who perform on a voluntary basis. Individuals are nominated for membership to the IAB by Internet community members and selected by the ISOC trustees for two-year, renewable terms [8]. The IAB may create task forces, committees, and working groups as required within the scope of the IAB's responsibility. The initial appointments are the following.

- The editor of the RFC publication series;
- The chairs of the IETF and the IRTF.

Members of the IAB appoint the chair of the IAB who then has the authority to organize and direct task forces as deemed necessary.

The *Internet Engineering Steering Group* (IESG) members are nominated by the Internet community and selected by the IAB. All terms are two years, renewable. The chairman and IESG members organize and manage the IETF.

There is an overlap of functions and membership between the IETF and the IRTF, with the major differences being viewpoint and sometimes time frame. The overlap is regarded as vital for cross-fertilization of technology transfer. Sections A.1.3–A.1.5 briefly describe the IETF, IRTF, and IANA.

A.1.3 Internet Engineering Task Force (IETF)

The IETF coordinates the operation, management, and evolution of the Internet protocols. It does *not* deal with the operation of any Internet network, nor does it set any operational policies. Its charter is to specify the protocols and architecture of the Internet and recommend standards for IAB approval. The IETF is currently (it changes periodically) organized around the following nine technical areas:

1. Applications area (APP);
2. Internet area (INT);
3. Network management area (MGT);
4. OSI integration area (OSI);

5. operational requirements area (OPS);

6. Routing area (RTG);

7. Security area (SEC);

8. Transport and services area (TSV);

9. User services area (USV).

The IETF Chairman and a technical area director from each area make up the IESG membership. Each technical area director has responsibility for a subset of all the IETF working groups. There are many working groups, each with a narrow focus and the goal of completing a specific task before moving to a new task (with some exceptions for tasks requiring ongoing support).

The IETF is the major source of proposed protocol standards for final approval by the IESG. The IETF meets three times annually, and extensive minutes of the plenary proceedings as well as reports from each of the working groups are issued by the IETF secretariat.

A.1.4 Internet Research Task Force (IRTF)

The IRTF is a community of network researchers that make up the following seven IRTF work groups.

1. End-to-end research group;

2. Autonomous networks research group;

3. Privacy and security research group;

4. Electronic libraries research group;

5. Internet architecture workshop;

6. Electronic communities research group;

7. Resource discovery research group.

The IRTF is concerned with understanding technologies and how they may be used by the Internet rather than products or standard protocols. However, specific experimental protocols may be developed, implemented, and tested to gain the required understanding.

The work of the IRTF is governed by its IRSG. The chairman of the IRTF and IRSG appoints a chair for each *research group* (RG). Each RG typically has 10 to 20 members and covers a broad area of research, which is determined by interests of the members and by recommendations from the IAB.

A.1.5 Internet Assigned Number Authority (IANA)

The IANA, located at the USC/Information Sciences Institute, is chartered by the IAB to coordinate the assigned values of protocol parameters—including type codes, protocol numbers, port numbers, Internet addresses, and Ethernet addresses. The IANA delegates the responsibility of assigning IP network numbers and autonomous system numbers to the *Network Information Center* (NIC) at Network Solutions, Inc.

All published documents maintained by the NIC are called RFCs. Older papers were called *Internet Experiment Notes* (IENs). About 150 of the RFCs are standards (now called *STDs*) and published in the RFC series [9].

The "Assigned Numbers" RFC indentifies the currently assigned values used in several network protocol implementations [10].

A.1.6 Standards Track

Each layer of the TCP/IP Internet architecture has multiple protocols, except the Internet layer. Protocols of the TCP/IP protocol suite were originally identified by a name and RFC number only. This caused problems since the latest RFC number, status, or maturity level was not known by implementers. The problem has been corrected with a process called the "standards track."

Each RFC providing a specification of a protocol is assigned a "Maturity Level" (state of standardization) and a "Requirement Level" (status). The "Maturity Level" of a standards track protocol begins with *proposed standard.* There are also protocols with a maturity level of *experimental.* Experimental protocols may remain experimental indefinitely, become *historic* or be reassigned as a *proposed standard* and enter the standards track. Protocols with a *proposed standard* maturity level must be implemented and reviewed for a minimum of six months before progressing to *draft standard.* Figure A.3 illustrates the standards track process.

Progressing from *draft standard* to *standard* requires evaluation for a minimum of four months and the blessing of the IESG and the IAB. (It takes action from both the IESG and the IAB to start or progress on the standards track.) A unique, standard (STD) number is assigned to each protocol reaching the *maturity level* of *standard.* The STD number identifies one or more RFCs that provide a specification for the protocol. Although the RFC identified by the STD number may change, the STD number is constant. A protocol may have the status *required, recommended, elective, limited use,* or *not recommended.* Protocols with the status of *required* must be implemented. See Section C.1 for examples of protocols with status and STD numbers. Standard Internet

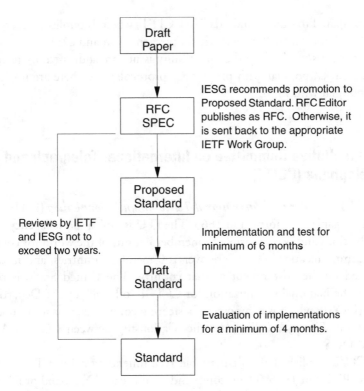

Figure A.3 Standards track process.

protocols that existed prior to the new process have been grandfathered from the standards track and assigned STD numbers (1–32) [11].

RFCs are reissued with new numbers regularly to update logic and correct errors. However, interim fixes for any RFC may be published in a special series of RFCs designed specifically to note corrections and recommended default values for multiple RFCs, including the most current. The first series of RFCs in this category is the "Requirements for Internet Hosts Application and Support" RFC (Current number RFC 1123). The second series of RFCs is the "Requirements for Internet Hosts Communication Layers Requirements for Internet Hosts" RFC (Current number RFC 1122). The third in the series is "Requirements for IP Version 4 Routers" RFC (Current number RFC 1812). Although you may have the latest standard number and the latest RFC number for that standard, it is still necessary to check the above special RFC series to obtain last-minute corrections.

So what is gained by this slow (relative to some proprietary protocols), elaborate, and time-consuming process of standards track? For starters, it is not slow if compared to other standards like CCITT, which is published every four years. Most importantly, however, the implementation and evaluation portion of the process assures backward compatibility issues are addressed, there are no system fallbacks (popular with proprietary protocols) and there are no "paper tigers" (popular with ISO).

A.2 Consultative Committee on International Telegraph and Telephone (CCITT)

The CCITT is under the *International Telecommunications Union* (ITU) which is a treaty organization formed in 1865. The ITU is now a special agency of the United Nations and is responsible to member nations of the world for compatible telecommunications. There are over 160 member countries, and most are represented by the government of that country. The United States is represented by the National Organization for CCITT (officially by the Department of State (DOS)), which is composed of a steering committee plus four domestic study groups. Figure A.4 illustrates the relationship between CCITT, ANSI, ISO, and DOS.

CCITT published the first, usable recommendation for a Public Data Network (PDN) in 1976. The recommendation is called X.25 and provides an interface specification between a network and a user. This is called DCE on the network side of the interface and called DTE on the user side of the interface. The specification identifies the lowest layer protocol of ISO's *high-level data link control* (HDLC) procedure as the link layer. The CCITT X.25 recommendation only specified layers 1, 2, and 3 because, it was believed, that with a minimal transport service, all requirements for a reliable network service were met.

A.3 Organization for International Standardization (ISO)

The ISO is a voluntary activity that represents nearly 100 participating countries, each with one vote. It was formed in 1947. The U.S. member body is ANSI, and the CCITT participates as a liaison member. Organization of the ISO standards is by subject material, and there is a near one-to-one correspondence to the organization of ANSI standards.

The ISO recognized the need for dynamic bandwidth allocation and a set of nonproprietary protocols to accomplish networking. In 1977, the ISO's Technical Committee 97 established a subcommittee (SC 16) to develop a

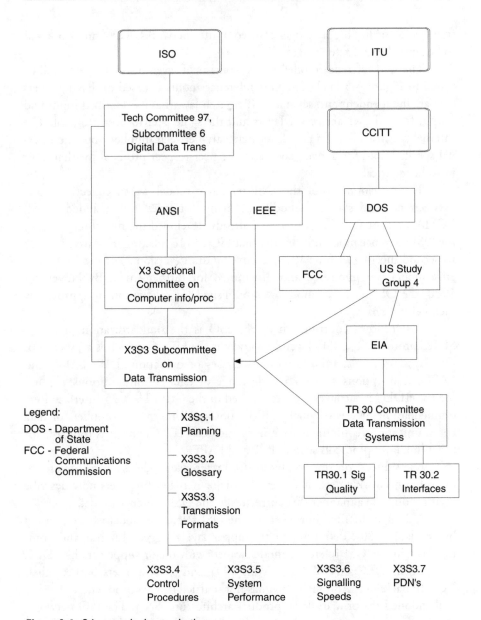

Figure A.4 Other standards organizations.

framework for linking heterogeneous computer networks. The framework was to become the OSI reference model.

The OSI reference model, which consists of seven distinct layers, is illustrated in Figure A.5 [12]. The OSI reference model is based on having layers that are independent and separable. (i.e., each layer has a prescribed input and output from the $n1$ and $n + 1$ layers and the same peer-to-peer protocol. The internal construction of a peer layer may vary between suppliers, but the external specifications for a given layer and the peer-to-peer protocol for that layer must be identical.

The development of the OSI reference model was a collective effort between national standards bodies of many countries and included ANSI, ECMA, and the CCITT (which had already developed the first three levels of the OSI reference model). From the first ISO SC 16 meeting in March 1978 to its second meeting in June 1979, a working draft (version 2) of the ISO reference model was published and distributed for further study. By November 1980, the OSI reference model was published for solicitation of approval by member nations.

A significant contribution of the ISO is the standardization of IBM's SDLC protocol. The ISO layer 2 protocol is the HDLC, which is a superset of link layer protocols. HDLC with its classes of operation (LAP, LAPB, and LAPD) and options satisfies a wide range of data communications requirements. HDLC is the layer 2 protocol used in the CCITT's X.25 Interface Recommendation (it was originally IBM's BSC). All the major suppliers of data communications equipment have implemented HDLC, and it is the basis for other link layer protocols such as PPP and LAPF.

A less direct contribution by the OSI reference model is a basis of describing other networking architectures. All major network providers now describe their products in a manner that correlates to the OSI reference model.

Although the first four layers of the OSI reference model were operating by the late 1980s, definition of the upper layers lagged. Interest and commitment in the OSI reference model waned with major supporters like DEC scuttling its DECnet/OSI Phase V in 1991 and sending users back to their TCP/IP backbone. Although IBM has not scuttled its OSI product, neither has it abandoned the primary IBM product architecture, SNA. The ISO networking standards were given another breath of life by the *National Institute of Standards and Technology* (NIST) issuing an edict, "Suppliers of information systems and networks to government agencies must comply with GOSIP by 1992." However, when 1992 ended and the ISO networking standards continued to be either incomplete or inadequate, the NIST mandate and GOSIP ultimately failed.

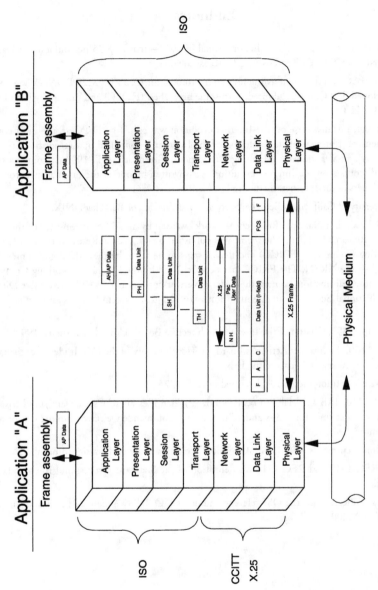

Figure A.5   OSI reference model.

Endnotes

[1] The line and computing cost became equal in 1972—that is, $.35 per million bits transported from New York to Los Angeles (coast to coast).

[2] IBM's BSC was first used as the link layer protocol for CCITT's X.25 PDN. When IBM developed SDLC, it was adopted, with some new labels (HDLC) and tuning, by both ISO and CCITT.

[3] Although this discourse on the Internet evolution was edited by Dr. Cerf, the most comprehensive treatment of the evolution of the Internet is contained in a recent book titled, *Internet System Handbook.* Manning Publications, Greenwich, CT, 1993. It contains several discourses on the Internet evolution, each written by one of the original inventors or developers of the Internet, including Dr. Cerf.

[4] University of California's Berkeley Software Distribution, or Berkeley UNIX.

[5] See "Towards a National Research Network" written by an ad hoc committee of the National Research Council. The report was published by the Federal Research Internet Coordinating Committee (FRICC) and details steps to be taken by the federal government to establish the NREN. The FRICC has been replaced by the Federal Networking Council. For more information, contact the Federal Networking Council, ATTN: Executive Director, National Science Foundation, Room 416, 1800 G Street, NW, Washington, DC, 22050.

[6] See RFC 1358, Charter of the Internet Architecture Board (IAB), dated August 1992.

[7] RFC 1160, the Internet Activities Board, provides a history of the IAB. It identifies the task forces, members, and goals of the IAB.

[8] There is a petition method as described in RFC 1325.

[9] RFC 2200, IAB Official Protocol Standards, identifies all RFCs that are standards. It associates a status with each standard (required, recommended, elective, limited use, or not recommended).

[10] The most current is RFC 1700. Appendix C contains many of the assigned numbers.

[11] See RFC 1310 and RFC 1311 for definition of the Standards Process and the Standards Track.

[12] The specification for the OSI reference model is ISO 7498 (Addenda 1, 2, 3 and 4), and CCITT X.200.

Appendix B: Numbering Systems

This appendix is directed to those who do not use different numbering systems on a regular basis. Reference material on this topic is not usually provided in data communications books because it is assumed that readers know it. From where? Those proficient in handling base 2, base 10, and base 16 conversions should skip this appendix. Those who are not, or who would like a refresher, please continue. It does not provide an "everything you ever wanted to know" treatment of numbering systems. It does provide everything you need to know about numbering systems to understand the concepts of data communications and the TCP/IP protocol suite.

Working with different numbering systems adds a degree of difficulty to understanding data communications. The term *digital computer* refers to its memory and *digital transmission* refers to the units being transmitted—and both refer to binary digits (either one or zero). For humans to make sense of the binary digits, they are converted in groups to decimal numbers that we can relate to. If the fundamentals of this conversion process are understood, a concept can be visualized at a glance. Without the ability to see the same picture as the inventor did (a binary picture), concepts will be evasive and difficult to comprehend.

The base 10 numbering system is presented first for definition. The same terms used to define the base 10 numbering system are used for the base 2 and base 16 numbering systems. By describing the base 10 numbering system first, the similarities should be clear. Converting to and from binary (base 2) is discussed next, followed by conversion to and from hexadecimal (base 16) directly from binary and from decimal.

B.1 Base 10 (Decimal)

The number 10 in base 10 represents the number of digits in the units position. The numbers in the units position of the base 10 system are 0, 1, 2, 3, 4, 5, 6, 7, 8, and 9, which is a total of 10 digits—hence, base 10.

The convention for stating a number in the base 10 system is to commence on the left side with the digit of the highest order. For example, with the number 6,309, the digit six is noted first because it is the thousands digit and so on. The digit three is next because it is in the hundredths position, digit 0 is next because it is in the tens position (second column from the right), and the nine digit is last because it is in the units position (first from the right.) This is illustrated as:

$$6\,(10^3) + 3(10^2) + 0(10^1) + 9(10^0) =$$

$$6(1000) + 3(100) + 0(10) + 9(1) = 6,309$$

The position (or column) within the number also determines the total value of that digit. For example a number in the units position means, that number times 10 raised to the zero power. But any number raised to the zero power is equal to one. Hence, $10^0 = 1$, $10^1 = 10$, $10^2 = 100$, $10^3 = 1000$, and so on. This can be illustrated as shown in Figure B.1 where X represents any number from 0 through 9. Applying this to the original number from above, it may be represented as illustrated in Figure B.2.

Notice that the column titles 10^0, 10^1, 10^2, 10^3, 10^4, and 10^5 were replaced with the power only (0, 1, 2, 3, 4, and 5)—the 10 is implied.

10^5	10^4	10^3	10^2	10^1	10^0
X	X	X	X	X	X

Figure B.1 Base 10 number representation.

5	4	3	2	1	0
0	0	6	3	0	9

Figure B.2 Base 10 number representation.

Since the numbers in each column are circular, if a number is added to a number in one column and the result is greater than nine, the next number commences with zero, or cycles through zero, *n* times. The number of times cycled through zero, called the carry, is added to the next higher digit. As in the example above, if one is added to the units position (9) resulting in a sum of 10, the units position is zero and the number of times cycled, or the carry (1), is added to the 10s digit. The new number is 6,310.

So far, nothing new has been presented. This is the system you were taught from time zero (though perhaps more formal than you have seen it since grade school)! With this as an understanding of base 10 numbers, let's go on to the base 2 number system.

B.2 Base 2 (Binary)

As the name suggests, the base 2 number system has two digits, zero and one. It is a boring system for humans because it is too easy. How long would it take first graders to learn to count (0, 1, 0, 1, 0, 1, 0, 1, , ,)? Just remember to start with zero. The string of zeros and ones gets too long to make sense to humans, but computers thrive on it.

The base 2 number system works the same as the base 10 number system, but with only 2 numbers instead of 10. To visualize this, observe the addition example in Figure B.3.

Notice that commencing with the value of zero, one is added to form a subtotal. Then one is added to the subtotal to form a new subtotal, and so on. With binary numbers this process would quickly result in a number too large to represent on one line.

Another observation is that the format is identical to the base 10 numbering system. That is, the value of each digit is equal to the base number (2) raised to the power of the bit position, as illustrated in Figure B.4.

In this illustration, X can only have the value of zero or one. If the value of X is equal to one, the value of the field is determined by raising 2 to the power indicated for that position. That is, the value of either zero or one is multiplied by the power of 2. From the previous addition example the decimal number 8 in binary format is illustrated in Figure B.5.

Notice that each digit position is labeled with only the power that the number two must be raised for this position—the character 2 is implied. Hence, raising 2 to the third power is equal to 8, which agrees with the previous addition example.

For example, consider bit positions 4 and 1 with each equal to the value one and all other bits equal to zero. Then the decimal value of the binary 2

Binary								Decimal
0	0	0	0	0	0	0	0	0
						+	1	
0	0	0	0	0	0	0	1	1
						+	1	
0	0	0	0	0	0	1	0	2
						+	1	
0	0	0	0	0	0	1	1	3
						+	1	
0	0	0	0	0	1	0	0	4
						+	1	
0	0	0	0	0	1	0	1	5
						+	1	
0	0	0	0	0	1	1	0	6
						+	1	
0	0	0	0	0	1	1	1	7
						+	1	
0	0	0	0	1	0	0	0	8

Figure B.3 Base 2 numbering.

2^7	2^6	2^5	2^4	2^3	2^2	2^1	2^0
X	X	X	X	X	X	X	X

Figure B.4 General octet of binary digits.

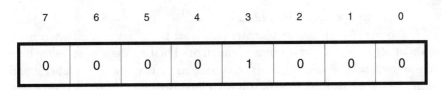

Figure B.5 Illustration of the decimal value 8 in binary.

number is equal to $2^4 + 2^1$, equal to $16 + 2$, equal to 18. This process, or technique, will work for any binary number. If all of the bits for a field of n bits are each equal to one, an easy technique of determining the decimal value is to determine the decimal value of 2^{n+1} and subtract one. This is similar to adding one to the decimal number 999 (equal to 1,000) and subtracting one.

It is straightforward to convert from binary to decimal, but how about going the other direction—from decimal to binary? Although not as straightforward, it is a mechanical process that is easy if you either memorize the powers of two or have a chart handy. First, select the digit position (power of 2) that produces a number either equal to or less than the decimal number to be converted to binary, and call it n for example. Subtract that number from the decimal number to be converted. If the remainder is not zero, repeat the same process with the remainder as the number to be converted until the remainder is zero. Each power of two used identifies a bit in the binary number equal to one, and all others are each equal to zero.

As an example, consider the decimal number 170. Looking at a chart of the powers of two, or from memory, the first power of two less than (or equal to) 170 is 2^7, which is equal to 128_{10}. Subtracting 128 from 170 leaves a remainder of 42. The next smaller power of two is 2^6 (64_{10}) is larger than 42, so evaluate the next smaller power of two, 2^5, or 32_{10}. Subtracting 32 from the remainder (42) leaves a new remainder of 10. The next power of two that is smaller than the remainder is 2^3, or 8, and the new remainder is 2 (10 - 8), which is equal to 2^1. Now to determine the binary value of the number is a simple matter of writing a one in the bit positions used for subtraction. In this example, the bit positions 7, 5, 3, and 1 were used and the binary number is illustrated in Figure B.6.

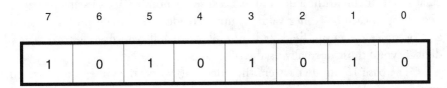

Figure B.6 Example of the decimal value 170 in binary.

To verify the conversion, add $2^7 + 2^5 + 2^3 + 2^1 = 128 + 32 + 8 + 2 = 170$.

Caution: Many (most) presentations label the digits of a binary number without regard for the real value of the digit. The first digit of an octet to be transmitted, with a value of 2^0, is often labeled as 1, while others label it with either 7 or 8 (as in RFC specifications). The mislabeling probably started with an editor looking for spelling errors and thought that zero was not a real number—once published it could not be corrected. Just remember, you must use the value 2^0 for the first bit transmitted, 2^1 for the second bit transmitted, and so on. Note that some people become proficient with binary conversions by associating only the correct decimal value of each bit position and do not understand that these values are derived from the powers of two. (They ignore the labeling.) If it works, this is okay. However, if you are not sure or lack an understanding of the conversion process, adopt these fundamental basics and you will not need a calculator to determine the correct value of a binary field.

B.3 Logic Operators

There are three basic logic operators used in binary arithmetic. They are the logical "and - AND," "inclusive or - OR," and "exclusive or - EOR." The EOR operator tests two bits (or logic states) and if each value is the same (one or zero), the result is zero (false). The OR operator tests two bits and if either bit is equal to one (true), the result is one (true). The AND operator tests two bits and makes the result equal to one (true) only if both bits are each equal to one.

If you are programming at the assembly language level, all three logic operators will be used. If you are configuring TCP/IP options, interfaces, and addresses, you will probably only need to use the AND operator. It is required for subnet and supernet addressing.

B.4 Base 16 (Hexadecimal)

The hexadecimal (hex) numbering system is primarily used by programmers or people required to work extensively with binary numbers. The binary numbers are cumbersome (too long) and can be converted directly to hex without a calculator or difficult arithmetic. This is because the number 16 is itself a power of two. That is, each of 16 hex values is uniquely identified in a 4-bit binary field and occupies the entire field. It is not all down hill, you must learn six new digits beyond those normally used (0–9).

The base 16 number system (hexadecimal) works the same as the base 10 or base 2 number systems, except it has 16 numbers in the units position instead of 10 or 2. The first 10 numbers are the same as base 10, but new expressions had to be invented for each of the six numbers above 9. This was a chance for someone back in an earlier era to be remembered for the new characters used (new sounds, new representations, etc.), but for ease, the first six characters of the alphabet were used. Hence, the numbers of the base 16 numbering system are; 0, 1, 2, 3, 4, 5, 6, 7, 8, 9, A, B, C, D, E, and F. The characters A, B, C, D, E, and F represent, respectively, 10, 11, 12, 13, 14, and 15. Since decimal to binary conversion has already been introduced, Figure B.7 illustrates base 10, base 2, and base 16.

Decimal	Binary				Hexidecimal
0	0	0	0	0	0
1	0	0	0	1	1
2	0	0	1	0	2
3	0	0	1	1	3
4	0	1	0	0	4
5	0	1	0	1	5
6	0	1	1	0	6
7	0	1	1	1	7
8	1	0	0	0	8
9	1	0	0	1	9
10	1	0	1	0	A
11	1	0	1	1	B
12	1	1	0	0	C
13	1	1	0	1	D
14	1	1	1	0	E
15	1	1	1	1	F

Figure B.7 The representation of base 16 numbers.

If you will be working with binary and hexadecimal numbers, the above chart should be memorized, which will make all conversions easy. It is relatively easy to visualize binary numbers from zero to seven. Now simply repeat the exact bit sequence with bit +3 (the 4th bit) equal to 1, which represents the numbers 8 through 15, or 8_{16} through F_{16}. With a little practice, binary numbers can be converted to hex and decimal without a calculator or pencil and paper.

Consider again the example using the decimal value 170. The binary value was determined to be "10101010_2." From the chart above the binary number "1010" is equal to "A" hex. Then the number "10101010_2" is equal to "AA" hex.

Can you convert hex numbers directly to decimal? The process is identical to the process of converting binary to decimal. Instead of working with powers of 2 or 10, use powers of 16. Observe the similarity between the illustration of a hexadecimal umber in Figure B.8 and the illustration of a base 10 number in Figure B.1.

Where the value of X ranges from 0 to F_{16}.

Again, from the example above, 170_{10} was determined to be equal to AA_{16}. Verify this by multiplying A_{16} (10_{10}) times 16^1, which is equal to 160_{10}. Now add to this 16^0 times A_{16}, which produces a sum equal to 170. More simply put, $(10 \times 16)+10 = 170$—and it checks.

Convert three or four numbers per day for a couple weeks, and it will be like typing or riding a bicycle. You will never forget.

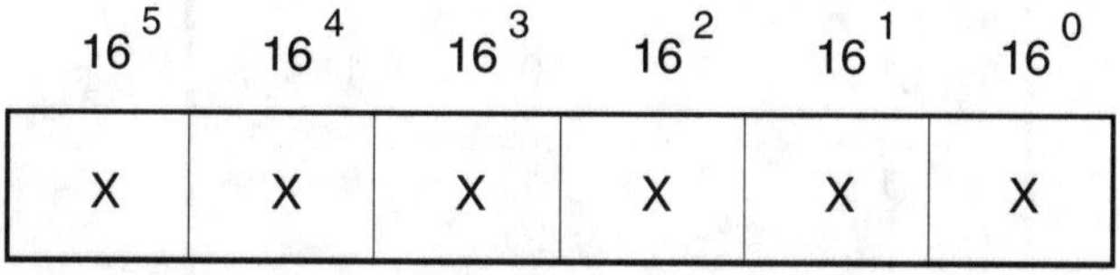

$$16^5 \qquad 16^4 \qquad 16^3 \qquad 16^2 \qquad 16^1 \qquad 16^0$$

X	X	X	X	X	X

Figure B.8 General illustration of base 16 number.

Appendix C: Official Internet Protocol Standards

C.1   Standard Protocols

Protocol	Name	Status	RFC	STD
—	Internet Official Protocol Standards	Req	2200	1
—	Assigned Numbers	Req	1700	2
—	Host Requirements—Communications	Req	1122	3
—	Host Requirements—Applications	Req	1123	3
IP	Internet Protocol as amended by —	Req	791	5
—	IP Subnet Extension	Req	950	5
—	IP Broadcast Datagrams	Req	919	5
—	IP Broadcast Datagrams with Subnets	Req	922	5
ICMP	Internet Control Message Protocol	Req	792	5
IGMP	Internet Group Multicast Protocol	Rec	1112	5
UDP	User Datagram Protocol	Rec	768	6
TCP	Transmission Control Protocol	Rec	793	7
TELNET	Telnet Protocol	Rec	854,855	8
FTP	File Transfer Protocol	Rec	959	9
SMTP	Simple Mail Transfer Protocol	Rec	821	10
SMTP-SIZE	SMTP Service Ext for Message Size	Rec	1870	10
SMTP-EXT	SMTP Service Extensions	Rec	1869	10

Protocol	Name	Status	RFC	STD
MAIL	Format of Electronic Mail Messages	Rec	822	11
CONTENT	Content Type Header Field	Rec	1049	11
NTPV2	Network Time Protocol (Version 2)	Rec	1119	12
DOMAIN	Domain Name System	Rec	1034,1035	13
DNS-MX	Mail Routing and the Domain System	Rec	974	14
SNMP	Simple Network Management Protocol	Rec	1157	15
SMI	Structure of Management Information	Rec	1155	16
Concise-MIB	Concise MIB Definitions	Rec	1212	16
MIB-II	Management Information Base-II	Rec	1213	17
NETBIOS	NetBIOS Service Protocols	Ele	1001,1002	19
ECHO	Echo Protocol	Rec	862	20
DISCARD	Discard Protocol	Ele	863	21
CHARGEN	Character Generator Protocol	Ele	864	22
QUOTE	Quote of the Day Protocol	Ele	865	23
USERS	Active Users Protocol	Ele	866	24
DAYTIME	Daytime Protocol	Ele	867	25
TIME	Time Server Protocol	Ele	868	26
TFTP	Trivial File Transfer Protocol	Ele	1350	33
TP-TCP	ISO Transport Service on top of the TCP	Ele	1006	35
ETHER-MIB	Ethernet MIB	Ele	1643	50
PPP	Point-to-Point Protocol (PPP)	Ele	1661	51
PPP-HDLC	PPP in HDLC Framing	Ele	1662	51
IP-SMDS	IP Datagrams over the SMDS Service	Ele	1209	52
POP3	Post Office Protocol, Version 3	Ele	1939	53

C.2 Network-Specific Standard Protocols

Protocol	Name	Status	RFC	STD
IP-ATM	Classical IP and ARP over ATM	Prop	1577	
IP-FR	Multiprotocol over Frame Relay	Draft	1490	
ATM-ENCAP	Multiprotocol Encapsulation over ATM	Prop	1483	
IP-TR-MC	IP Multicast over Token-Ring LANs	Prop	1469	

Protocol	Name	Status	RFC	STD
IP-FDDI	Transmission of IP and ARP over FDDI Net	Std	1390	36
IP-X.25	X.25 and ISDN in the Packet Mode	Draft	1356	
ARP	Address Resolution Protocol	Std	826	37
RARP	A Reverse Address Resolution Protocol	Std	903	38
IP-ARPA	Internet Protocol on ARPANET	Std	BBN1822	39
IP-WB	Internet Protocol on Wideband Network	Std	907	40
IP-E	Internet Protocol on Ethernet Networks	Std	894	41
IP-EE	Internet Protocol on Exp. Ethernet Nets	Std	895	42
IP-IEEE	Internet Protocol on IEEE 802	Std	1042	43
IP-DC	Internet Protocol on DC Networks	Std	891	44
IP-HC	Internet Protocol on Hyperchannel	Std	1044	45
IP-ARC	Transmitting IP Traffic over ARCNET Nets	Std	1201	46
IP-SLIP	Transmission of IP over Serial Lines	Std	1055	47
IP-NETBIOS	Transmission of IP over NETBIOS	Std	1088	48
IP-IPX	Transmission of 802.2 over IPX Networks	Std	1132	49

C.3 Draft Standard Protocols

Protocol	Name	Status	RFC
BOOTP	DHCP Options and BOOTP Extensions	Recommended	2132
DHCP	Dynamic Host Configuration Protocol	Elective	2131
—	Clarifications and Extensions BOOTP	Elective	1542
DHCP-BOOTP	Interoperation Between DHCP and BOOTP	Elective	1534
MIME-CONF	MIME Conformance Criteria	Elective	2049
MIME-MSG	MIME Msg Header Ext for Non-ASCII	Elective	2047
MIME-MEDIA	MIME Media Types	Elective	2046
MIME	Multipurpose Internet Mail Extensions	Elective	2045
PPP-CHAP	PPP Challenge Handshake Authentication	Elective	1994
PPP-MP	PPP Multilink Protocol	Elective	1990
PPP-LINK	PPP Link Quality Monitoring	Elective	1989
COEX-MIB	Coexistence between SNMPV1 & SNMPV2	Elective	1908
SNMPv2-MIB	MIB for SNMPv2	Elective	1907

Protocol	Name	Status	RFC
TRANS-MIB	Transport Mappings for SNMPv2	Elective	1906
OPS-MIB	Protocol Operations for SNMPv2	Elective	1905
CONF-MIB	Conformance Statements for SNMPv2	Elective	1904
CONV-MIB	Textual Conventions for SNMPv2	Elective	1903
SMIV2	SMI for SNMPv2	Elective	1902
CON-MD5	Content-MD5 Header Field	Elective	1864
OSPF-MIB	OSPF Version 2 MIB	Elective	1850
STR-REP	String Representation...	Elective	1779
X.500syn	X.500 String Representation...	Elective	1778
X.500lite	X.500 Lightweight...	Elective	1777
BGP-4-APP	Application of BGP-4	Elective	1772
BGP-4	Border Gateway Protocol 4	Elective	1771
PPP-DNCP	PPP DECnet Phase IV Control Protocol	Elective	1762
RMON-MIB	Remote Network Monitoring MIB	Elective	1757
802.5-MIB	IEEE 802.5 Token Ring MIB	Elective	1748
BGP-4-MIB	BGP-4 MIB	Elective	1657
RIP2-MIB	RIP Version 2 MIB Extension	Elective	1724
RIP2	RIP Version 2-Carrying Additional Info.	Elective	1723
RIP2-APP	RIP Version 2 Protocol App. Statement	Elective	1722
SIP-MIB	SIP Interface Type MIB	Elective	1694
—	Def Man Objs Parallel-printer-like	Elective	1660
—	Def Man Objs RS-232-like	Elective	1659
—	Def Man Objs Character Stream	Elective	1658
SMTP-8BIT	SMTP Service Ext or 8bit-MIMEtransport	Elective	1652
OSI-NSAP	Guidelines for OSI NSAP Allocation	Elective	1629
OSPF2	Open Shortest Path First Routing V2	Elective	1583
ISO-TS-ECHO	Echo for ISO-8473	Elective	1575
DECNET-MIB	DECNET MIB	Elective	1559
BRIDGE-MIB	BRIDGE-MIB	Elective	1493
NTPV3	Network Time Protocol (Version 3)	Elective	1305
IP-MTU	Path MTU Discovery	Elective	1191
FINGER	Finger Protocol	Elective	1288
NICNAME	WhoIs Protocol	Elective	954

C.4 Proposed Standard Protocols

Protocol	Name	Status	RFC
IPv6-Jumbo	TCP and UDP over IPv6 Jumbograms	Elective	2147
MAIL-SERV	Mailbox Names for Common Services	Elective	2142
URN-SYNTAX	URN Syntax	Elective	2141
RADIUS	Remote Authentication Dial In Service	Elective	2138
SDNSDU	Secure Domain Name System Dynamic Update	Elective	2137
DNS-UPDATE	Dynamic Updates in the DNS	Elective	2136
DC-MIB	Dial Control MIB using SMIv2	Elective	2128
ISDN-MIB	ISDN MIB using SMIv2	Elective	2127
ITOT	ISO Transport Service on top of TCP	Elective	2126
BAP-BACP	PPP-BAP, PPP-BACP	Elective	2125
VEMMI-URL	VEMMI URL Specification	Elective	2122
ROUT-ALERT	IP Router Alert Option	Elective	2113
MIME-RELAT	MIME Multipart/Related Content-type	Elective	2112
CIDMID-URL	Content-ID and Message-ID URLs	Elective	2111
MHTML	MIME E-mail Encapsulation	Elective	2110
HTTP-STATE	HTTP State Management Mechanism	Elective	2109
802.3-MIB	802.3 Repeater MIB using SMIv2	Elective	2108
PPP-NBFCP	PPP NetBIOS Frames Control Protocol	Elective	2097
TABLE-MIB	IP Forwarding Table MIB	Elective	2096
IMAPPOPAU	IMAP/POP AUTHorize Extension	Elective	2095
RIP-TRIG	Trigger RIP	Elective	2091
IMAP4-LIT	IMAP4 non-synchronizing literals	Elective	2088
IMAP4-QUO	IMAP4 QUOTA extension	Elective	2087
IMAP4-ACL	IMAP4 ACL Extension	Elective	2086
HMAC-MD5	HMAC-MD5 IP Auth. with Replay Prevention	Elective	2085
RIP2-MD5	RIP-2 MD5 Authentication	Elective	2082
RIPNG-IPV6	RIPng for IPv6	Elective	2080
URI-ATT	URI Attribute Type and Object Class	Elective	2079
GSSAP	Generic Security Service Application	Elective	2078
MIME-MODEL	Model Primary MIME Types	Elective	2077
RMON-MIB	Remote Network Monitoring MIB	Elective	2074
IPV6-UNI	IPv6 Provider-Based Unicast Address	Elective	2073
HTML-INT	HTML Internationalization	Elective	2070

Protocol	Name	Status	RFC
DAA	Digest Access Authentication	Elective	2069
HTTP-1.1	Hypertext Transfer Protocol — HTTP/1.1	Elective	2068
DNS-SEC	Domain Name System Security Extensions	Elective	2065
IMAPV4	Internet Message Access Protocol v4rev1	Elective	2060
URLZ39.50	Uniform Resource Locators for Z39.50	Elective	2056
SNANAU-APP	SNANAU APPC MIB using SMIv2	Elective	2051
PPP-SNACP	PPP SNA Control Protocol	Elective	2043
RTP-MPEG	RTP Payload Format for MPEG1/MPEG2	Elective	2038
ENTITY-MIB	Entity MIB using SMIv2	Elective	2037
RTP-JPEG	RTP Payload Format for JPEG-compressed	Elective	2035
SMTP-ENH	SMTP Enhanced Error Codes	Elective	2034
RTP-H.261	RTP Payload Format for H.261	Elective	2032
RTP-CELLB	RTP Payload Format of Sun's CellB	Elective	2029
SPKM	Simple Public-Key GSS-API Mechanism	Elective	2025
DLSW-MIB	DLSw MIB using SMIv2	Elective	2024
IPV6-PPP	IP Version 6 over PPP	Elective	2023
MULTI-UNI	Multicast over UNI 3.0/3.1 based ATM	Elective	2022
RMON-MIB	RMON MIB using SMIv2	Elective	2021
802.12-MIB	IEEE 802.12 Interface MIB	Elective	2020
IPV6-FDDI	Transmission of IPv6 Packets Over FDDI	Elective	2019
TCP-ACK	TCP Selective Acknowledgement Options	Elective	2018
URL-ACC	URL Access-Type	Elective	2017
MIME-PGP	MIME Security with PGP	Elective	2015
MIB-UDP	SNMPv2 MIB for UDP	Elective	2013
MIB-TCP	SNMPv2 MIB for TCP	Elective	2012
MIB-IP	SNMPv2 MIB for IP	Elective	2011
MOBILEIPMIB	Mobile IP MIB Definition using SMIv2	Elective	2006
MOBILEIPAPP	Applicability Statement for IP Mobility	Elective	2005
MINI-IP	Minimal Encapsulation within IP	Elective	2004
IPENCAPIP	IP Encapsulation within IP	Elective	2003
MOBILEIPSUPIP	Mobility Support	Elective	2002
TCPSLOWSRT	TCP Slow Start, Congestion Avoidance...	Elective	2001
BGP-COMM	BGP Communities Attribute	Elective	1997
DNS-NOTIFY	Mech. for Notification of Zone Changes	Elective	1996
DNS-IZT	Incremental Zone Transfer in DNS	Elective	1995

Protocol	Name	Status	RFC
SMTP-ETRN	SMTP Service Extension ETRN	Elective	1985
SNA	Serial Number Arithmetic	Elective	1982
MTU-IPV6	Path MTU Discovery for IP version 6	Elective	1981
PPP-FRAME	PPP in Frame Relay	Elective	1973
IPV6-ETHER	Transmission IPv6 Packets Over Ethernet	Elective	1972
IPV6-AUTO	IPv6 Stateless Address Autoconfiguration	Elective	1971
IPV6-ND	Neighbor Discovery for IP Version 6	Elective	1970
PPP-ECP	PPP Encryption Control Protocol	Elective	1968
GSSAPI-KER	Kerberos Version 5 GSS-API Mechanism	Elective	1964
PPP-CCP	PPP Compression Control Protocol	Elective	1962
GSSAPI-SOC	GSS-API Auth for SOCKS Version 5	Elective	1961
LDAP-STR	String Rep. of LDAP Search Filters	Elective	1960
LDAP-URL	LDAP URL Format	Elective	1959
ONE-PASS	One-Time Password System	Elective	1938
TRANS-IPV6	Transition Mechanisms IPv6 Hosts/Routers	Elective	1933
AUTH-SOCKS	Username Authentication for SOCKS V5	Elective	1929
SOCKSV5	SOCKS Protocol Version 5	Elective	1928
WHOIS++M	How to Interact with a Whois++ Mesh	Elective	1914
WHOIS++A	Architecture of Whois++ Index Service	Elective	1913
DSN	Delivery Status Notifications	Elective	1894
EMS-CODE	Enhanced Mail System Status Codes	Elective	1893
MIME-RPT	Multipart/Report	Elective	1892
SMTP-DSN	SMTP Delivery Status Notifications	Elective	1891
RTP-AV	RTP Audio/Video Profile	Elective	1890
RTP	Transport Protocol for Real-Time Apps	Elective	1889
DNS-IPV6	DNS Extensions to support IPv6	Elective	1886
ICMPv6	ICMPv6 for IPv6	Elective	1885
IPV6-Addr	IPv6 Addressing Architecture	Elective	1884
IPV6	IPv6 Specification	Elective	1883
HTML	Hypertext Markup Language - 2.0	Elective	1866
SMTP-Pipe	SMTP Serv. Ext. for Command Pipelining	Elective	1854
MIME-Sec	MIME Object Security Services	Elective	1848
MIME-Encyp	MIME: Signed and Encrypted	Elective	1847
WHOIS++	Architecture of the WHOIS++ service	Elective	1835
—	Binding Protocols for ONC RPC Version 2	Elective	1833

Protocol	Name	Status	RFC
XDR	External Data Representation Standard	Elective	1832
RPC	Remote Procedure Call Protocol V. 2	Elective	1831
—	ESP DES-CBC Transform	Ele/Req	1829
—	IP Authentication using Keyed MD5	Ele/Req	1828
ESP	IP Encapsulating Security Payload	Ele/Req	1827
IPV6-AH	IP Authentication Header	Ele/Req	1826
—	Security Architecture for IP	Ele/Req	1825
RREQ	Requirements for IP Version 4 Routers	Elective	1812
URL	Relative Uniform Resource Locators	Elective	1808
CLDAP	Connection-less LDAP	Elective	1798
OSPF-DC	Ext. OSPF to Support Demand Circuits	Elective	1793
TMUX	Transport Multiplexing Protocol	Elective	1692
TFTP-Opt	TFTP Options	Elective	1784
TFTP-Blk	TFTP Blocksize Option	Elective	1783
TFTP-Ext	TFTP Option Extension	Elective	1782
OSI-Dir	OSI User Friendly Naming...	Elective	1781
MIME-EDI	MIME Encapsulation of EDI Objects	Elective	1767
Lang-Tag	Tags for Identification of Languages	Elective	1766
XNSCP	PPP XNS IDP Control Protocol	Elective	1764
BVCP	PPP Banyan Vines Control Protocol	Elective	1763
Print-MIB	Printer MIB	Elective	1759
ATM-SIG	ATM Signaling Support for IP over ATM	Elective	1755
IPNG	Recommendation for IP Next Generation	Elective	1752
802.5-SSR	802.5 SSR MIB using SMIv2	Elective	1749
SDLCSMIv2	SNADLC SDLC MIB using SMIv2	Elective	1747
BGP4/IDRP	BGP4/IDRP for IP/OSPF Interaction	Elective	1745
AT-MIB	Appletalk MIB	Elective	1742
MacMIME	MIME Encapsulation of Macintosh files	Elective	1740
URL	Uniform Resource Locators	Elective	1738
POP3-AUTH	POP3 AUTHentication command	Elective	1734
IMAP4-AUTH	IMAP4 Authentication Mechanisms	Elective	1731
RDBMS-MIB	RDMS MIB - using SMIv2	Elective	1697
MODEM-MIB	Modem MIB - using SMIv2	Elective	1696
ATM-MIB	ATM Management Version 8.0 using SMIv2	Elective	1695
SNANAU-MIB	SNA NAUs MIB using SMIv2	Elective	1666

Protocol	Name	Status	RFC
PPP-TRANS	PPP Reliable Transmission	Elective	1663
—	Postmaster Convention X.400 Operations	Elective	1648
TN3270-En	TN3270 Enhancements	Elective	1647
PPP-BCP	PPP Bridging Control Protocol	Elective	1638
UPS-MIB	UPS Management Information Base	Elective	1628
AAL5-MTU	Default IP MTU for use over ATM AAL5	Elective	1626
PPP-SONET	PPP over SONET/SDH	Elective	1619
PPP-ISDN	PPP over ISDN	Elective	1618
DNS-R-MIB	DNS Resolver MIB Extensions	Elective	1612
DNS-S-MIB	DNS Server MIB Extensions	Elective	1611
FR-MIB	Frame Relay Service MIB	Elective	1604
PPP-X25	PPP in X.25	Elective	1598
OSPF-NSSA	The OSPF NSSA Option	Elective	1587
OSPF-Multi	Multicast Extensions to OSPF	Elective	1584
SONET-MIB	MIB SONET/SDH Interface Type	Elective	1595
RIP-DC	Extensions to RIP to Support Demand Cir.	Elective	1582
—	Evolution of the Interfaces Group of MIB-II	Elective	1573
PPP-LCP	PPP LCP Extensions	Elective	1570
X500-MIB	X.500 Directory Monitoring MIB	Elective	1567
MAIL-MIB	Mail Monitoring MIB	Elective	1566
NSM-MIB	Network Services Monitoring MIB	Elective	1565
CIPX	Compressing IPX Headers Over WAM Media	Elective	1553
IPXCP	PPP Internetworking Packet Exchange Control	Elective	1552
SRB-MIB	Source Routing Bridge MIB	Elective	1525
CIDR-STRA	CIDR Address Assignment...	Elective	1519
CIDR-ARCH	CIDR Architecture...	Elective	1518
CIDR-APP	CIDR Applicability Statement	Elective	1517
—	802.3 MAU MIB	Elective	1515
HOST-MIB	Host Resources MIB	Elective	1514
—	Token Ring Extensions to RMON MIB	Elective	1513
FDDI-MIB	FDDI Management Information Base	Elective	1512
KERBEROS	Kerberos Network Authentication Ser (V5)	Elective	1510
GSSAPI	Generic Security Service API: C-bindings	Elective	1509
DASS	Distributed Authentication Security...	Elective	1507
—	X.400 Use of Extended Character Sets	Elective	1502

Protocol	Name	Status	RFC
HARPOON	Rules for Downgrading Messages...	Elective	1496
Mapping	MHS/RFC-822 Message Body Mapping	Elective	1495
Equiv	X.400/MIME Body Equivalences	Elective	1494
IDPR	Inter-Domain Policy Routing Protocol	Elective	1479
IDPR-ARCH	Architecture for IDPR	Elective	1478
PPP/Bridge	MIB Bridge PPP MIB	Elective	1474
PPP/IP MIB	IP Network Control Protocol of PPP MIB	Elective	1473
PPP/SEC MIB	Security Protocols of PPP MIB	Elective	1472
PPP/LCP MIB	Link Control Protocol of PPP MIB	Elective	1471
X25-MIB	Multiprotocol Interconnect on X.25 MIB	Elective	1461
SNMPv2	Introduction to SNMPv2	Elective	1441
PEM-KEY	PEM—Key Certification	Elective	1424
PEM-ALG	PEM—Algorithms, Modes, and Identifiers	Elective	1423
PEM-CKM	PEM—Certificate-Based Key Management	Elective	1422
PEM-ENC	PEM—Message Encryption and Auth	Elective	1421
SNMP-IPX	SNMP over IPX	Elective	1420
SNMP-AT	SNMP over AppleTalk	Elective	1419
SNMP-OSI	SNMP over OSI	Elective	1418
FTP-FTAM	FTP-FTAM Gateway Specification	Elective	1415
IDENT-MIB	Identification MIB	Elective	1414
IDENT	Identification Protocol	Elective	1413
DS3/E3-MIB	DS3/E3 Interface Type	Elective	1407
DS1/E1-MIB	DS1/E1 Interface Type	Elective	1406
BGP-OSPF	BGP OSPF Interaction	Elective	1403
—	Route Advertisement In BGP2 And BGP3	Elective	1397
SNMP-X.25	SNMP MIB Extension for X.25 Packet Layer	Elective	1382
SNMP-LAPB	SNMP MIB Extension for X.25 LAPB	Elective	1381
PPP-ATCP	PPP AppleTalk Control Protocol	Elective	1378
PPP-OSINLCP	PPP OSI Network Layer Control Protocol	Elective	1377
SNMP-PARTY-MIB	Administration of SNMP	Elective	1353
SNMP-SEC	SNMP Security Protocols	Elective	1352
SNMP-ADMIN	SNMP Administrative Model	Elective	1351
TOS	Type of Service in the Internet	Elective	1349
PPP-IPCP	PPP Control Protocol	Elective	1332
—	X.400 1988 to 1984 downgrading	Elective	1328

Protocol	Name	Status	RFC
—	Mapping between X.400 (1988)	Elective	1327
TCP-EXT	TCP Extensions for High Performance	Elective	1323
FRAME-MIB	Management Information Base for Frame	Elective	1315
NETFAX	File Format for the Exchange of Images	Elective	1314
IARP	Inverse Address Resolution Protocol	Elective	1293
FDDI-MIB	FDDI-MIB	Elective	1285
—	Encoding Network Addresses	Elective	1277
—	Replication and Distributed Operations	Elective	1276
—	COSINE and Internet X.500 Schema	Elective	1274
BGP-MIB	Border Gateway Protocol MIB (Version 3)	Elective	1269
ICMP-ROUT	ICMP Router Discovery Messages	Elective	1256
OSI-UDP	OSI TS on UDP	Elective	1240
STD-MIBs	Reassignment of Exp MIBs to Std MIBs	Elective	1239
IPX-IP	Tunneling IPX Traffic through IP Nets	Elective	1234
IS-IS	OSI IS-IS for TCP/IP Dual Environments	Elective	1195

C.5 Telnet Options

Protocol	Name	Number	State	Status	RFC	STD
TOPT-BIN	Binary Transmission	0	Std	Rec	856	27
TOPT-ECHO	Echo	1	Std	Rec	857	28
TOPT-RECN	Reconnection	2	Prop	Ele	...	
TOPT-SUPP	Suppress Go Ahead	3	Std	Rec	858	29
TOPT-APRX	Approx Message Size Negotiation	4	Prop	Ele	...	
TOPT-STAT	Status	5	Std	Rec	859	30
TOPT-TIM	Timing Mark	6	Std	Rec	860	31
TOPT-REM	Remote Controlled Trans and Echo	7	Prop	Ele	726	
TOPT-OLW	Output Line Width	8	Prop	Ele	...	
TOPT-OPS	Output Page Size	9	Prop	Ele	...	
TOPT-OCRD	Output Carriage-Return Disposition	10	Prop	Ele	652	
TOPT-OHT	Output Horizontal Tabstops	11	Prop	Ele	653	
TOPT-OHTD	Output Horizontal Tab Disposition	12	Prop	Ele	654	

Protocol	Name	Number	State	Status	RFC	STD
TOPT-OFD	Output Formfeed Disposition	13	Prop	Ele	655	
TOPT-OVT	Output Vertical Tabstops	14	Prop	Ele	656	
TOPT-OVTD	Output Vertical Tab Disposition	15	Prop	Ele	657	
TOPT-OLD	Output Linefeed Disposition	16	Prop	Ele	658	
TOPT-EXT	Extended ASCII	17	Prop	Ele	698	
TOPT-LOGO	Logout	18	Prop	Ele	727	
TOPT-BYTE	Byte Macro	19	Prop	Ele	735	
TOPT-DATA	Data Entry Terminal	20	Prop	Ele	1043	
TOPT-SUP	SUPDUP	21	Prop	Ele	736	
TOPT-SUPO	SUPDUP Output	22	Prop	Ele	749	
TOPT-SNDL	Send Location	23	Prop	Ele	779	
TOPT-TERM	Terminal Type	24	Prop	Ele	1091	
TOPT-EOR	End of Record	25	Prop	Ele	885	
TOPT-TACACS	TACACS User Identification	26	Prop	Ele	927	
TOPT-OM	Output Marking	27	Prop	Ele	933	
TOPT-TLN	Terminal Location Number	28	Pro	Ele	946	
TOPT-3270	Telnet 3270 Regime	29	Prop	Ele	1041	
TOPT-X.3	X.3 PAD	30	Prop	Ele	1053	
TOPT-NAWS	Negotiate About Window Size	31	Prop	Ele	1073	
TOPT-TS	Terminal Speed	32	Prop	Ele	1079	
TOPT-RFC	Remote Flow Control	33	Prop	Ele	1372	
TOPT-LINE	Linemode	34	Draft	Ele	1184	
TOPT-XDL	X Display Location	35	Prop	Ele	1096	
TOPT-ENVIR	Telnet Environment Option	36	Hist	Not	1408	
TOPT-AUTH	Telnet Authentication Option	37	Exp	Ele	1416	
TOPT-ENVIR	Telnet Environment Option	39	Prop	Ele	1572	
TOPT-TN3270E	TN3270 Enhancements	40	Prop	Ele	1647	
TOPT-AUTH	Telnet XAUTH	41	Exp			

C.6 Experimental Protocols

Protocol	Name	RFC
IP-SCSI	Encapsulating IP with the SCSI	2143
X.500-NAME	Managing the X.500 Root Naming Context	2120
TFTP-MULTI	TFTP Multicast Option	2090
IP-Echo	IP Echo Host Service	2075
METER-MIB	Traffic Flow Measurement Meter MIB	2064
TFM-ARCH	Traffic Flow Measurement Architecture	2063
DNS-SRV	Location of Services in the DNS	2052
URAS	Uniform Resource Agents	2016
GPS-AR	GPS-Based Addressing and Routing	2009
ETFTP	Enhanced Trivial File Transfer Protocol	1986
BGP-RR	BGP Route Reflection	1966
BGP-ASC	Autonomous System Confederations for BGP	1965
SMKD	Scalable Multicast Key Distribution	1949
HTML-TBL	HTML Tables	1942
MIME-VP	Voice Profile for Internet Mail	1911
SNMPV2SM	User-based Security Model for SNMPv2	1910
SNMPV2AI	SNMPv2 Administrative Infrastructure	1909
SNMPV2CB	Introduction to Community-based SNMPv2	1901
—	IPv6 Testing Address Allocation	1897
DNS-LOC	Location Information in the DNS	1876
SGML-MT	SGML Media Types	1874
CONT-MT	Access Type Content-ID	1873
UNARP	ARP Extension - UNARP	1868
—	Form-based File Upload in HTML	1867
—	BGP/IDRP Route Server Alternative	1863
—	IP Authentication using Keyed SHA	1852
ESP3DES	ESP Triple DES Transform	1851
—	SMTP 521 Reply Code	1846
—	SMTP Serv. Ext. for Checkpoint/Restart	1845
—	X.500 Mapping X.400 and RFC 822 Addresses	1838
—	Tables and Subtrees in the X.500 Directory	1837
—	O/R Address hierarchy in X.500	1836
—	SMTP Serv. Ext. Large and Binary MIME Msgs.	1830

Protocol	Name	RFC
ST2	Stream Protocol Version 2	1819
—	Content-Disposition Header	1806
—	Schema Publishing in X.500 Directory	1804
—	X.400-MHS use X.500 to support X.400-MHS Routing	1801
—	Class A Subnet Experiment	1797
TCP/IPXMIB	TCP/IPX Connection Mib Specification	1792
—	TCP And UDP Over IPX Networks With Fixed Path MTU	1791
ICMP-DM	ICMP Domain Name Messages	1788
CLNP-MULT	Host Group Extensions for CLNP Multicasting	1768
OSPF-OVFL	OSPF Database Overflow	1765
RWP	Remote Write ProtocolL - Version 1.0	1756
NARP	NBMA Address Resolution Protocol	1735
DNS-DEBUG	Tools for DNS debugging	1713
DNS-ENCODE	DNS Encoding of Geographical Location	1712
TCP-POS	An Extension to TCP: Partial Order Service	1693
—	DNS to Distribute RFC1327 Mail Address Mapping Tables	1664
T/TCP	TCP Extensions for Transactions	1644
MIME-UNI	Using Unicode with MIME	1641
FOOBAR	FTP Operation Over Big Address Records	1639
X500-CHART	Charting Networks in the X.500 Directory	1609
X500-DIR	Representing IP Information in the X.500 Directory	1608
SNMP-DPI	SNMP Distributed Protocol Interface	1592
CLNP-TUBA	Use of ISO CLNP in TUBA Environments	1561
REM-PRINT	TPC.INT Subdomain Remote Printing - Technical	1528
EHF-MAIL	Encoding Header Field for Internet Messages	1505
RAP	Internet Route Access Protocol	1476
TP/IX	TP/IX: The Next Internet	1475
X400	Routing Coordination for X.400 Services	1465
DNS	Storing Arbitrary Attributes in DNS	1464
IRCP	Internet Relay Chat Protocol	1459
TOS-LS	Link Security TOS	1455
SIFT/UFT	Sender-Initiated/Unsolicited File Transfer	1440
DIR-ARP	Directed ARP	1433
TEL-SPX	Telnet Authentication: SPX	1412
TEL-KER	Telnet Authentication: Kerberos V4	1411

Protocol	Name	RFC
MAP-MAIL	X.400 Mapping and Mail-11	1405
TRACE-IP	Traceroute Using an IP Option	1393
DNS-IP	Experiment in DNS Based IP Routing	1383
RMCP	Remote Mail Checking Protocol	1339
TCP-HIPER	TCP Extensions for High Performance	1323
MSP2	Message Send Protocol 2	1312
DSLCP	Dynamically Switched Link Control	1307
—	X.500 and Domains	1279
IN-ENCAP	Internet Encapsulation Protocol	1241
CLNS-MIB	CLNS-MIB	1238
CFDP	Coherent File Distribution Protocol	1235
IP-AX.25	IP Encapsulation of AX.25 Frames	1226
ALERTS	Managing Asynchronously Generated Alerts	1224
MPP	Message Posting Protocol	1204
SNMP-BULK	Bulk Table Retrieval with the SNMP	1187
DNS-RR	New DNS RR Definitions	1183
IMAP2	Interactive Mail Access Protocol	1176
NTP-OSI	NTP over OSI Remote Operations	1165
DMF-MAIL	Digest Message Format for Mail	1153
RDP	Reliable Data Protocol	908,1151
TCP-ACO	TCP Alternate Checksum Option	1146
IP-DVMRP	IP Distance Vector Multicast Routing	1075
VMTP	Versatile Message Transaction Protocol	1045
COOKIE-JAR	Authentication Scheme	1004
NETBLT	Bulk Data Transfer Protocol	998
IRTP	Internet Reliable Transaction Protocol	938
LDP	Loader Debugger Protocol	909
RLP	Resource Location Protocol	887
NVP-II	Network Voice Protocol	ISI-memo
PVP	Packet Video Protocol	ISI-memo

C.7 Informational Protocols

Protocol	Name	RFC
PPP-EXT	PPP Vendor Extensions	2153
UTF-7	UTF-7	2152
CAST-128	CAST-128 Encryption Algorithm	2144
DLSCAP	Data Link Switching Client Access Protocol	2114
PNG	Portable Network Graphics Version 1.0	2083
RC5	RC5	RC5-CBC
SNTP	Simple Network Time Protocol v4 for IPv4	IPv6 and OSI
PGP-MEF	PGP Message Exchange Formats	1991
PPP-DEFL	PPP Deflate Protocol	1979
PPP-PRED	PPP Predictor Compression Protocol	1978
PPP-BSD	PPP BSD Compression Protocol	1977
PPP-DCE	PPP for Data Compression in DCE	1976
PPP-MAG	PPP Magnalink Variable Resource Compression	1975
PPP-STAC	PPP Stac LZS Compression Protocol	1974
GZIP	GZIP File Format Specification Version 4.3	1952
DEFLATE	DEFLATE Compressed Data Format Specification V. 1.3	1951
ZLIB	ZLIB Compressed Data Format Specification V. 3.3	1950
HTTP-1.0	Hypertext Transfer Protocol — HTTP/1.0	1945
—	Text/enriched MIME Content-type	1896
—	Application/CALS-1840 Content-type	1895
—	PPP IPCP Extensions for Name Server Addresses	1877
SNPP	Simple Network Paging Protocol - Version 2	1861
—	ISO Transport Class 2 Non-use Explicit Flow Control over TCP RFC1006 extension	1859
—	IP in IP Tunneling	1853
—	PPP Network Control Protocol for LAN Extension	1841
TESS	The Exponential Security System	1824
NFSV3	NFS Version 3 Protocol Specification	1813
—	A Format for Bibliographic Records	1807
—	Data Link Switching: Switch-to-Switch Protocol	1795
BGP-4	Experience with the BGP-4 Protocol	1773
SDMD	IPv4 Option for Sender Directed MD Delivery	1770
SNOOP	Snoop Version 2 Packet Capture File Format	1761

Protocol	Name	RFC
BINHEX	MIME Content Type for BinHex Encoded Files	1741
RWHOIS	Referral Whois Protocol	1714
DNS-NSAP	DNS NSAP Resource Records	1706
RADIO-PAGE	TPC.INT Subdomain: Radio Paging—Technical Procedures	1703
GRE-IPv4	Generic Routing Encapsulation over IPv4	1702
GRE	Generic Routing Encapsulatio	1701
ADSNA-IP	Advanced SNA/IP: A Simple SNA Transport Protocol	1538
TACACS	Terminal Access Control Protocol	1492
MD4	MD4 Message Digest Algorithm	1320
SUN-NFS	Network File System Protocol	1094
SUN-RPC	Remote Procedure Call Protocol Version 2	1057
GOPHER	The Internet Gopher Protocol	1436
LISTSERV	Listserv Distribute Protocol	1429
—	Replication Requirements	1275
PCMAIL	Pcmail Transport Protocol	1056
MTP	Multicast Transport Protocol	1301
BSD Login	BSD Login	1282
DIXIE	DIXIE Protocol Specification	1249
IP-X.121	IP to X.121 Address Mapping for DDN	1236
OSI-HYPER	OSI and LLC1 on HYPERchannel	1223
HAP2	Host Access Protocol	1221
SUBNETASGN	On the Assignment of Subnet Numbers	1219
SNMP-TRAPS	Defining Traps for use with SNMP	1215
DAS	Directory Assistance Service	1202
LPDP	Line Printer Daemon Protocol	1179

Appendix D: Important Assigned Numbers

D.1 Protocol Numbers

There is a field called protocol in the IP header that identifies the transport layer protocol to receive the datagram. The following table summarizes the protocol numbers.

Decimal	Keyboard	Protocol
0		Reserved
1	ICMP	Internet Control Message
2	IGMP	Internet Group Management
3	GGP	Gateway-to-Gateway
4	IP	IP in IP (encapsulation)
5	ST	Stream
6	TCP	Transmission Control
7	UCL	UCL
8	EGP	Exterior Gateway Protocol
9	IGP	any private interior gateway
10	BBN-RCC-MON	BBN RCC Monitoring
11	NVP-II	Network Voice Protocol
12	PUP	PUP
13	ARGUS	ARGUS
14	EMCON	EMCON

Decimal	Keyboard	Protocol
15	XNET	Cross Net Debugger
16	CHAOS	Chaos
17	UDP	User Datagram
18	MUX	Multiplexing
19	DCN-MEAS	DCN Measurement Subsystems
20	HMP	Host Monitoring
21	PRM	Packet Radio Measurement
22	XNS-IDP	XEROX NS IDP
23	TRUNK-1	Trunk-1
24	TRUNK-2	Trunk-2
25	LEAF-1	Leaf-1
26	LEAF-2	Leaf-2
27	RDP	Reliable Data Protocol
28	IRTP	Internet Reliable Transaction
29	ISO-TP4	ISO Transport Protocol Class 4
30	NETBLT	Bulk Data Transfer Protocol
31	MFE-NSP	MFE Network Services Protocol
32	MERIT-INP	MERIT Internodal Protocol
33	SEP	Sequential Exchange Protocol
34	3PC	Third Party Connect Protocol
35	IDPR	Inter-Domain Policy Routing Protocol
36	XTP	XTP
37	DDP	Datagram Delivery Protocol
38	IDPR-CMTP	IDPR Control Message Transport Protocol
39	TP++	TP++ Transport Protocol
40	IL	IL Transport Protocol
41	SIP	Simple Internet Protocol
42	SDRP	Source Demand Routing Protocol
43	SIP-SR	SIP Source Route
44	SIP-FRAG	SIP Fragment
45	IDRP	Inter-Domain Routing Protocol
46	RSVP	Reservation Protocol
47	GRE	General Routing Encapsulation
48	MHRP	Mobile Host Routing Protocol
49	BNA	BNA

Decimal	Keyboard	Protocol
50	SIPP-ESP	SIPP Encap Security Payload
51	SIPP-AH	SIPP Authentication Header
52	I-NLSP	Integrated Net Layer Security TUBA
53	SWIPE	IP with Encryption
54	NHRP	NBMA Next Hop Resolution Protocol
55-60		Unassigned
61		any host internal protocol
62	CFTP	CFTP
63		any local network
64	SAT-EXPAK	SATNET and Backroom EXPAK
65	KRYPTOLAN	Kryptolan
66	RVD	MIT Remote Virtual Disk Protocol
67	IPPC	Internet Pluribus Packet Core
68		any distributed file system
69	SAT-MON	SATNET Monitoring
70	VISA	VISA Protocol
71	IPCV	Internet Packet Core Utility
72	CPNX	Computer Protocol Network Executive
73	CPHB	Computer Protocol Heart Beat
74	WSN	Wang Span Network
75	PVP	Packet Video Protocol
76	BR-SAT-MON	Backroom SATNET Monitoring
77	SUN-ND	SUN ND PROTOCOL-Temporary
78	WB-MON	WIDEBAND Monitoring
79	WB-EXPAK	WIDEBAND EXPAK
80	ISO-IP	ISO Internet Protocol
81	VMTP	VMTP
82	SECURE-VMTP	SECURE-VMTP
83	VINES	VINES
84	TTP	TTP
85	NSFNET-IGP	NSFNET-IGP
86	DGP	Dissimilar Gateway Protocol
87	TCF	TCF
88	IGRP	IGRP
89	OSPFIGP	OSPFIGP

Decimal	Keyboard	Protocol
90	Sprite-RPC	Sprite RPC Protocol
91	LARP	Locus Address Resolution Protocol
92	MTP	Multicast Transport Protocol
93	AX.25	AX.25 Frames
94	IPIP	IP-within-IP Encapsulation Protocol
95	MICP	Mobile Internetworking Control Pro.
96	SCC-SP	Semaphore Communications Sec. Pro.
97	ETHERIP	Ethernet-within-IP Encapsulation
98	ENCAP	Encapsulation Header
99		any private encryption scheme
100	GMTP	GMTP
101-254		Unassigned
255		Reserved

D.2 Well-Known Port Numbers

Ports are used in the TCP to name the ends of logical connections that carry long-term conversations. For the purpose of providing services to unknown callers, a service contact port is defined. This list specifies the port used by the server process as its contact port. The contact port is sometimes called the *well-known port* (WKP). The numbers below 1023 are managed by the IANA.

Keyword	Number	Description
	0/tcp	Reserved
tcpmux	1/tcp	TCP Port Service Multiplexer
compressnet	2/tcp	Management Utility
compressnet	3/tcp	Compression Process
#	4/tcp	Unassigned
rje	5/tcp	Remote Job Entry
#	6/tcp	Unassigned
echo	7/tcp	Echo
#	8/tcp	Unassigned

Keyword	Number	Description
discard	9/tcp	Discard
#	10/tcp	Unassigned
systat	11/tcp	Active Users
#	12/tcp	Unassigned
daytime	13/tcp	Daytime
#	14/tcp	Unassigned
#	15/tcp	Unassigned [was netstat]
#	16/tcp	Unassigned
qotd	17/tcp	Quote of the Day
msp	18/tcp	Message Send Protocol
chargen	19/tcp	Character Generator
ftp-data	20/tcp	File Transfer [Default Data]
ftp	21/tcp	File Transfer [Control]
#	22/tcp	Unassigned
telnet	23/tcp	Telnet
	24/tcp	any private mail system
smtp	25/tcp	Simple Mail Transfer
#	26/tcp	Unassigned
nsw-fe	27/tcp	NSW User System FE
#	28/tcp	Unassigned
msg-icp	29/tcp	MSG ICP
#	30/tcp	Unassigned
msg-auth	31/tcp	MSG Authentication
#	32/tcp	Unassigned
dsp	33/tcp	Display Support Protocol
#	34/tcp	Unassigned
	35/tcp	any private printer server
#	36/tcp	Unassigned
time	37/tcp	Time
rap	38/tcp	Route Access Protocol
rlp	39/tcp	Resource Location Protocol
#	40/tcp	Unassigned
graphics	41/tcp	Graphics
nameserver	42/tcp	Host Name Server
nicname	43/tcp	Who Is

Keyword	Number	Description
mpm-flags	44/tcp	MPM FLAGS Protocol
mpm	45/tcp	Message Processing Module [recv]
mpm-snd	46/tcp	MPM [default send] ni-ftp
	47/tcp	NI FTP
auditd	48/tcp	Digital Audit Daemon
login	49/tcp	Login Host Protocol
re-mail-ck	50/tcp	Remote Mail Checking Protocol
la-maint	51/tcp	IMP Logical Address Maintenance
xns-time	52/tcp	XNS Time Protocol
domain	53/tcp	Domain Name Server
xns-ch	54/tcp	XNS Clearinghouse
isi-gl	55/tcp	ISI Graphics Language
xns-auth	56/tcp	XNS Authentication
	57/tcp	any private terminal access
xns-mail	58/tcp	XNS Mail
	59/tcp	any private file service
	60/tcp	Unassigned
ni-mail	61/tcp	NI MAIL
acas	62/tcp	ACA Services
#	63/tcp	Unassigned
covia	64/tcp	Communications Integrator (CI)
tacacs-ds	65/tcp	TACACS-Database Service
sql*net	66/tcp	Oracle SQL*NET
bootps	67/tcp	Bootstrap Protocol Server
bootpc	68/tcp	Bootstrap Protocol Client
tftp	69/tcp	Trivial File Transfer
gopher	70/tcp	Gopher
netrjs-1	71/tcp	Remote Job Service
netrjs-2	72/tcp	Remote Job Service
netrjs-3	73/tcp	Remote Job Service
netrjs-4	74/tcp	Remote Job Service
	75/tcp	any private dial out service
deos	76/tcp	Distributed External Object Store
	77/tcp	any private RJE service
vettcp	78/tcp	vettcp

Keyword	Number	Description
finger	79/tcp	Finger
www-http	80/tcp	World Wide Web HTTP
hosts2-ns	81/tcp	HOSTS2 Name Server
xfer	82/tcp	XFER Utility
mit-ml-dev	83/tcp	MIT ML Device
ctf	84/tcp	Common Trace Facility
mit-ml-dev	85/tcp	MIT ML Device
mfcobol	86/tcp	Micro Focus Cobol
	87/tcp	any private terminal link
kerberos	88/tcp	Kerberos
su-mit-tg	89/tcp	SU/MIT Telnet Gateway
dnsix	90/tcp	DNSIX Securit Attribute Token Map
mit-dov	91/tcp	MIT Dover Spooler
npp	92/tcp	Network Printing Protocol
dcp	93/tcp	Device Control Protocol
objcall	94/tcp	Tivoli Object Dispatcher
supdup	95/tcp	SUPDUP
dixie	96/tcp	DIXIE Protocol Specification
swift-rvf	97/tcp	Swift Remote Vitural File swift-rvf
tacnews	98/tcp	TAC News
metagram	99/tcp	Metagram Relay
newacct	100/tcp	[unauthorized use]
hostname	101/tcp	NIC Host Name Server
iso-tsap	102/tcp	ISO-TSAP
gppitnp	103/tcp	Genesis Point-to-Point Trans Net
acr-nema	104/tcp	ACR-NEMA Digital Imag. & Comm.
csnet-ns	105/tcp	Mailbox Name Nameserver
3com-tsmux	106/tcp	3COM-TSMUX
rtelnet	107/tcp	Remote Telnet Service
snagas	108/tcp	SNA Gateway Access Server
pop2	109/tcp	Post Office Protocol - Version 2
pop3	110/tcp	Post Office Protocol - Version 3
sunrpc	111/tcp	SUN Remote Procedure Call
mcidas	112/tcp	McIDAS Data Transmission Protocol
auth	113/tcp	Authentication Service

Keyword	Number	Description
audionews	114/tcp	Audio News Multicast
sftp	115/tcp	Simple File Transfer Protocol
ansanotify	116/tcp	ANSA REX Notify
uucp-path	117/tcp	UUCP Path Service
sqlserv	118/tcp	SQL Services
nntp	119/tcp	Network News Transfer Protocol
cfdptkt	120/tcp	CFDPTKT
erpc	121/tcp	Encore Expedited Remote Pro.Call
smakynet	122/tcp	SMAKYNET
ntp	123/tcp	Network Time Protocol
ansatrader	124/tcp	ANSA REX Trader
locus-map	125/tcp	Locus PC-Interface Net Map Ser
unitary	126/tcp	Unisys Unitary Login
locus-con	127/tcp	Locus PC-Interface Conn Server
gss-xlicen	128/tcp	GSS X License Verification
pwdgen	129/tcp	Password Generator Protocol
cisco-fna	130/tcp	cisco FNATIVE
cisco-tna	131/tcp	cisco TNATIVE
cisco-sys	132/tcp	cisco SYSMAINT
statsrv	133/tcp	Statistics Service
ingres-net	134/tcp	INGRES-NET Service
loc-srv	135/tcp	Location Service
profile	136/tcp	PROFILE Naming System
netbios-ns	137/tcp	NETBIOS Name Service
netbios-dgm	138/tcp	NETBIOS Datagram Service
netbios-ssn	139/tcp	NETBIOS Session Service
emfis-data	140/tcp	EMFIS Data Service
emfis-cntl	141/tcp	EMFIS Control Service
bl-idm	142/tcp	Britton-Lee IDM
imap2	143/tcp	Interim Mail Access Protocol v2
news	144/tcp	NewS
uaac	145/tcp	UAAC Protocol
iso-tp0	146/tcp	ISO-IP0
iso-ip	147/tcp	ISO-IP
cronus	148/tcp	CRONUS-SUPPORT

Keyword	Number	Description
aed-512	149/tcp	AED 512 Emulation Service
sql-net	150/tcp	SQL-NET
hems	151/tcp	HEMS
bftp	152/tcp	Background File Transfer Program
sgmp	153/tcp	SGMP
netsc-prod	154/tcp	NETSC
netsc-dev	155/tcp	NETSC
sqlsrv	156/tcp	SQL Service
knet-cmp	157/tcp	KNET/VM Command/Message Protocol
pcmail-srv	158/tcp	PCMail Server
nss-routing	159/tcp	NSS-Routing
sgmp-traps	160/tcp	SGMP-TRAPS
snmp	161/tcp	SNMP
snmptrap	162/tcp	SNMPTRAP
cmip-man	163/tcp	CMIP/TCP Manager
cmip-agent	164/tcp	CMIP/TCP Agent
xns-courier	165/tcp	Xerox
s-net	166/tcp	Sirius Systems
namp	167/tcp	NAMP
rsvd	168/tcp	RSVD
send	169/tcp	SEND
print-srv	170/tcp	Network PostScript
multiplex	171/tcp	Network Innovations Multiplex
cl/1	172/tcp	Network Innovations CL/1
xyplex-mux	173/tcp	Xyplex
mailq	174/tcp	MAILQ
vmnet	175/tcp	VMNET
genrad-mux	176/tcp	GENRAD-MUX
xdmcp	177/tcp	X Display Manager Control Protocol
nextstep	178/tcp	NextStep Window Server
bgp	179/tcp	Border Gateway Protocol
ris	180/tcp	Intergraph
unify	181/tcp	Unify
audit	182/tcp	Unisys Audit SITP
ocbinder	183/tcp	OCBinder

Keyword	Number	Description
ocserver	184/tcp	OCServer
remote-kis	185/tcp	Remote-KIS
kis	186/tcp	KIS Protocol
aci	187/tcp	Appl Comm Interface
mumps	188/tcp	Plus Five's MUMPS
qft	189/tcp	Queued File Transport
gacp	190/tcp	Gateway Access Control Protocol
prospero	191/tcp	Prospero Directory Service
osu-nms	192/tcp	OSU Network Monitoring System
srmp	193/tcp	Spider Remote Monitoring Protocol
irc	194/tcp	Internet Relay Chat Protocol
dn6-nlm-aud	195/tcp	DNSIX Network Level Module Audit
dn6-smm-red	196/tcp	DNSIX Session Mgt Module Audit
dls	197/tcp	Directory Location Service
dls-mon	198/tcp	Directory Location Service
smux	199/tcp	SMUX
src	200/tcp	IBM System Resource Controller
at-rtmp	201/tcp	AppleTalk Routing Maintenance
at-nbp	202/tcp	AppleTalk Name Binding
at-3	203/tcp	AppleTalk Unused
at-echo	204/tcp	AppleTalk Echo
at-5	205/tcp	AppleTalk Unused
at-zis	206/tcp	AppleTalk Zone Information
at-7	207/tcp	AppleTalk Unused
at-8	208/tcp	AppleTalk Unused
tam	209/tcp	Trivial Authenticated Mail
z39.50	210/tcp	ANSI Z39.50
914c/g	211/tcp	Texas Instruments 914C/G Terminal
anet	212/tcp	ATEXSSTR
ipx	213/tcp	IPX
vmpwscs	214/tcp	VM PWSCS
softpc	215/tcp	Insignia Solutions
atls	216/tcp	Access Technology License Server
dbase	217/tcp	dBASE Unix
mpp	218/tcp	Netix Message Posting Protocol

Keyword	Number	Description
uarps	219/tcp	Unisys ARPs
imap3	220/tcp	Interactive Mail Access Protocol
fln-spx	221/tcp	Berkeley rlogin with SPX auth
rsh-spx	222/tcp	Berkeley rshd with SPX auth
cdc	223/tcp	Certificate Distribution Center
#	224-241	Reserved
#	242/tcp	Unassigned
sur-meas	243/tcp	Survey Measurement
#	244/tcp	Unassigned
link	245/tcp	LINK
dsp3270	246/tcp	Display Systems Protocol
#	247-255	Reserved

D.3 Internet Multicast Addresses

Host Extensions for IP multicasting specifies the extensions required of a host implementation of the IP to support multicasting. Current addresses are listed below.

Address	Purpose
224.0.0.0	Base Address (Reserved)
224.0.0.1	All Systems on this Subnet
224.0.0.2	All Routers on this Subnet
224.0.0.3	Unassigned
224.0.0.4	DVMRP Routers
224.0.0.5	OSPFIGP OSPFIGP All Routers
224.0.0.6	OSPFIGP OSPFIGP Designated Routers
224.0.0.7	ST Routers
224.0.0.8	ST Hosts
224.0.0.9	RIP2 Routers
224.0.0.10	IGRP Routers
224.0.0.11	Mobile-Agents

Address	Purpose
224.0.0.12-224.0.0.255	Unassigned
224.0.1.0	VMTP Managers Group
224.0.1.1	NTP Network Time Protocol
224.0.1.2	SGI-Dogfight
224.0.1.3	Rwhod
224.0.1.4	VNP
224.0.1.5	Artificial Horizons - Aviator
224.0.1.6	NSS - Name Service Server
224.0.1.7	AUDIONEWS - Audio News Multicast
224.0.1.8	SUN NIS+ Information Service
224.0.1.9	MTP Multicast Transport Protocol
224.0.1.10	IETF-1-LOW-AUDIO
224.0.1.11	IETF-1-AUDIO
224.0.1.12	IETF-1-VIDEO
224.0.1.13	IETF-2-LOW-AUDIO
224.0.1.14	IETF-2-AUDIO
224.0.1.15	IETF-2-VIDEO
224.0.1.16	MUSIC-SERVICE
224.0.1.17	SEANET-TELEMETRY
224.0.1.18	SEANET-IMAGE
224.0.1.19	MLOADD
224.0.1.20	any private experiment
224.0.1.21	DVMRP on MOSPF
224.0.1.22	SVRLOC
224.0.1.23	XINGTV
224.0.1.24	microsoft-ds
224.0.1.25	nbc-pro
224.0.1.26	nbc-pfn
224.0.1.27-224.0.1.255	Unassigned
224.0.2.1	"rwho" Group (BSD) (unofficial)
224.0.2.2	SUN RPC PMAPPROC_CALLIT
224.0.3.000-224.0.3.255	RFE Generic Service
224.0.4.000-224.0.4.255	RFE Individual Conferences
224.0.5.000-224.0.5.127	CDPD Groups
224.0.5.128-224.0.5.255	Unassigned

Address	Purpose
224.0.6.000-224.0.6.127	Cornell ISIS Project
224.0.6.128-224.0.6.255	Unassigned
224.1.0.0-224.1.255.255	ST Multicast Groups
224.2.0.0-224.2.255.255	Multimedia Conference Calls
224.252.0.0-224.255.255.255	DIS transient groups
232.0.0.0-232.255.255.255	VMTP transient groups

D.4 ICMP Type Numbers

ICMP messages are identified by the type field. The following is a summary of the type field values.

Value	Function
0	Echo Reply
1	Unassigned
2	Unassigned
3	Destination Unreachable
4	Source Quench
5	Redirect
6	Alternate Host Address
7	Unassigned
8	Echo
9	Router Advertisement
10	Router Selection
11	Time Exceeded
12	Parameter Problem
13	Timestamp
14	Timestamp Reply
15	Information Request
16	Information Reply
17	Address Mask Request
18	Address Mask Reply
19	Reserved (for Security)

Value	Function
20-29	Reserved (for Robustness Experiment)
30	Traceroute
31	Datagram Conversion Error
32	Mobile Host Redirect
33	IPv6 Where-Are-You
34	IPv6 I-Am-Here
35	Mobile Registration Request
36	Mobile Registration Reply
37-255	Reserved

D.5 BOOTP and DHCP Parameters

Tag	Name	Data Length	Meaning
0	Pad	0	None
1	Subnet Mask	4	Subnet Mask Value
2	Time Offset	4	Seconds offset from UTC
3	Gateways	N	N/4 Gateway addresses
4	Time Server	N	N/4 Timeserver addresses
5	Name Server	N	N/4 Server addresses
6	Domain Server	N	N/4 DNS Server addresses
7	Log Server	N	N/4 Log Server addresses
8	Quotes Server	N	N/4 Quotes Server addresses
9	LPR Server	N	N/4 Printer Server addresses
10	Impress Server	N	N/4 Impress Server addresses
11	RLP Server	N	N/4 RLP Server addresses
12	Hostname	N	Hostname string
13	Boot File Size	2	Boot size in 512 byte chunks
14	Merit Dump File		Client dump file name
15	Domain Name	N	DNS domain name of the client
16	Swap Server	N	Swap Server addeess
17	Root Path	N	Path name for root disk
18	Extension File	N	Path name for more BOOTP info

Tag	Name	Data Length	Meaning
19	Forward On/Off	1	Enable/Disable IP Forwarding
20	SrcRte On/Off	1	Enable/Disable Source Routing
21	Policy Filter	N	Routing Policy Filters
22	Max DG Assembly	2	Max Datagram Reassembly Size
23	Default IP TTL	1	Default IP Time to Live
24	MTU Timeout	4	Path MTU Aging Timeout
25	MTU Plateau	N	Path MTU Plateau Table
26	MTU Interface	2	Interface MTU Size
27	MTU Subnet	1	All Subnets are Local
28	Broadcast Address	4	Broadcast Address
29	Mask Discovery	1	Perform Mask Discovery
30	Mask Supplier	1	Provide Mask to Others
31	Router Discovery	1	Perform Router Discovery
32	Router Request	4	Router Solicitation Address
33	Static Route	N	Static Routing Table
34	Trailers	1	Trailer Encapsulation
35	ARP Timeout	4	ARP Cache Timeout
36	Ethernet	1	Ethernet Encapsulation
37	Default TCP TTL	1	Default TCP Time to Live
38	Keepalive Time	4	TCP Keepalive Interval
39	Keepalive Data	1	TCP Keepalive Garbage
40	NIS Domain	N	NIS Domain Name
41	NIS Servers	N	NIS Server Addresses
42	NTP Servers	N	NTP Server Addresses
43	Vendor Specific	N	Vendor Specific Information
44	NETBIOS Name Srv	N	NETBIOS Name Servers
45	NETBIOS Dist Srv	N	NETBIOS Datagram Distribution
46	NETBIOS Note Type	1	NETBIOS Note Type
47	NETBIOS Scope	N	NETBIOS Scope
48	X Window Font	N	X Window Font Server
49	X Window Manmager	N	X Window Display Manager
50	Address Request	4	Requested IP Address
51	Address Time	4	IP Address Lease Time
52	Overload	1	Overloaf "sname" or "file"
53	DHCP Msg Type	1	DHCP Message Type

Tag	Name	Data Length	Meaning
54	DHCP Server Id	4	DHCP Server Identification
55	Parameter List	N	Parameter Request List
56	DHCP Message	N	DHCP Error Message
57	DHCP Max Msg Size	2	DHCP Maximum Message Size
58	Renewal Time	4	DHCP Renewal (T1) Time
59	Rebinding Time	4	DHCP Rebinding (T2) Time
60	Class Id	N	Class Identifier
61	Client Id	N	Client Identifier
62	Netware/IP Domain	N	Netware/IP Domain Name
63	Netware/IP Option	N	Netware/IP sub Options
64-127	Unassigned		
128-154	Reserved		
255	End	0	None

D.6 Address Resolution Protocols (ARP, RARP, DARP, and IARP)

D.6.1 Operation Codes

Number	Hardware Type (hrd)
1	REQUEST
2	REPLY
3	request Reverse
4	reply Reverse
5	DRARP-Request
6	DRARP-Reply
7	DRARP-Error
8	InARP-Request
9	InARP-Reply
10	ARP-NAK

D.6.2 Hardware Types

Number	Hardware Type (hrd)
1	Ethernet (10Mb)
2	Experimental Ethernet (3Mb)
3	Amateur Radio AX.25
4	Proteon ProNET Token Ring
5	Chaos
6	IEEE 802 Networks
7	ARCNET
8	Hyperchannel
9	Lanstar
10	Autonet Short Address
11	LocalTalk
12	LocalNet (IBM PCNet or SYTEK LocalNET)
13	Ultra link
14	SMDS
15	Frame Relay
16	Asynchronous Transmission Mode (ATM)
17	HDLC
18	Fiber Channel
19	Asynchronous Transmission Mode (ATM)
20	Serial Line
21	Asynchronous Transmission Mode (ATM)

D.7 Address Resolution Protocol Parameters

D.7.1 Operation Codes

No	Operation Code (op)
1	REQUEST
2	REPLY
3	request Reverse
4	reply Reverse

No	Operation Code (op)
5	DRARP-Request
6	DRARP-Reply
7	DRARP-Error
8	InARP-Request
9	InARP-Reply
10	ARP-NAK

D.7.2 Hardware Types

No	Hardware Type (hrd)
1	Ethernet (10Mb)
2	Experimental Ethernet (3Mb)
3	Amateur Radio AX.25
4	Proteon ProNET Token Ring
5	Chaos
6	IEEE 802 Networks
7	ARCNET
8	Hyperchannel
9	Lanstar
10	Autonet Short Address
11	LocalTalk
12	LocalNet (IBM PCNet or SYTEK LocalNET)
13	Ultra link
14	SMDS
15	Frame Relay
16	Asynchronous Transmission Mode (ATM)
17	HDLC
18	Fibre Channel
19	Asynchronous Transmission Mode (ATM)
20	Serial Line
21	Asynchronous Transmission Mode (ATM)

D.8 IEEE 802 Numbers of Interest

Link Service	Access Point		Description
IEEE	Internet	decimal	
00000000	00000000	0	Null LSAP
01000000	00000010	2	Indiv LLC Sublayer Mgt
11000000	00000011	3	Group LLC Sublayer Mgt
00100000	00000100	4	SNA Path Control
01100000	00000110	6	Reserved (DOD IP)
01110000	00001110	14	PROWAY-LAN
01110010	01001110	78	EIA-RS 511
01111010	01011110	94	ISI IP
01110001	10001110	142	PROWAY-LAN
01010101	10101010	170	SNAP
01111111	11111110	254	ISO CLNS IS 8473
11111111	11111111	255	Global DSAP

D.9 Ethernet Type Codes

Ethernet		Exp. Ethernet		
Dec	Hex	decimal	octal	Description
000	0000-05DC	-	-	IEEE802.3 Length Field
257	0101-01FF	-	-	Experimental
512	0200	512	1000	XEROX PUP (see 0A00)
513	0201	-	-	PUP Addr Trans
	0400			Nixdorf
1536	0600	1536	3000	XEROX NS IDP
	0660			DLOG
	0661			DLOG
2048	0800	513	1001	Internet IP (IPv4)
2049	0801	-	-	X.75 Internet
2050	0802	-	-	NBS Internet

	Ethernet		Exp. Ethernet		
Dec	**Hex**	**decimal**	**octal**	**Description**	
2051	0803	-	-	ECMA Internet	
2052	0804	-	-	Chaosnet	
2053	0805	-	-	X.25 Level 3	
2054	0806	-	-	ARP	
2055	0807	-	-	XNS Compatability	
2076	081C	-	-	Symbolics Private	
2184	0888-088A	-	-	Xyplex	
2304	0900	-	-	Ungermann-Bass net debugr	
2560	0A00	-	-	Xerox IEEE802.3 PUP	
2561	0A01	-	-	PUP Addr Trans	
2989	0BAD	-	-	Banyan Systems	
4096	1000	-	-	Berkeley Trailer nego	
4097	1001-100F	-	-	Berkeley Trailer encap/IP	
5632	1600	-	-	Valid Systems	
16962	4242	-	-	PCS Basic Block Protocol	
21000	5208	-	-	BBN Simnet	
24576	6000	-	-	DEC Unassigned (Exp.)	
24577	6001	-	-	DEC MOP Dump/Load	
24578	6002	-	-	DEC MOP Remote Console	
24579	6003	-	-	DEC DECNET Phase IV Route	
24580	6004	-	-	DEC LAT	
24581	6005	-	-	DEC Diagnostic Protocol	
24582	6006	-	-	DEC Customer Protocol	
24583	6007	-	-	DEC LAVC	
24584	6008-6009	-	-	DEC Unassigned	
24586	6010-6014	-	-	3Com Corporation	
28672	7000	-	-	Ungermann-Bass download	
28674	7002	-	-	Ungermann-Bass dia/loop	
28704	7020-7029	-	-	LRT	
28720	7030	-	-	Proteon	
28724	7034	-	-	Cabletron	
32771	8003	-	-	Cronus VLN	
32772	8004	-	-	Cronus Direct	
32773	8005	-	-	HP Probe	

Ethernet		Exp. Ethernet		
Dec	**Hex**	**decimal**	**octal**	**Description**
32774	8006	-	-	Nestar
32776	8008	-	-	AT&T
32784	8010	-	-	Excelan
32787	8013	-	-	SGI diagnostics
32788	8014	-	-	SGI network games
32789	8015	-	-	SGI reserved
32790	8016	-	-	SGI bounce server
32793	-8019	-	-	Apollo Computers
32815	802E	-	-	Tymshare
32816	802F	-	-	Tigan
32821	8035	-	-	Reverse ARP
32822	8036	-	-	Aeonic Systems
32824	8038	-	-	DEC LANBridge
32825	8039-803C	-	-	DEC Unassigned
32829	803D	-	-	DEC Ethernet Encryption
32830	803E	-	-	DEC Unassigned
32831	803F	-	-	DEC LAN Traffic Monitor
32832	8040-8042	-	-	DEC Unassigned
32836	8044	-	-	Planning Research Corp.
32838	8046	-	-	AT&T
32839	8047	-	-	AT&T
32841	8049	-	-	ExperData
32859	805B	-	-	Stanford V Kernel exp.
32860	805C	-	-	Stanford V Kernel prod.
32861	805D	-	-	Evans & Sutherland
32864	8060	-	-	Little Machines
32866	8062	-	-	Counterpoint Computers
32869	8065	-	-	Univ. of Mass. @ Amherst
32870	8066	-	-	Univ. of Mass. @ Amherst
32871	8067	-	-	Veeco Integrated Auto.
32872	8068	-	-	General Dynamics
32873	8069	-	-	AT&T
32874	806A	-	-	Autophon
32876	806C	-	-	ComDesign

Ethernet		Exp. Ethernet		
Dec	Hex	decimal	octal	Description
32877	806D	-	-	Computgraphic Corp.
32878	806E-8077	-	-	Landmark Graphics Corp.
32890	807A	-	-	Matra
32891	807B	-	-	Dansk Data Elektronik
32892	807C	-	-	Merit Internodal
32893	807D-807F	-	-	Vitalink Communications
32896	8080	-	-	Vitalink TransLAN III
32897	8081-8083	-	-	Counterpoint Computers
32923	809B	-	-	Appletalk
32924	809C-809E	-	-	Datability
32927	809F	-	-	Spider Systems Ltd.
32931	80A3	-	-	Nixdorf Computers
32932	80A4-80B3	-	-	Siemens Gammasonics Inc.
32960	80C0-80C3	-	-	DCA Data Exchange Cluster
	80C4			Banyan Systems
	80C5			Banyan Systems
32966	80C6	-	-	Pacer Software
32967	80C7	-	-	Applitek Corporation
32968	80C8-80CC	-	-	Intergraph Corporation
32973	80CD-80CE	-	-	Harris Corporation
32975	80CF-80D2	-	-	Taylor Instrument
32979	80D3-80D4	-	-	Rosemount Corporation
32981	80D5	-	-	IBM SNA Service on Ether
32989	80DD	-	-	Varian Associates
32990	80DE-80DF	-	-	Integrated Solutions TRFS
32992	80E0-80E3	-	-	Allen-Bradley
32996	80E4-80F0	-	-	Datability
33010	80F2	-	-	Retix
33011	80F3	-	-	AppleTalk AARP (Kinetics)
33012	80F4-80F5	-	-	Kinetics
33015	80F7	-	-	Apollo Computer
33023	80FF-8103	-	-	Wellfleet Communications
33031	8107-8109	-	-	Symbolics Private
33072	8130	-	-	Hayes Microcomputers

Ethernet		Exp. Ethernet		
Dec	**Hex**	**decimal**	**octal**	**Description**
33073	8131	-	-	VG Laboratory Systems
	8132-8136			Bridge Communications
33079	8137-8138	-	-	Novell
33081	8139-813D	-	-	KTI
	8148			Logicraft
	8149			Network Computing Devices
	814A			Alpha Micro
33100	814C	-	-	SNMP
	814D			BIIN
	814E			BIIN
	814F			Technically Elite Concept
	8150			Rational Corp
	8151-8153			Qualcomm
	815C-815E			Computer Protocol Pty Ltd[
	8164-8166			Charles River Data System
	817D-818C			Protocol Engines
	818D			Motorola Computer
	819A-81A3			Qualcomm
	81A4			ARAI Bunkichi
	81A5-81AE			RAD Network Devices
	81B7-81B9			Xyplex
	81CC-81D5			Apricot Computers
	81D6-81DD			Artisoft
	81E6-81EF			Polygon
	81F0-81F2			Comsat Labs
	81F3-81F5			SAIC
	81F6-81F8			VG Analytical
	8203-8205			Quantum Software
	8221-8222			Ascom Banking Systems
	823E-8240			Advanced Encryption Syste
	827F-8282			Athena Programming
	8263-826A			Charles River Data System
	829A-829B			Inst Ind Info Tech
	829C-82AB			Taurus Controls

	Ethernet		Exp. Ethernet		
Dec	Hex	decimal	octal		Description
	82AC-8693				Walker Richer & Quinn
	8694-869D				Idea Courier
	869E-86A1				Computer Network Tech
	86A3-86AC				Gateway Communications
	86DB				SECTRA
	86DE				Delta Controls
34543	86DF	-	-		ATOMIC
	86E0-86EF				Landis & Gyr Powers
	8700-8710				Motorola
	8A96-8A97				Invisible Software
36864	9000	-	-		Loopback
36865	9001	-	-		3Com(Bridge) XNS Sys Mgmt
36866	9002	-	-		3Com(Bridge) TCP-IP Sys
36867	9003	-	-		3Com(Bridge) loop detect
65280	FF00	-	-		BBN VITAL-LanBridge cache
	FF00-FF0F				ISC Bunker Ramo

D.10 PPP Assigned Numbers

D.10.1 PPP DLL Protocol Numbers

Value (in hex)	Protocol Name
0001	Padding Protocol
0003 to 001f	reserved (transparency inefficient)
0021	Internet Protocol
0023	OSI Network Layer
0025	Xerox NS IDP
0027	DECnet Phase IV
0029	Appletalk
002b	Novell IPX
002d	Van Jacobson Compressed TCP/IP

Value (in hex)	Protocol Name
002f	Van Jacobson Uncompressed TCP/IP
0031	Bridging PDU
0033	Stream Protocol (ST-II)
0035	Banyan Vines
0037	reserved (until 1993)
0039	AppleTalk EDDP
003b	AppleTalk SmartBuffered
003d	Multi-Link
003f	NETBIOS Framing
0041	Cisco Systems
0043	Ascom Timeplex
0045	Fujitsu Link Backup and Load Balancing (LBLB)
0047	DCA Remote Lan
0049	Serial Data Transport Protocol (PPP-SDTP)
004b	SNA over 802.2
004d	SNA
004f	IP6 Header Compression
006f	Stampede Bridging
007d	reserved (Control Escape) [RFC1661]
007f	reserved (compression inefficient) [RFC1662]
00cf	reserved (PPP NLPID)
00fb	compression on single link in multilink group
00fd	1st choice compression
00ff	reserved (compression inefficient)
0201	802.1d Hello Packets
0203	IBM Source Routing BPDU
0205	DEC LANBridge100 Spanning Tree
0231	Luxcom
0233	Sigma Network Systems
8001-801f	Not Used - reserved
8021	Internet Protocol Control Protocol
8023	OSI Network Layer Control Protocol
8025	Xerox NS IDP Control Protocol
8027	DECnet Phase IV Control Protocol
8029	Appletalk Control Protocol

Value (in hex)	Protocol Name
802b	Novell IPX Control Protocol
802d	reserved
802f	reserved
8031	Bridging NCP
8033	Stream Protocol Control Protocol
8035	Banyan Vines Control Protocol
8037	reserved till 1993
8039	reserved
803b	reserved
803d	Multi-Link Control Protocol
803f	NETBIOS Framing Control Protocol
807d	Not Used - reserved
8041	Cisco Systems Control Protocol
8043	Ascom Timeplex
8045	Fujitsu LBLB Control Protocol
8047	DCA Remote Lan Network Control Protocol
8049	Serial Data Control Protocol
804b	SNA over 802.2 Control Protocol
804d	SNA Control Protocol
804f	IP6 Header Compression Control Protocol
006f	Stampede Bridging Control Protocol
80cf	Not Used - reserved
80fb	compression on single link in multilink group control
80fd	Compression Control Protocol
80ff	Not Used - reserved
c021	Link Control Protocol
c023	Password Authentication Protocol
c025	Link Quality Report
c027	Shiva Password Authentication Protocol
c029	CallBack Control Protocol
c081	Container Control Protocol
c223	Challenge Handshake Authentication Protocol
c281	Proprietary Authentication Protocol

D.10.2 PPP LCP Callback Operation Fields

Operation	Description
0	Location determined by user authentication
1	Dialing string
2	Location identifier
3	E.164 number
4	X.500 distinguished name
5	unassigned
6	Location is determined during CBCP negotiation

Glossary

This glossary contains some originally defined terms, but primarily the terms were collected from various sources such as RFCs (e.g., RFC1583 and InterNIC FYI 18, , ,), ACC's Acronyms, *IBM's Dictionary of Computing A-Z,* and the *Communications Standard Dictionary* by Weik. It contains only the subset of Internet terms that are directly related to the TCP/IP protocol suite.

10Base2 The cable type specified for IEEE 802.3 operation over a coaxial cable at 10 Mbps with a maximum segment length of 185m. The number 10 represents the millions of bps, the word *Base* denotes baseband signaling, and the number 2 represents, to the nearest 100, the maximum cable segment length in meters divided by 100. (Also called thinnet.)

10Base5 The cable type specified for IEEE 802.3 operation over a coaxial cable at 10 Mbps with a maximum segment length of 500m. The number 10 represents the millions of bps, the word *Base* denotes baseband signaling, and the number 5 represents, to the nearest 100, the maximum cable segment length in meters divided by 100. (Also called thicknet.)

10BaseF The cable type specified for operation over a fiber optic cable at 10 Mbps. The number 10 represents the millions of bps, the word *Base* denotes baseband signaling, and the character F denotes fiber optic cable.

10BaseT The type medium specified for IEEE 802.3 operation over twisted pair copper wire at 10 Mbps. The number 10 represents the millions of bps, the word *Base* denotes baseband signaling, and the character T denotes twisted pair wire.

802.x   The family of IEEE standards for LAN protocols.

Abstract Syntax Notation One (ASN.1)   The language used by the OSI protocols for describing abstract syntax. This language is also used to encode SNMP packets. ASN.1 is defined in ISO documents 8824, and the BER are defined in ISO 8825.

access line   The part of a circuit between a user and a switch.

ACK   Acknowledgment

acknowledgment (ACK)   A type of message sent to give a positive indication of the receipt of data in the form of commands, requests, etc. *See also* negative acknowledgment.

address   A value used to identify endpoints. Common address types used in the Internet are e-mail, IP, MAC (hardware), and uniform resource locator (URL).

address resolution   Conversion of an Internet layer address (e.g., IP address) into the corresponding physical address (e.g., MAC address). *See also* IP address, MAC address.

address resolution protocol (ARP)   Used to dynamically discover the low-level physical network hardware address that corresponds to the Internet layer IP address for a given host. ARP is limited to physical network systems that support broadcast packets (e.g., Ethernet and SMDS). *See also* proxy ARP, RARP.

adjacency   A relationship between selected neighboring routers for the purpose of exchanging routing information. Physically connected routers are not necessarily logically adjacent.

administrative domain (AD)   A collection of hosts and routers and the interconnecting network(s), managed by a single administrative authority.

Advanced Research Projects Agency (ARPA)   An agency of the U.S. Department of Defense responsible for the development of new technology for use by the military. ARPA was responsible for funding much of the development of the Internet we know today, including the Berkeley version of UNIX and TCP/IP.

Advanced Research Projects Agency Network (ARPANET) A pioneering longhaul network funded by ARPA. Now retired, it served as the basis for early networking research as well as a central backbone during the development of the Internet. The ARPANET consisted of individual packet switching computers interconnected by leased lines. *See also* Advanced Research Projects Agency.

agent In the client-server model, the part of the system that performs information preparation and exchange on behalf of a client or server application.

alias A name, usually short and easy to remember, that is translated into another name, usually long and difficult to remember.

alternate mark inversion (AMI) A pseudo-ternary signal, in which successive marks are of alternating polarity. Synonymous with bipolar signal.

American National Standards Institute (ANSI) The organization responsible for approving U.S. standards in many areas, including computers and communications. Standards approved by this organization are often called ANSI standards.

American Standard Code for Information Interchange (ASCII) A standard character-to-number encoding system widely used in the computer industry.

analog Pertaining to the representation of data or physical quantities in the form of a continuous signal.

anonymous FTP Anonymous FTP allows a user to retrieve documents, files, programs, and other archived data from anywhere in the Internet without having to establish a userid and password. By using the special userid of "anonymous" the network user will bypass local security checks and have access to publicly accessible files on the remote system. *See also* archive site, file transfer protocol, World Wide Web.

ANSI *See* American National Standards Institute

AppleTalk A networking protocol developed by Apple Computer for communication between Apple Computer products and other computers. This protocol is independent of the network layer on which it is run. Current implementations exist for LocalTalk, a 235-Kbps local area network; and Ether-Talk, a 10-Mbps local area network.

application   A program that performs a function directly for a user. FTP, mail, and Telnet clients are examples of network applications.

application layer   The top layer of the ISO network protocol stack. The application layer is concerned with the semantics of work (e.g. formatting electronic mail messages). How to represent that data and how to reach the foreign node are issues for lower layers of the network. The application layer translates to the Internet upper layer.

application program interface (API)   A set of calling conventions that define how a service is invoked through a software package.

archie   A system to automatically gather, index, and serve information on the Internet. The initial implementation of archie provided an indexed directory of filenames from all anonymous FTP archives on the Internet. Later versions provide other collections of information. *See also* archive site, Gopher, Prospero, and wide area information servers.

archive site   A machine that provides access to a collection of files across the Internet. For example, an anonymous FTP archive site provides access to archived material via the FTP protocol. WWW servers can also serve as archive sites. *See also* anonymous FTP, archie, Gopher, Prospero, wide area information servers, World Wide Web.

ARP   *See* address resolution protocol

ARPA   *See* Advanced Research Projects Agency

ARPANET   *See* Advanced Research Projects Agency Network

AS   *See* autonomous system

ASCII   *See* American Standard Code for Information Interchange

ASN.1   *See* Abstract Syntax Notation One

assigned numbers   The RFC [STD2] that documents the currently assigned values from several series of numbers used in network protocol implementations. This RFC is updated periodically and, in any case, current information can be obtained from the Internet Assigned Numbers Authority (IANA). If you are developing a protocol or application that will require the use of a link, socket, port, protocol, etc., please contact the IANA to receive a number assignment. *See also* Internet Assigned Numbers Authority, STD.

asynchronous operation   An operation that occurs without a regular or predictable time relationship to a specified event.

asynchronous transfer mode (ATM)   A standard that defines high-load, high-speed (1.544 Mbps 1.2Gbps), fixed-size packet (cell) switching with dynamic bandwidth allocation and two levels of multiplexing.

ATM   *See* asynchronous transfer mode

authentication   The verification of the identity of a person or process.

autonomous system (AS)   A collection of routers under a single administrative authority using a common interior gateway protocol (e.g., OSPF) for routing packets.

backbone   The top level in a hierarchical network. Stub and transit networks that connect to the same backbone are guaranteed to be interconnected. *See also* stub network, transit network.

bandwidth   Technically, the difference, in Hertz (Hz), between the highest and lowest frequencies of a transmission channel. However, as typically used, the amount of data that can be sent through a given communications circuit (expressed in bits per second [bps]).

baseband   A transmission medium through which digital signals are sent without complicated frequency shifting. In general, only one communication channel is available at any given time. Ethernet is an example of a baseband network. *See also* broadband, Ethernet.

basic encoding rules (BER)   Standard rules for encoding data units described in ASN.1. Sometimes incorrectly lumped under the term ASN.1, which properly refers only to the abstract syntax description language, not the encoding technique. *See also* Abstract Syntax Notation One, bit error rate.

BBS   *See* bulletin board system

BCP   The newest subseries of RFCs that are written to describe best current practices in the Internet. Rather than specifying a protocol, these documents specify the best ways to use the protocols and the best ways to configure options to ensure interoperability between various vendors' products. BCPs carry the endorsement of the IESG. *See also* Request For Comments, Internet Engineering Steering Group.

BER   *See* basic encoding rules

Berkeley Software Distribution (BSD)   Implementation of the UNIX operating system and its utilities developed and distributed by the University of California at Berkeley. "BSD" is usually preceded by the version number of the distribution, e.g., "4.3 BSD" is version 4.3 of the Berkeley UNIX distribution. Many Internet hosts run BSD software, and it is the ancestor of many commercial UNIX implementations.

BGP   *See* border gateway protocol

big-endian   A format for storage or transmission of binary data in which the most significant bit (or byte) comes first. The term comes from *Gulliver's Travels* by Jonathan Swift. The Lilliputians, being very small, had correspondingly small political problems. In the book, the Big-Endian and Little-Endian parties debated over whether soft-boiled eggs should be opened at the big end or the little end. *See also* little-endian.

binary   The base 2 numbering system.

bipolar   Pertaining to a method of transmitting a signal with the use of both positive and negative voltages in the signal.

bit error rate (BER)   Bits that are received in error divided by the total number of bits transmitted. The BER is typically expressed in bits per million. e.g., 25 bits per million or simply 25×10^{-6}. Also see basic encoding rules.

Bitnet   An academic computer network that provides interactive electronic mail and file transfer services, using a store-and-forward protocol, based on IBM Network Job Entry protocols. Bitnet-II encapsulates the Bitnet protocol within IP packets and depends on the Internet to route them.

Blocking   In a telephone switched system, the inability to make a connection or obtain a service because the devices needed are not available. The telephone system is basically a blocking system. When there are emergencies that tie up trunk lines, users trying to use the service are blocked (busy tone), or worse yet they are unable to obtain a dial tone. Blocking is the result of insufficient committed resources. In an ATM network the result of blocking is lost cells.

BOOTP The Bootstrap Protocol, described in RFCs 2131 an 2132, is used for booting diskless nodes. *See also* dynamic host configuration protocol, reverse address resolution protocol, and inverse address resolution protocol.

border gateway protocol (BGP) The BGP is an exterior gateway protocol defined in RFC 1771. Its design is based on experience gained with EGP, as defined in RFC 904, and EGP usage in the NSFNET Backbone, as described in RFCs 1092 and 1093. *See also* exterior gateway protocol.

bridge A device that forwards traffic between network segments based on link layer information. These segments would have a common network layer address, if configured. That is, they would form a single network. *See also* gateway, router.

broadband A transmission medium capable of supporting a wide range of frequencies. It can carry multiple signals by dividing the total capacity of the medium into multiple, independent bandwidth channels where each channel operates only on a specific range of frequencies. *See also* baseband.

broadcast A special type of multicast packet that all nodes on the network are always willing to receive. *See also* multicast, unicast.

broadcast storm An incorrect packet broadcast onto a network that causes multiple hosts to respond all at once, typically with equally incorrect packets that cause the storm to grow exponentially in severity. *See also* Ethernet meltdown.

brouter A device that bridges some packets (i.e. forwards based on datalink layer information) and routes other packets (i.e., forwards based on network layer information). The bridge/route decision is based on configuration information. *See also* bridge, router.

BSD *See* Berkeley Software Distribution

bulletin board system (BBS) A computer, and associated software, that typically provides electronic messaging services, archives of files, and any other services or activities of interest to the BBS's operator. Although BBSs have traditionally been the domain of hobbyists, an increasing number of BBSs are connected directly to the Internet, and many BBSs are currently operated by government, educational, and research institutions. *See also* electronic mail, Internet, Usenet.

carrier   A wave, pulse train, or other signal suitable for modulation by an information bearing signal.

CCITT   *See* Comite Consultatif International de Telegraphique et Telephonique

CERT   *See* Computer Emergency Response Team

channel   A connection between two nodes in a network.

channel associated signaling   Signaling in which the signals necessary for the traffic carried by a single channel are transmitted in the channel itself or in a signaling channel permanently associated with it.

characteristic impedance   The impedance that a transmission line would present at its input terminals if it were infinitely long.

checksum   A computed value that is dependent upon the contents of a packet. This value is sent along with the packet when it is transmitted. The receiving system computes a new checksum based upon the received data and compares this value with the one sent with the packet. If the two values are the same, the receiver has a high degree of confidence that the data was received correctly. *See also* cyclic redundancy check.

CIDR   *See* classless interdomain routing

circuit   An electrical path between two or more points capable of providing a number of channels.

circuit switching   A process, usually on demand, connecting two or more users by permitting the exclusive use of a circuit between them until the connection is released. The telephone system is an example of a circuit switched network. *See also* connection-oriented, connectionless, packet switching.

classless interdomain routing (CIDR)   A proposal, set forth in RFC 1519, to allocate IP addresses so as to allow the addresses to be aggregated when advertised as routes. It is based on the elimination of intrinsic IP network addresses, that is, the determination of the network address based on the first few bits of the IP address. *See also* IP address, network address, supernet.

client   A computer system or process that requests a service of another computer system or process. A workstation requesting the contents of a file from a file server is a client of the file server. *See also* client-server model, server.

client-server model   A common way to describe the paradigm of many network protocols. Examples include the name-server/name-resolver relationship in DNS and the file-server/file-client relationship in NFS. *See also* client, server, domain name system, network file system.

clock   A device that generates periodic signals used for synchronization.

Comite Consultatif International de Telegraphique et Telephonique (CCITT)   This organization is now part of the International Telecommunications Union and is responsible for making technical recommendations about telephone and data communications systems. Every four years CCITT holds plenary sessions where it adopts new standards; the most recent was in 1992. Recently, the ITU reorganized and CCITT was renamed the ITU-TSS. *See also* International Telecommunications Union—Telecommunications Standards Sector.

Computer Emergency Response Team (CERT)   The CERT was formed by ARPA in November 1988 in response to the needs exhibited during the Internet worm incident. The CERT charter is to work with the Internet community to facilitate its response to computer security events involving Internet hosts, to take proactive steps to raise the community's awareness of computer security issues, and to conduct research targeted at improving the security of existing systems. CERT products and services include 24-hour technical assistance for responding to computer security incidents, product vulnerability assistance, technical documents, and tutorials. In addition, the team maintains a number of mailing lists (including one for CERT Advisories), and provides an anonymous FTP server, at "cert.org," where security-related documents and tools are archived. The CERT may be reached by e-mail at "cert@cert.org" and by telephone at +1-412-268-7090 (24-hour hotline). *See also* Advanced Research Projects Agency, worm.

congestion   Congestion occurs when the offered load exceeds the capacity of a data communication path.

connection-oriented   The data communication method in which communication proceeds through three well-defined phases: connection establishment, data transfer, and connection release. TCP is a connection-oriented protocol. *See also* circuit switching, connectionless, packet switching, transmission control protocol.

connectionless   The data communication method in which communication occurs between hosts with no previous setup. Packets between two hosts may take different routes, as each is independent of the other. UDP is a connectionless protocol. *See also* circuit switching, connection-oriented, packet switching, user datagram protocol.

Coordinating Committee for Intercontinental Research Networks (CCIRN) A committee that includes the U.S. FNC and its counterparts in North America and Europe. Co-chaired by the executive directors of the FNC and the European Association of Research Networks (RARE), the CCIRN provides a forum for cooperative planning among the principal North American and European research networking bodies. *See also* Federal Networking Council, RARE.

core gateway   Historically, one of a set of gateways (routers) operated by the Internet Network Operations Center at Bolt, Beranek and Newman (BBN). The core gateway system formed a central part of Internet routing in that all groups must advertise paths to their networks from a core gateway.

Corporation for Research and Educational Networking (CREN)   This organization was formed in October 1989, when Bitnet and CSNET (Computer + Science NETwork) were combined under one administrative authority. CSNET is no longer operational, but CREN still runs Bitnet. *See also* Bitnet.

cracker   A cracker is an individual who attempts to access computer systems without authorization. These individuals are often malicious, as opposed to hackers, and have many means at their disposal for breaking into a system. See also hacker, Computer Emergency Response Team, Trojan Horse, virus, worm.

CRC   *See* cyclic redundancy check

cyclic redundancy check (CRC)   A number derived from a set of data that will be transmitted. By recalculating the CRC at the remote end and comparing it to the value originally transmitted, the receiving node can detect some types of transmission errors. *See also* checksum.

DANTE A nonprofit company founded in July 1993 to help the European research community enhance its networking facilities. It focuses on the establishment of a high-speed computer network infrastructure.

DARPA Defense Advanced Research Projects Agency. *See* Advanced Research Projects Agency

data encryption key (DEK) Used for the encryption of message text and for the computation of message integrity checks (signatures). *See also* encryption.

data encryption standard (DES) A popular, standard encryption scheme. *See also* encryption, pretty good privacy, RSA.

datagram A self-contained, independent entity of data carrying sufficient information to be routed from the source to the destination computer without reliance on earlier exchanges between this source and destination computer and the transporting network. *See also* frame, packet.

DCE data circuit-terminating equipment (Internet layer)

DCE data computer equipment (physical layer)

DDN *See* Defense Data Network

DDN NIC *See* Defense Data Network Network Information Center

DECnet A proprietary network protocol designed by Digital Equipment Corporation. The functionality of each phase of the implementation, such as phase IV and phase V, is different.

default route A routing table entry that is used to direct packets addressed to networks not explicitly listed in the routing table.

defense data network (DDN) A global communications network serving the U.S. Department of Defense composed of MILNET, other portions of the Internet, and classified networks that are not part of the Internet. The DDN is used to connect military installations and is managed by the Defense Information Systems Agency. *See also* Defense Information Systems Agency.

Defense Data Network Network Information Center (DDN NIC) Previously called the *NIC,* the DDN NIC's primary responsibility was the assignment of Internet network addresses and AS numbers and the administration of the root domain and providing information and support services to the Internet for the DDN. Since the creation of the InterNIC, the DDN NIC performs these functions only for the DDN. *See also* autonomous system, network address, Internet Registry, InterNIC, Network Information Center, request For comments.

Defense Information Systems Agency (DISA) Formerly called the Defense Communications Agency (DCA), this is the government agency responsible for managing the DDN portion of the Internet, including the MILNET. Currently, DISA administers the DDN and supports the user assistance services of the DDN NIC. *See also* Defense Data Network.

DEK See data encryption key

demodulation To undo or reverse the effects of modulation.

DES *See* data encryption standard

dial-up A temporary, as opposed to dedicated, connection between machines established over a phone line (analog or ISDN). *See also* integrated services digital network.

distributed database A collection of several different data repositories that looks like a single database to the user. A prime example in the Internet is the domain name system.

DIX Ethernet *See* Ethernet

DNS *See* domain name system

domain Domain is a heavily overused term in the Internet. It can be used in the administrative dDomain context, or the domain name context. *See also* administrative domain, domain name system.

domain name system (DNS) The DNS is a general purpose distributed, replicated, data query service. The principal use is the lookup of host IP addresses based on host names. The style of host names now used in the Internet is called *domain name,* because they are the style of names used to look up anything in the DNS. Some important domains are: .COM (commercial), .EDU (educational), .NET (network operations), .GOV (U.S. government), and .MIL (U.S. military). Most countries also have a domain. The country domain names are based on ISO 3166. For example, .US (United States), .UK (United Kingdom), .AU (Australia). *See also* Fully Qualified Domain Name, Mail Exchange Record.

dot address (dotted decimal notation)   Dot address refers to the common notation for IP addresses of the form A.B.C.D; where each letter represents, in decimal, one byte of a four byte IP address. *See also* IP address.

DTE   data terminal equipment

dynamic adaptive routing   Automatic rerouting of traffic based on a sensing and analysis of current actual network conditions. *Note:* this does not include cases of routing decisions taken on predefined information.

E1   The basic building block for European multimegabit data rates, with a bandwidth of 2.048 Mbps.

E3   A European standard for transmitting data at 57.344 Mbps. *See also* T3.

EARN   European Academic and Research Network. *See* Trans-European Research and Education Networking Association.

EBCDIC   *See* extended binary coded decimal interchange code

Ebone   A pan-European backbone service.

EGP   See exterior gateway protocol

electronic mail (e-mail)   A system whereby a computer user can exchange messages with other computer users (or groups of users) via a communications network. E-mail is one of the most popular uses of the Internet.

e-mail   *See* electronic mail

e-mail address   The domain-based address that is used to send e-mail to a specified destination. The two parts of an Internet e-mail address are separated by an (@). The first is the local part (e.g., Jim) and the second is the domain. For example, Jim@ubmr.com.

encapsulation   The technique used by layered protocols in which a layer adds header information to the protocol data unit (PDU) from the layer above. For example, in Internet terminology, a packet would contain a header from the physical layer, followed by a header from the datalink layer (e.g. Ethernet), followed by a header from the network layer (IP), followed by a header from the transport layer (e.g. TCP), followed by the application protocol data.

encryption   Encryption is the manipulation of a packet's data in order to prevent any but the intended recipient from reading that data. There are many types of data encryption, and they are the basis of network security. *See also* data encryption standard.

error checking   The examination of received data for transmission errors. *See also* checksum, cyclic redundancy check.

Ethernet   A 10-Mbps standard for LANs, initially developed by Xerox, and later refined by Digital, Intel, and Xerox (DIX). All hosts are connected to a coaxial cable where they contend for network access using a carrier sense multiple access with collision detection (CSMA/CD) paradigm. *See also* 802.x, local area network, token ring.

Ethernet meltdown   An event that causes saturation, or near saturation, on an Ethernet. It usually results from illegal or misrouted packets and typically lasts only a short time. *See also* broadcast storm.

extended binary coded decimal interchange code (EBCDIC)   A standard character-to-number encoding used primarily by IBM computer systems. *See also* ASCII.

exterior gateway protocol (EGP)   A protocol that distributes routing information to the routers that connect autonomous systems. The term *gateway* is historical, as *router* is currently the preferred term. There is also a routing protocol called EGP defined in RFC 904. *See also* autonomous system, border gateway protocol, interior gateway protocol.

external data representation (XDR)   A standard for machine independent data structures developed by Sun Microsystems and defined in RFCs 1014 and 1832. It is similar to ASN.1. *See also* Abstract Syntax Notation One.

FAQ   frequently asked question

FDDI   *See* fiber distributed data interface

Federal Networking Council (FNC)   The coordinating group of representatives from those federal agencies involved in the development and use of federal networking, especially those networks using TCP/IP and the Internet. Current members include representatives from DOD, DOE, ARPA, NSF, NASA, and HHS. *See also* Advanced Research Projects Agency, National Science Foundation.

fiber distributed data interface (FDDI)   A high-speed (100-Mbps) LAN standard. The underlying medium is fiber optics, and the topology is a dual-attached, counter-rotating token ring. *See also* local area network, token ring.

file transfer   The copying of a file from one computer to another over a computer network. *See also* file transfer protocol, Kermit, Gopher, World Wide Web.

file transfer protocol (FTP)   A protocol that allows a user on one host to access, and transfer files to and from, another host over a network. Also, FTP is usually the name of the program the user invokes to execute the protocol. *See also* anonymous FTP.

finger   A protocol, defined in RFC 1288, that allows information about a system or user on a system to be retrieved. Finger also refers to the commonly used program that retrieves this information. Information about all logged in users, as well as information about specific users may be retrieved from local or remote systems. Some sites consider finger to be a security risk and have either disabled it, or replaced it with a simple message.

FNC   *See* Federal Networking Council

FQDN   See fully qualified domain name

fragment   A piece of a packet. When a router is forwarding an IP packet to a network that has a maximum transmission unit smaller than the packet size, it is forced to break up that packet into multiple fragments. These fragments will be reassembled by the IP layer at the transmission unit.

fragmentation   The IP process in which a packet is broken into smaller pieces to fit the requirements of a physical network over which the packet must pass. *See also* reassembly.

frame   A frame is a datalink layer *packet* that contains the header and trailer information required by the physical medium. That is, network layer packets are encapsulated to become frames. *See also* datagram, encapsulation, packet.

frame-based user-to-network interface (FUNI)   The ATM Forum Specification V1, September 1995.

FTP   *See* file transfer protocol

fully qualified domain name (FQDN)   The FQDN is the full name of a system, rather than just its hostname. For example, "Jim" is a hostname and "jim.ubmr.com" is an FQDN. *See also* hostname, domain name system.

FUNI   *See* frame-based user-to-network interface

funneling   A form of implied routing. Funneling is one form of forwarding used to keep a virtual private network's traffic separate. *See also* tunneling.

FYI   A subseries of RFCs that are not technical standards or descriptions of protocols. FYIs convey general information about topics related to TCP/IP or the Internet. *See also* request for comments.

gated   Gatedaemon. A program that supports multiple routing protocols and protocol families. It may be used for routing and makes an effective platform for routing protocol research. The software is freely available by anonymous FTP from "gated.cornell.edu." Pronounced "gate-dee." *See also* exterior gateway protocol, open shortest-path first, routing information protocol, routed.

gateway   The term router is now used in place of the original definition of gateway. Currently, a gateway is a communications device/program that passes data between networks having similar functions but dissimilar implementations. This should not be confused with a protocol converter. By this definition, a router is a layer 3 (network layer) gateway, and a mail gateway is a layer 7 (application layer) gateway. *See also* mail gateway, router, protocol converter.

Gopher   A distributed information service, developed at the University of Minnesota, that makes hierarchical collections of information available across the Internet. Gopher uses a simple protocol, defined in RFC 1436, that allows a single Gopher client to access information from any accessible Gopher server, providing the user with a single "Gopher space" of information. Public domain versions of the client and server are available. *See also* archie, archive site, Prospero, wide area information servers.

GOSIP   See Government OSI Profile

Government OSI Profile (GOSIP)   A subset of OSI standards specific to U.S. government procurements, designed to maximize interoperability in areas where plain OSI standards are ambiguous or allow excessive options.

hacker   A person who delights in having an intimate understanding of the internal workings of a system, computers, and computer networks in particular. The term is often misused in a pejorative context, where *cracker* would be the correct term. See also cracker.

header   The portion of a packet, preceding the actual data, containing source and destination information. It may also contain error checking and other fields. A header is also the part of an electronic mail message that precedes the body of a message and contains, among other things, the message originator, date, and time. *See also* electronic mail, packet, error checking.

heterogeneous network   A network running multiple network layer protocols. *See also* DECnet, IP, IPX, XNS.

hierarchical routing   The complex problem of routing on large networks can be simplified by reducing the size of the networks. This is accomplished by breaking a network into a hierarchy of networks, where each level is responsible for its own routing. The Internet has, basically, three levels: the backbones, the midlevels, and the stub networks. The backbones know how to route between the midlevels, the midlevels know how to route between the sites, and each site (being an autonomous system) knows how to route internally. *See also* autonomous system, exterior gateway protocol, interior gateway protocol, stub network, transit network.

high-performance computing and communications (HPCC)   High-performance computing encompasses advanced computing, communications, and information technologies, including scientific workstations, supercomputer systems, high-speed networks, special purpose and experimental systems, the new generation of large-scale parallel systems, and application and systems software with all components well-integrated and linked over a high-speed network.

high-performance parallel interface (HIPPI)   An emerging ANSI standard that extends the computer bus over fairly short distances at speeds of 800 and 1600 Mbps. HIPPI is often used in a computer room to connect a super-computer to routers, frame buffers, mass-storage peripherals, and other computers. *See also* American National Standards Institute.

HIPPI   *See* high-performance parallel interface

HTML   *See* hypertext markup language

network layer protocol   *See also* DECnet, IP, IPX, XNS, heterogeneous network.

hop   A term used in routing. A path to a destination on a network is a series of hops, through routers, away from the origin.

host   A computer that allows users to communicate with other host computers on a network. Individual users communicate by using application programs, such as electronic mail, Telnet, and FTP.

host address   See internet address

hostname   The name given to a machine. *See also* fully qualified domain name.

host number   *See* host address

HPCC   See high-performance computing and communications

HTTP   See hypertext transfer protocol

hub   A device connected to several other devices. In ARCnet, a hub is used to connect several computers. In a message handling service, a hub is used for the transfer of messages across the network.

hyperlink   A pointer within a hypertext document that points (links) to another document, which may or may not also be a hypertext document. *See also* hypertext.

hypertext   A document, written in HTML, that contains hyperlinks to other documents, which may or may not also be hypertext documents. Hypertext documents are usually retrieved using WWW. *See also* hyperlink, hypertext markup language, World Wide Web.

hypertext markup language (HTML)   The language used to create hypertext documents. It is a subset of SGML and includes the mechanisms to establish hyperlinks to other documents. *See also* hypertext, hyperlink, standardized general markup language.

hypertext transfer protocol (HTTP)   The protocol used by WWW to transfer HTML files. A formal standard is still under development in the IETF. *See also* hyperlink, hypertext, hypertext markup language, World Wide Web.

I-D   *See* Internet-Draft

IAB   *See* Internet Architecture Board

IANA   *See* Internet Assigned Numbers Authority

ICMP   *See* Internet control message protocol

IEEE   Institute of Electrical and Electronics Engineers

IEN   *See* Internet Experiment Note

IEPG   *See* Internet Engineering Planning Group

IESG   *See* Internet Engineering Steering Group

IETF   *See* Internet Engineering Task Force

IGP   *See* Interior Gateway Protocol

interface   The connection between a router and one of its attached networks. An interface to a network has associated with it a single IP address and subnet mask, unless the network is an unnumbered point-to-point network. An interface is sometimes referred to as a link.

integrated services digital network (ISDN)   An emerging technology that is beginning to be offered by the telephone carriers of the world. ISDN combines voice and digital network services in a single medium, making it possible to offer customers digital data services as well as voice connections through a single wire. The standards that define ISDN are specified by ITU-T. *See also* CCITT.

interior gateway protocol (IGP) A protocol that distributes routing information to the routers within an autonomous system. The term *gateway* is historical, as router is currently the preferred term. *See also* autonomous system, exterior gateway protocol, open shortest-path first, routing information protocol.

International Organization for Standardization (ISO) A voluntary, non-treaty organization founded in 1946 that is responsible for creating international standards in many areas, including computers and communications. Its members are the national standards organizations of the 89 member countries, including ANSI for the United States. *See also* American National Standards Institute, Open Systems Interconnection.

International Telecommunications Union (ITU) An agency of the United Nations that coordinates the various national telecommunications standards so that people in one country can communicate with people in another country.

International Telecommunications Union Telecommunications Standards Sector (ITU-TSS)—The new name for CCITT since the ITU reorganization. The function is the same; only the name has been changed. It is abbreviated as ITU-T.

internet While an internet is a network, the term *internet* is usually used to refer to a collection of networks interconnected with routers. The latest word to describe internet is intranet. *See also* network.

Internet The Internet (capitalized) is the largest internet in the world. It is a three-level hierarchy composed of backbone networks (e.g. Ultranet), mid-level networks (e.g., NEARnet), and stub networks. The Internet is a multiprotocol internet. *See also* backbone, mid-level network, stub network, transit network, Internet protocol.

internet address An IP address that uniquely identifies a node on an internet. An Internet address uniquely identifies a node on the Internet. *See also* internet, Internet, IP address.

Internet Architecture Board (IAB) The IAB has been many things over the years. Originally, as the Internet Activities Board, it was responsible for the development of the protocols that make up the Internet. It later changed its name and charter to become the group most responsible for the architecture of the Internet, leaving the protocol details to the IESG. In June of 1992, it was chartered as a component of the Internet Society; this is the charter it holds today. The IAB is responsible for approving nominations to the IESG, architectural oversight for Internet Standard Protocols, IETF standards process oversight and appeals, IANA and RFC activities, and liaison to peer standards groups (e.g., ISO). *See also* Internet Engineering Task Force, Internet Research Task Force, Internet Engineering Steering Group, Internet Assigned Numbers Authority, request for comments.

Internet Assigned Numbers Authority (IANA)   The central registry for various Internet protocol parameters, such as port, protocol and enterprise numbers, and options, codes, and types. The currently assigned values are listed in the "Assigned Numbers" document [STD2]. To request a number assignment, contact the IANA at "iana@isi.edu". *See also* assigned numbers, STD.

Internet control message protocol (ICMP)   ICMP is an extension to the Internet protocol. It allows for the generation of error messages, test packets, and informational messages related to IP.

Internet-Draft (I-D)   Internet-Drafts are working documents of the IETF, its areas, and its working groups. As the name implies, Internet-Drafts are draft documents. They are valid for a maximum of six months and may be updated, replaced, or obsoleted by other documents at any time. Very often, I-Ds are precursors to RFCs. *See also* Internet Engineering Task Force, request For comments.

Internet Engineering Planning Group (IEPG)   A group, primarily composed of Internet service operators, whose goal is to promote a globally coordinated Internet operating environment. Membership is open to all.

Internet Engineering Steering Group (IESG)   The IESG is composed of the IETF area directors and the IETF chair. It provides the first technical review of Internet standards and is responsible for day-to-day management of the IETF. *See also* Internet Engineering Task Force.

Internet Engineering Task Force (IETF)   The IETF is a large, open community of network designers, operators, vendors, and researchers whose purpose is to coordinate the operation, management, and evolution of the Internet, and to resolve short-range and mid-range protocol and architectural issues. It is a major source of proposals for protocol standards that are submitted to the IAB for final approval. The IETF meets three times a year and extensive minutes are included in the IETF Proceedings. *See also* Internet, Internet Architecture Board.

Internet Experiment Note (IEN) A series of reports pertinent to the Internet. IENs were published in parallel to RFCs and were intended to be working documents. They have been replaced by Internet-Drafts and are currently of historic value only. *See also* Internet-Draft, Request For Comments.

Internet Monthly Report (IMR) Published monthly, the purpose of the *Internet Monthly Report* is to communicate to the Internet Research Group the accomplishments, milestones reached, or problems discovered by the participating organizations.

internet number *See* internet address

Internet Protocol (IP, IPv4) The Internet protocol (version 4), defined in RFC 791, is the network layer for the TCP/IP protocol suite. It is a connectionless, best-effort packet switching protocol. *See also* packet switching, TCP/IP protocol suite, Internet protocol version 6.

Internet protocol version 6 (IPng, IPv6) IPv6 (version 5 is a stream protocol used for special applications) is a new version of the Internet protocol that is designed to be an evolutionary step from its predecessor, version 4. There are many RFCs defining various portions of the protocol, its auxiliary protocols, and the transition plan from IPv4. The core RFCs are 1883 through 1886. The name IPng (IP next generation) is a nod to STNG (Star Trek Next Generation).

Internet Registry (IR) The IANA has the discretionary authority to delegate portions of its responsibility and, with respect to network address and autonomous system identifiers, has lodged this responsibility with an IR. The IR function is performed by the DDN NIC. *See also* autonomous system, network address, Defense Data Network, Internet Assigned Numbers Authority.

Internet relay chat (IRC) A worldwide "party line" protocol that allows one to converse with others in real time. IRC is structured as a network of servers, each of which accepts connections from client programs, one per user. *See also* talk.

Internet Research Steering Group (IRSG)   The *governing body* of the IRTF. *See also* Internet Research Task Force.

Internet Research Task Force (IRTF)   The IRTF is chartered by the IAB to consider long-term Internet issues from a theoretical point of view. It has research groups, similar to IETF working groups, that are each tasked to discuss different research topics. Multicast audio/video conferencing and privacy enhanced mail are samples of IRTF output. *See also* Internet Architecture Board, Internet Engineering Task Force, privacy enhanced mail.

Internet Society (ISOC)   The ISOC is a nonprofit, professional membership organization that facilitates and supports the technical evolution of the Internet; stimulates interest in and educates the scientific and academic communities, industry, and the public about the technology, uses, and applications of the Internet; and promotes the development of new applications for the system. The society provides a forum for discussion and collaboration in the operation and use of the global Internet infrastructure. The ISOC publishes a quarterly newsletter, the *Internet Society News* and holds an annual conference, INET. The development of Internet technical standards takes place under the auspices of the Internet Society with substantial support from the Corporation for National Research Initiatives under a cooperative agreement with the U.S. federal government.

Internetwork Packet eXchange (IPX)   Novell's protocol used by Netware. A router with IPX routing can interconnect LANs so that Novell Netware clients and servers can communicate. *See also* local area network.

InterNIC   A five-year project, partially supported by the National Science Foundation, to provide network information services to the networking community. The InterNIC began operations in April of 1993 and is now a collaborative project of two organizations: AT&T, which provides directory and database services from South Plainsfield, NJ; and Network Solutions, Inc., which provides registration services from its headquarters in Herndon, VA. Services are provided via the Internet and by telephone, fax, and hardcopy.

interoperability   The ability of software and hardware on multiple machines from multiple vendors to communicate meaningfully.

intranet   A private internet. Also called intranet.

IP (IPv4)   *See* Internet protocol

IPng (IPv6)   *See* Internet protocol version 6

IP address   The 32-bit address defined by the Internet protocol in RFC 791. It is usually represented in dotted decimal notation. *See also* dot address, internet address, Internet Protocol, network address, subnet address, host address.

IP datagram   *See* datagram

IPX   *See* Internetwork Packet eXchange

IR   *See* Internet Registry

IRC   *See* Internet relay chat

IRSG   *See* Internet Research Steering Group

IRTF   *See* Internet Research Task Force

ISDN   *See* integrated services digital network

ISO   *See* International Organization for Standardization

ISO Development Environment (ISODE)   Software that allows OSI services to use a TCP/IP network. Pronounced eye-so-dee-eee. *See also* Open Systems Interconnection, TCP/IP protocol suite.

ISOC   *See* Internet Society

ISODE   *See* ISO Development Environment

ITU   *See* International Telecommunications Union–Telecommunications Standards Sector

ITU-TSS   *See* International Telecommunications Union

KA9Q   A popular implementation of TCP/IP and associated protocols for amateur packet radio systems. *See also* TCP/IP Protocol Suite.

Kerberos   Kerberos is the security system of MIT's Project Athena. It is based on symmetric key cryptography. *See also* encryption.

Kermit   A popular file transfer protocol developed by Columbia University. Because Kermit runs in most operating environments, it provides an easy method of file transfer. Kermit is *not* the same as FTP. *See also* file transfer protocol

LAN   *See* local area network

layer   Communication networks for computers may be organized as a set of more or less independent protocols, each in a different layer (also called level). The lowest layer governs direct host-to-host communication between the hardware at different hosts; the highest consists of user applications. Each layer builds on the layer beneath it. For each layer, programs at different hosts use protocols appropriate to the layer to communicate with each other. TCP/IP has five layers of protocols; OSI has seven. The advantages of different layers of protocols is that the methods of passing information from one layer to another are specified clearly as part of the protocol suite, and changes within a protocol layer are prevented from affecting the other layers. This greatly simplifies the task of designing and maintaining communication programs. *See also* Open Systems Interconnection, TCP/IP protocol suite.

link   A pointer that may be used to retrieve the file or data to which the pointer points.

little-endian   A format for storage or transmission of binary data in which the least significant byte (bit) comes first. *See also* big-endian.

LLC   See logical link control

local area network (LAN)   A data network intended to serve an area of only a few square kilometers or less. Because the network is known to cover only a small area, optimizations can be made in the network signal protocols that permit data rates up to 100 Mbps. *See also* Ethernet, fiber distributed data interface, token ring, metropolitan area network, wide area network.

logical link control (LLC)   The upper portion of the data link layer, as defined in IEEE 802.2. The LLC sublayer presents a uniform interface to the user of the datalink service, usually the network layer. Beneath the LLC sublayer is the MAC sublayer. *See also* 802.x, layer, media access control.

MAC   *See* media access control

MAC address   The hardware address of a device connected to a shared media. *See also* media access control, Ethernet, token ring.

mail bridge   A mail gateway that forwards electronic mail between two or more networks while ensuring that the messages it forwards meet certain administrative criteria. A mail bridge is simply a specialized form of mail gateway that enforces an administrative policy with regard to what mail it forwards. *See also* electronic mail, mail gateway.

mail exchange record (MX record)   A DNS resource record type indicating which host can handle mail for a particular domain. *See also* domain name system, electronic mail.

mail exploder   Part of an electronic mail delivery system that allows a message to be delivered to a list of addresses. Mail exploders are used to implement mailing lists. Users send messages to a single address and the mail exploder takes care of delivery to the individual mailboxes in the list. *See also* electronic mail, e-mail address, mailing list.

mail gateway   A machine that connects two or more electronic mail systems (including dissimilar mail systems) and transfers messages between them. Sometimes the mapping and translation can be quite complex, and it generally requires a store-and-forward scheme whereby the message is received from one system completely before it is transmitted to the next system, after suitable translations. *See also* electronic mail.

mail server   A software program that distributes files or information in response to requests sent via e-mail. Internet examples include Almanac and netlib. Mail servers have also been used in Bitnet to provide FTP-like services. *See also* Bitnet, Electronic Mail, FTP.

mailing list   A list of e-mail addresses, used by a mail exploder, to forward messages to groups of people. Generally, a mailing list is used to discuss certain set of topics, and different mailing lists discuss different topics. A mailing list may be moderated. This means that messages sent to the list are actually sent to a moderator who determines whether or not to send the messages on to everyone else. Requests to subscribe to, or leave, a mailing list should *always* be sent to the list's "-request" address (e.g. ietf-request@cnri.reston.va.us for the IETF mailing list) or majordomo server.

MAN   *See* metropolitan area network

management information base (MIB)   The set of parameters an SNMP management station can query or set in the SNMP agent of a network device (e.g. router). Standard, minimal MIBs have been defined, and vendors often have private enterprise MIBs. In theory, any SNMP manager can talk to any SNMP agent with a properly defined MIB. *See also* client-server model, simple network management protocol.

Martian   A humorous term applied to packets that turn up unexpectedly on the wrong network because of bogus routing entries. Also used as a name for a packet that has an altogether bogus (unregistered or ill-formed) internet address.

maximum transmission onit (MTU)   The largest frame length that may be sent on a physical medium. *See also* frame, fragment, fragmentation.

MAU   *See* medium attachment unit (IEEE 802.3).

MAU   *See* multistation access unit (IBM Token Ring).

mbone   The multicast backbone is based on IP multicasting using class-D addresses. The mbone concept was adopted at the March 1992 IETF in San Diego, during which it was used to audiocast to 40 people throughout the world. At the following meeting, in Cambridge, the name mbone was adopted. Since then the audiocast has become full two-way audio/video conferencing using two video channels and four audio channels and involving hundreds of remote users. *See also* multicast, Internet Engineering Task Force.

media access control (MAC, IEEE 802.3) or medium access control (MAC, IEEE 802.5)   The lower portion of the datalink layer. The MAC differs for various physical media. *See also* MAC address, Ethernet, logical link control, token ring.

message digest (MD-2, MD-4, MD-5)   Message digests are algorithmic operations, generally performed on text, that produce a unique signature for that text. MD-2, described in RFC 1319; MD-4, described in RFC 1320; and MD-5, described in RFC 1321 all produce a 128-bit signature. They differ in their operating speed and resistance to crypto-analytic attack. Generally, one must be traded off for the other.

message switching   *See* packet switching

metropolitan area network (MAN) A data network intended to serve an area approximating that of a large city. Such networks are being implemented by innovative techniques, such as running fiber cables through subway tunnels. A popular example of a MAN is SMDS. *See also* local area network, switched multimegabit data service, wide area network.

MIB *See* management information base

microcom networking protocol (MNP) A series of protocols built into most modems that error-check or compress data being transmitted over a phone line.

mid-level network Mid-level networks (a.k.a. regionals) make up the second level of the Internet hierarchy. They are the transit networks that connect the stub networks to the backbone networks. *See also* backbone, Internet, stub network, transit network.

MIME See multipurpose Internet mail extensions

MNP See microcom networking protocol

modulation The variation of a characteristic of a carrier wave in accordance with a characteristic of another wave. The carrier is used as a means of propagation.

modulation rate The reciprocal of the measure of the shortest nominal unit interval between successive significant instants of the modulated signal. If this measure is expressed in seconds, the modulation rate in baud (bauds) is obtained.

MOSPF multicast open shortest-path first. See open shortest-path first.

MTU *See* maximum transmission unit

multicast A packet with a special destination address that multiple nodes on the network may be willing to receive. *See also* broadcast, unicast.

multifrequency (MF) signaling A signaling method in which a combination of frequencies for signaling purposes is used. For example, two-out-of-six (MF 2/6) audio frequencies are used to indicate telephone number, precedence, and control signals, such as line-busy or trunk-busy signals.

multihomed host   A host that has more than one connection to a network. The host may send and receive data over any of the links but will not route traffic for other nodes. *See also* host, router.

multipurpose Internet mail extensions (MIME)   An extension to Internet e-mail that provides the ability to transfer nontextual data, such as graphics, audio, and fax. *See also* electronic mail

multistation access unit (MAU)   IBM wiring concentrator (hub) used to control workstations in a token ring LAN.

MX Record   *See* mail exchange record

NAK   *See* negative acknowledgment

name resolution   The process of mapping a name into its corresponding address. *See also* domain name system.

namespace   A commonly distributed set of names in which all names are unique.

National Institute of Standards and Technology (NIST)   U.S. governmental body that provides assistance in developing standards. Formerly the National Bureau of Standards.

National Research and Education Network (NREN)   The NREN is the realization of an interconnected gigabit computer network devoted to high-performance computing and communications. *See also* HPCC.

National Science Foundation (NSF)   A U.S. government agency whose purpose is to promote the advancement of science. NSF funds science researchers, scientific projects, and infrastructure to improve the quality of scientific research. The NSFNET, funded by NSF, was once an essential part of academic and research communications. It was a high-speed, hierarchical "network of networks." At the highest level, it had a backbone network of nodes, interconnected with T3 (45-Mbps) facilities that spanned the continental United States. Attached to that were mid-level networks, and attached to the mid-levels were campus and local networks. *See also* backbone network, mid-level network.

negative acknowledgment (NAK)   Response to the receipt of either a corrupted or unexpected packet of information. *See also* acknowledgment.

neighboring routers   Two routers that have interfaces to a common network are neighboring routers. On multi-access networks, neighbors are dynamically discovered by the OSPF hello Protocol.

network   A computer network is a data communications system that inter-connects computer systems at various different sites. A network may be composed of any combination of LANs, MANs, or WANs. *See also* local area network, metropolitan area network, wide area network, internet.

network address   The network portion of an IP address. For a class A network, the network address is the first byte of the IP address. For a class B network, the network address is the first two bytes of the IP address. For a class C network, the network address is the first three bytes of the IP address. In each case, the remainder is the host address. In the Internet, assigned network addresses are globally unique. *See also* Internet, IP address, subnet address, host address, Internet Registry.

network file system (NFS)   A protocol developed by Sun Microsystems and defined in RFC 1094 (RFC 1813 defines Version 3) that allows a computer system to access files over a network as if they were on its local disks. This protocol has been incorporated in products by more than 200 companies and is now a de facto Internet standard.

network information center (NIC)   A NIC provides information, assistance and services to network users. *See also* network operations center.

network information services (NIS)   A set of services, generally provided by a NIC, to assist users in using the network. *See also* network information center.

network news transfer protocol (NNTP)   A protocol, defined in RFC 977, for the distribution, inquiry, retrieval, and posting of news articles. *See also* Usenet.

network operations center (NOC)   A location from which the operation of a network or internet is monitored. Additionally, this center usually serves as a clearinghouse for connectivity problems and efforts to resolve those problems. *See also* network information center.

network time protocol (NTP)   A protocol that assures accurate local time keeping with reference to radio and atomic clocks located on the Internet. This protocol is capable of synchronizing distributed clocks within milliseconds over long time periods. *See also* Internet.

NFS   *See* network file system

NIC   *See* network information center

NIC.DDN.MIL   This is the domain name of the DDN NIC. See also Defense Data Network, domain name system, network information center.

NIS   *See* network information services

NIST   *See* National Institute of Standards and Technology

NNTP   *See* network news transfer protocol

NOC   *See* network operations center

nodal switching system (NSS)   Main routing nodes in the NSFnet backbone. *See also* backbone, National Science Foundation.

node   An addressable device attached to a computer network. *See also* host, router.

nonblocking   Pertaining to a communications system in which 100% of all access attempts (calls) are completed. *See* blocking.

non-return-to-zero (NRZ)   Pertaining to a signal that represents a sequence of digits in which the signal level does not return to a zero amplitude, or to a value that represents a zero level, between digits.

NREN   *See* National Research and Education Network

NSF   *See* National Science Foundation

NSS   *See* nodal switching system

NTP   *See* network time protocol

octet   An octet is eight contiguous bits. This term is used in networking, rather than byte, because some systems have bytes that are not eight bits long.

open shortest-path first (OSPF) A link state, as opposed to distance vector, routing protocol. It is an Internet standard IGP defined in RFC 2178. The multicast version, MOSPF, is defined in RFC 1584. *See also* interior gateway protocol, routing information protocol.

Open Systems Interconnection (OSI) A suite of protocols, designed by ISO committees, to be the international standard computer network architecture. *See also* International Organization for Standardization.

OSI *See* Open Systems Interconnection

OSI reference model A seven-layer structure designed to describe computer network architectures and the way that data passes through them. This model was developed by the ISO in 1978 to clearly define the interfaces in multi-vendor networks and to provide users of those networks with conceptual guide-lines in the construction of such networks. *See also* International Organization for Standardization.

OSPF See open shortest-path first

packet The unit of data sent across a network. The expression *packet* was first used by Donald Davies with the British National Physical Laboratory (NPL) to describe the data transferred by a packet switched exchange. Packet is now a generic term used to describe a unit of data at all levels of the protocol stack, but it is most correctly used to describe application data units. *See also* datagram, frame.

Packet InterNet Groper (PING) A program used to test reachability of desti-nations by sending them an ICMP echo request and waiting for a reply. The term is used as a verb: "Ping host X to see if it is up!" *See also* Internet control message protocol. (Caution, spell checkers will change this to grouper.)

packet switch node (PSN) A dedicated computer whose purpose is to accept, route, and forward packets in a packet switched network. *See also* packet switch-ing, router.

packet switching A communications paradigm in which a message is seg-mented in packets and transmitted individually between logically connected source and destination. This paradigm generally replaced message switching, which transmitted the entire message without segmentation. *See also* circuit switching, connection-oriented, connectionless.

PD   public domain

PDU   *See* protocol data unit

PEM   *See* privacy enhanced mail

PGP   *See* pretty good privacy

PING   *See* packet Internet grouper

point of presence (POP)   A site where there exists a collection of telecommunications equipment, usually digital leased lines and multiprotocol routers.

point-to-point protocol (PPP)   The point-to-point protocol, defined in RFC 1661, provides a method for transmitting packets over serial point-to-point links. There are many other RFCs that define extensions to the basic protocol. *See also* serial line IP.

POP   *See* post office protocol and point of presence

port   A port is a transport layer demultiplexing value. Each application has a unique port number associated with it. *See also* transmission control protocol, user datagram protocol.

post office protocol (POP)   A protocol designed to allow single user hosts to read electronic mail from a server. Version 3, the most recent and most widely used, is defined in RFC 1725. *See also* electronic mail.

Postal Telegraph and Telephone (PTT)   Outside the United States, PTT refers to a telephone service provider, which is usually a monopoly, in a particular country.

postmaster   The person responsible for taking care of electronic mail problems, answering queries about users, and other related work at a site. *See also* electronic mail.

PPP   *See* point-to-point protocol

pretty good privacy (PGP)   A program, developed by Phil Zimmerman, that cryptographically protects files and electronic mail from being read by others. It may also be used to digitally sign a document or message, thus authenticating the creator. *See also* encryption, data encryption standard, RSA.

privacy enhanced mail (PEM)   Internet e-mail that provides confidentiality, authentication, and message integrity using various encryption methods. *See also* electronic mail, encryption.

Prospero   A distributed file system that provides the user with the ability to create multiple views of a single collection of files distributed across the Internet. Prospero provides a file naming system, and file access is provided by existing access methods (e.g. anonymous FTP and NFS). The Prospero protocol is also used for communication between clients and servers in the archie system. *See also* anonymous FTP, archie, archive site, Gopher, network file system, wide area information servers.

protocol   A formal description of message formats and the rules two computers must follow to exchange those messages. Protocols can describe low-level details of machine-to-machine interfaces (e.g., the order in which bits and bytes are sent across a wire) or high-level exchanges between allocation programs (e.g., the way in which two programs transfer a file across the Internet).

protocol converter   A device or program that translates between different protocols. Also called a gateway function.

protocol data unit (PDU)   PDU is international standards committee speak for packet. *See also* packet.

protocol stack   A layered set of protocols that work together to provide a set of network functions. *See also* layer, protocol.

proxy ARP   The technique in which one machine, usually a router, answers ARP requests intended for another machine. By "faking" its identity, the router accepts responsibility for routing packets to the real destination. Proxy ARP allows a site to use a single IP address with two physical networks. Subnetting would normally be a better solution. *See also* address resolution protocol.

PSN   *See* packet switch node.

PTT   *See* Postal, Telegraph and Telephone

pulse-code modulation (PCM)   Modulation involving the conversion of a waveform from analog to digital by means of coding.

quantization   The process of digital encoding that involves the conversion of the exact instantaneous sample value of a continuous signal to the nearest equivalent digital value from a finite set of discrete values.

queue   A backup of packets waiting to be processed.

RARE   Reseaux Associes pour la Recherche Europeenne. *See* Trans-European Research and Education Networking Association.

RARP   *See* reverse address resolution protocol

RBOC   regional Bell operating company

reassembly   The IP process in which a previously fragmented packet is re-assembled before being passed to the transport layer. *See also* fragmentation.

recursive   See recursive

regional   *See* mid-level network

remote login   Operating on a remote computer, using a protocol over a computer network, as though locally attached. *See also* Telnet.

Remote Procedure Call (RPC)   An easy and popular paradigm for implementing the client-server model of distributed computing. In general, a request is sent to a remote system to execute a designated procedure, using arguments supplied, and the result returned to the caller. There are many variations and subtleties in various implementations, resulting in a variety of different (incompatible) RPC protocols.

repeater   A device that propagates electrical signals from one cable to another. *See also* bridge, gateway, router.

request for comments (RFCs)   The document series, begun in 1969, that describes the Internet suite of protocols and related experiments. Not all (in fact very few) RFCs describe Internet standards, but all Internet standards are written up as RFCs. The RFC series of documents is unusual in that the proposed protocols are forwarded by the Internet research and development community, acting on their own behalf, as opposed to the formally reviewed and stand-ardized protocols that are promoted by organizations such as CCITT and ANSI. *See also* BCP, FYI, STD.

Reseaux IP Europeens (RIPE)   A collaboration between European networks that use the TCP/IP protocol suite.

reverse address resolution protocol (RARP)   A protocol, defined in RFC 903, that provides the reverse function of ARP. RARP maps a hardware (MAC) address to an internet address. It is used primarily by diskless nodes when they first initialize to find their internet address. *See also* address resolution protocol, BOOTP, internet address, MAC address.

RFC   *See* request for comments

RFC 822   The Internet standard format for electronic mail message headers. Mail experts often refer to "822 messages." The name comes from RFC 822, which contains the specification. 822 format was previously known as 733 for-mat. *See also* electronic mail.

RIP   *See* routing information protocol

RIPE   *See* Reseaux IP Europeenne

round-trip time (RTT)   A measure of the current delay on a network.

route   The path that network traffic takes from its source to its destination. Also, a possible path from a given host to another host or destination.

routed   Route Daemon. A program that runs under 4.2BSD/4.3BSD UNIX systems (and derived operating systems) to propagate routes among machines on a local area network, using the RIP protocol. Pronounced "route-dee." *See also* routing information protocol, gated.

router   A device that forwards traffic between networks. The forwarding deci-sion is based on Internet layer information and routing tables, often constructed by dynamic routing protocols. *See also* bridge, gateway, exterior gateway proto-col, interior gateway protocol.

router ID   A 32-bit number assigned to each router running the OSPF proto-col. This number uniquely identifies the router within an autonomous system.

routing   The process of selecting the correct interface and next hop for a packet being forwarded. *See also* hop, router, exterior gateway protocol, interior gateway protocol.

routing domain   A set of routers exchanging routing information within an administrative domain. *See also* administrative domain, router.

routing information protocol (RIP)   A distance vector, as opposed to link state, routing protocol. It is an Internet standard IGP defined in RFC 1058. *See also* interior gateway protocol, open shortest-path first.

RPC   *See* remote procedure call

RSA   A public-key cryptographic system that may be used for encryption and authentication. It was invented in 1977 and named for its inventors, Ron Rivest, Adi Shamir, and Leonard Adleman. *See also* encryption, data encryption standard, pretty good privacy.

RTT   *See* round-trip time

SDH   *See* synchronous digital hierarchy

serial line IP (SLIP)   A protocol used to run IP over serial lines, such as telephone circuits or RS-232 cables, interconnecting two systems. SLIP is defined in RFC 1055, but is not an Internet standard. It is being replaced by PPP. *See also* point-to-point protocol.

server   A provider of resources (e.g. file servers and name servers). *See also* client, domain name system, network file system.

SGML   *See* standardized generalized markup language

signature   The three- or four-line message at the bottom of a piece of e-mail or a Usenet article that identifies the sender. Large signatures (over five lines) are generally frowned upon. *See also* electronic mail, Usenet.

simple mail transfer protocol (SMTP)   A protocol used to transfer electronic mail between computers. It is specified in RFC 821, with extensions specified in many other RFCs. It is a server-to-server protocol, so other protocols are used to access the messages. *See also* electronic mail, post office protocol, RFC 822.

simple network management protocol (SNMP)   The Internet standard protocol developed to manage nodes on an IP network. The first version is defined in RFC 1157 (STD 15). SNMPv2 (version 2) is defined in too many RFCs to list. It is currently possible to manage wiring hubs, toasters, jukeboxes, etc. *See also* Management Information Base.

SLIP   *See* serial line IP

SMDS   *See* switched multimegabit data service

SMI   *See* structure of management information

SMTP   *See* simple mail transfer protocol

SNA   *See* systems network architecture

SNAP   *See* subnetwork attachment point

snail mail   A pejorative term referring to the U.S. postal service

SNMP   *See* simple network management protocol

SONET   *See* synchronous optical network

standardized generalized markup language (SGML)   An international standard for the definition of system-independent, device-independent methods of representing text in electronic form. *See also* hypertext markup language.

STD   A subseries of RFCs that specify Internet standards. The official list of Internet standards is in STD 1. *See also* request for comments.

stream-oriented   A type of transport service that allows its client to send data in a continuous stream. The transport service will guarantee that all data will be delivered to the other end in the same order as sent and without duplicates. *See also* transmission control protocol.

structure of management information (SMI)   The rules used to define the objects that can be accessed via a network management protocol. These rules are defined in RFC 1155 (STD 17). The acronym is pronounced "Ess Em Eye." *See also* management information base.

stub network   A stub network only carries packets to and from local hosts. Even if it has paths to more than one other network, it does not carry traffic for other networks. *See also* backbone, transit network.

subnet   A portion of a network, that may be a physically independent network segment, which shares a network address with other portions of the network and is distinguished by a subnet number. A subnet is to a network what a network is to an internet. *See also* internet, network.

subnet address   The subnet portion of an IP address. In a subnetted network, the host portion of an IP address is split into a subnet portion and a host portion using a subnet mask.

subnet mask   A bit mask used to identify which bits in an IP address correspond to the network and subnet portions of the address.

subnet number   *See* subnet address

subnetwork attachment point (SNAP)   The name given to IEEE 802.1b, which is a header used to provide multiprotocol support to several link layer protocols.

supernet   An aggregation of IP network addresses advertised as a single classless network address. For example, given four Class C IP networks, 192.0.8.0, 192.0.9.0, 192.0.10.0, and 192.0.11.0, each having the intrinsic network mask of 255.255.255.0, one can advertise the address 192.0.8.0 with a subnet mask of 255.255.252.0. *See also* IP address, network address, network mask, classless interdomain routing.

switched multimegabit data service (SMDS)   An emerging high-speed datagram-based public data network service developed by Bellcore and expected to be widely used by telephone companies as the basis for its data networks. *See also* metropolitan area network.

synchronous   Pertaining to events that occur at the same time or at the same rate.

synchronous digital hierarchy (SDH)   The European standard for high-speed data communications over fiber-optic media. The transmission rates range from 155.52 Mbps to 2.5 Gbps.

synchronous optical network (SONET)   SONET is an international standard for high-speed data communications over fiber-optic media. The transmission rates range from 51.84 Mbps to 2.5 Gbps.

systems network architecture (SNA)   A proprietary networking architecture used by IBM and IBM-compatible mainframe computers.

T1   A term for a digital carrier facility used to transmit a DS-1 formatted digital signal at 1.544 Mbps.

T3   A term for a digital carrier facility used to transmit a DS-3 formatted digital signal at 44.746 Mbps.

TAC   See terminal access controller (TAC)

TAC   technical assistance center (TAC for ACC)

talk   A protocol that allows two people on remote computers to communicate in a real-time fashion. *See also* Internet relay chat.

TCP   See transmission control protocol

TCP/IP protocol suite   transmission control protocol over Internet protocol. This is a common shorthand that refers to the suite of transport and upper layer protocols that runs over IP. *See also* IP, ICMP, TCP, UDP, FTP, Telnet, SMTP, SNMP.

TELENET   The original name for what is now SprintNet. It should not be confused with the Telnet protocol or application program.

telephony   A system of telecommunications in which voice or other data originally in the form of sounds are transmitted over long distances.

Telnet   Telnet is the Internet standard protocol for remote terminal connection service. It is defined in RFC 854 and extended with options by many other RFCs.

terminal emulator   A program that allows a computer to emulate a terminal. The workstation thus appears as a terminal to the remote host.

terminal server   A device that connects many terminals to a LAN through one network connection. A terminal server can also connect many network users to its asynchronous ports for dial-out capabilities and printer access. *See also* local area network.

time to live (TTL)   A field in the IP header that indicates how long this packet should be allowed to survive before being discarded. It is primarily used as a hop count. *See also* Internet protocol.

TN3270   A variant of the Telnet program that allows one to attach to IBM mainframes and use the mainframe as if you had a 3270 or similar terminal.

token ring   A token ring is a type of LAN with nodes wired into a ring. Each node constantly passes a control message (token) on to the next; whichever node has the token can send a message. Often, token ring, as named by IBM, is used to refer to the IEEE 802.5 token ring standard, which is the most common type of token ring. *See also* 802.x, local area network.

topology   A network topology shows the computers and the links between them. A network layer must stay abreast of the current network topology to be able to route packets to their final destination.

traceroute   A program available on many systems that traces the path a packet takes to a destination. It is mostly used to debug routing problems between hosts. There is also a traceroute protocol defined in RFC 1393.

Trans-European Research and Education Networking Association (TERENA)   TERENA was formed in October 1994 by the merger of RARE and EARN to promote and participate in the development of a high-quality international information and telecommunications infrastructure for the benefit of research and education. *See also* European Academic and Research Network (EARN).

transceiver   transmitter-receiver. The physical device that connects a host interface to a local area network, such as Ethernet. Ethernet transceivers contain electronics that apply signals to the cable and sense collisions. The equivalent IEEE 802.3 name is medium attachment unit (MAU).

transit network   A transit network passes traffic between networks in addition to carrying traffic for its own hosts. It must have paths to at least two other networks. *See also* backbone, stub network.

transmission control protocol (TCP)   An Internet standard transport layer protocol defined in RFC 793. It is connection-oriented and stream-oriented, as opposed to UDP. *See also* connection-oriented, stream-oriented, user datagram protocol.

Trojan horse   A computer program that carries within itself a means to allow the creator of the program access to the system using it. *See also* virus, worm.

trunk   A transmission channel, or group of channels, connected between two nodes in a communications system.

TTL   See time to live

tunneling   Tunneling refers to encapsulation of protocol A within protocol B, such that A treats B as though it were a datalink layer. Tunneling is used to get data between administrative domains that use a protocol that is not supported by the internet connecting those domains. *See also* administrative domain.

twisted pair   A type of cable in which pairs of conductors are twisted together to produce certain electrical properties.

UDP   See user datagram protocol

unicast   An address that only one host will recognize. *See also* broadcast, multicast.

uniform resource locator (URL)   A URL is a compact (most of the time) string representation for a resource available on the Internet. URLs are primarily used to retrieve information using WWW. The syntax and semantics for URLs are defined in RFC 1738. *See also* World Wide Web.

universal time coordinated (UTC)   This was previously called Greenwich mean time (GMT).

URL   *See* uniform resource locators

Usenet   A collection of thousands of topically named newsgroups, the computers that run the protocols, and the people who read and submit Usenet news. Not all Internet hosts subscribe to Usenet, and not all Usenet hosts are on the Internet. *See also* network news transfer protocol.

user datagram protocol (UDP) An Internet standard transport layer protocol defined in RFC 768. It is a connectionless protocol that adds a level of reliability and multiplexing to IP. *See also* connectionless, transmission control protocol.

UTC *See* universal time coordinated

Veronica A Gopher utility that effectively searches Gopher servers based on a user's list of keywords. The name was chosen to be a "mate" to another utility named "Archie." It later became an acronym for very easy rodent-oriented net-wide index to cComputer archives. *See also* archie, Gopher.

virtual circuit A network service that provides connection-oriented service without necessarily doing circuit-switching. *See also* connection-oriented.

virus A program that replicates itself on computer systems by incorporating itself into other programs that are shared among computer systems. *See also* Trojan horse, worm.

W3 *See* World Wide Web

WAIS *See* wide area information servers

WAN *See* wide area network

WebCrawler A WWW search engine. The aim of the WebCrawler Project is to provide a high-quality, fast, and free Internet search service. The WebCrawler may be reached at "http://webcrawler.com/".

WG *See* working group

whois An Internet program that allows users to query a database of people and other Internet entities, such as domains, networks, and hosts. The primary database is kept at the InterNIC. The information stored includes a person's company name, address, phone number, and e-mail address. The latest version of the protocol, WHOIS++, is defined in RFCs 1834 and 1835. See also InterNIC.

wide area information servers (WAIS) A distributed information service that offers simple natural language input, indexed searching for fast retrieval, and a "relevance feedback" mechanism that allows the results of initial searches to influence future searches. Public domain implementations are available. *See also* archie, Gopher, Prospero.

wide area network (WAN) A network, usually constructed with serial lines, that covers a large geographic area. *See also* local area network, metropolitan area network.

working group (WG) A working group, within the IETF, is a group of people who work under a charter to achieve a certain goal. That goal may be the creation of an informational document, the creation of a protocol specification, or the resolution of problems in the Internet. Most working groups have a finite lifetime. That is, once a working group has achieved its goal, it disbands. There is no official membership for a working group. Unofficially, a working group member is somebody who is on that working group's mailing list; however, anyone may attend a working group meeting. *See also* Internet Engineering Task Force.

World Wide Web (WWW, W3) A hypertext-based, distributed information system created by researchers at CERN in Switzerland. Users may create, edit, or browse hypertext documents. The clients and servers are freely available.

worm A computer program that replicates itself and is self-propagating. Worms, as opposed to viruses, are meant to spawn in network environments. Network worms were first defined by Shoch & Hupp of Xerox in ACM Communications (March 1982). The Internet worm of November 1988 is perhaps the most famous; it successfully propagated itself on over 6,000 systems across the Internet. *See also* Trojan horse, virus.

X X is the name for TCP/IP based network-oriented window systems. Network window systems allow a program to use a display on a different computer. The most widely-implemented window system is X11—a component of MIT's Project Athena.

X.25 A data communications interface specification developed to describe how data passes into and out of public data communications networks. The CCITT and ISO approved protocol suite defines protocol layers 1 through 3.

X.400 The CCITT and ISO standard for electronic mail. It is widely used in Europe and Canada.

X.500 The CCITT and ISO standard for electronic directory services. *See also* whois.

XDR *See* external data representation

Xerox Network System (XNS) A protocol suite developed by Xerox Corporation to run on LAN and WAN networks, where the LANs are typically Ethernet. Implementations exist for both Xerox's workstations and 4.3BSD, and 4.3BSD-derived, systems. XNS denotes not only the protocol stack, but also an architecture of standard programming interfaces, conventions, and service functions for authentication, directory, filing, e-mail, and remote procedure call. XNS is also the name of Xerox's implementation. *See also* Ethernet, Berkeley Software Distribution, local area network, wide area network.

XNS See Xerox Network System

Yahoo! Yahoo! is a hierarchical subject-oriented guide for the World Wide Web and Internet. Yahoo! lists sites and categorizes them into appropriate subject categories. Yahoo! may be reached at http://www.yahoo.com/.

zone A logical group of network devices.

Bibliography

Black, Uyless, *OSI: A Model for Computer Communications Standards,* Englewood Cliffs, NJ: Prentice Hall, 1991.

Black, Uyless, *ATM: Foundation for Broadband Networks,* Englewood Cliffs, NJ: Prentice Hall, 1995.

Cerf, Vinton, *Internet Technology Handbook,* Menlo Park, CA: SRI International, Network Information Service Center, 1991 (Available on CD-ROM).

Comer, Douglas E., *Internetworking With TCP/IP: Volume I, Principles, Protocols, and Architecture,* Englewood Cliffs, NJ: Prentice Hall, 1988.

Comer, Douglas E., *Internetworking With TCP/IP: Volume II, Design Implementation and Internals,* Englewood Cliffs, NJ: Prentice Hall, 1991.

Davidson, John, *Introduction to TCP/IP,* Ungermann-Bass Corporation.

Doll, Dixon R., *Data Communications Facilities, Networks and Systems Design,* New York, NY: John Wiley and Sons, 1978.

Feinler, Elizabeth J., Ole J. Jacobsen, Mary K. Stahl, and Carol A. Ward, *DDN Protocol Handbook* (3 volumes), Menlo Park, CA: SRI International, DDN Network Information Center, December 1985. (No longer sold—find a library copy.)

Folts, Harold C., *McGraw-Hill Compilation of Open Systems Standards, Edition IV* (six volumes), New York, NY: McGraw-Hill Inc., 1990.

Freeman, Roger L., *Telecommunication System Engineering,* New York, NY: John Wiley and Sons, 1996.

Garcia-Luna-Aceves, Jose J., Mary K. Stahl, and Carol A. Ward, *Internet Protocol Handbook: The Domain Name System (DNS) Handbook,* Menlo Park, CA: SRI International, Network Information Systems Center, August 1989. (No longer sold—find library copy.)

Graham, Buck, T*CP/IP Addressing,* Orlando, FL: AP Professional, 1977.

Griffiths, John M., *ISDN Explained,* Chichester, West Sussex, England: John Wiley and Sons, 1992.

Hedrick, Charles L., *Introduction to Administration of an Internet-Base Local Network,* Piscataway, NJ: Rutgers University Computer Science Facilities Group, July 24, 1988.

Held, Gilbert, *Understanding Data Communications,* Indianapolis, IN: Sams Publishing, 1996.

Kessler, Gary C., and Peter Southwick, *ISDN Concepts, Facilities and Services,* McGraw-Hill, 1997.

Lynch, Daniel C., and Marshall T. Rose, *Internet System Handbook,* Greenwich, CT: Manning Publications Co., 1993.

Manterfield, Richard J., *Common Channel Signalling,* Steverage, Herts, United Kingdom: Peter Peregrinus Ltd., 1991.

McNamara, John E., *Technical Aspects of Data Communications,* Bedford, MA: Digital Press, Digital Equipment Corporation, 1977.

Muller, Nathan J., *Intelligent Hubs,* Norwood, MA: Artech House, Inc., 1993.

McConnell, John, *Internetworking Computer Systems: Interconnecting Networks and Systems,* Englewood Cliffs, NJ: Prentice Hall, 1988.

Naugle, Matthew, *Network Protocol Handbook,* McGraw-Hill, Inc., 1994.

Rose, Marshall T., *The Simple Book, An Introduction to Management of TCP/IP-Based Internets,* Englewood Cliffs, NJ: Prentice Hall, 1991.

Rose, Marshall T., *The Open Book: A Practical Perspective on OSI,* Englewood Cliffs, NJ: Prentice Hall, 1989.

Santifaller, Michael, *TCP/IP and NFS: Internetworking in a UNIX Environment,* Reading, MA: Addison-Wesley Publishing Company, 1991.

Stallings, William, *Handbook of Computer-Communications Standards Volume 1: The Open System (OSI) Model and OSI-Related Standards,* New York, NY: Macmillan, 1990.

Stallings, William, *Handbook of Computer-Communications Standards Volume 2: Local Area Network Standards,* New York, NY: Macmillan, 1990.

Stallings, William, *Handbook of Computer-Communications Standards Volume 3: The TCP/IP Protocol Suite,* New York, NY: Macmillan, 1990.

Tanenbaum, Andrew S., *Computer Networks, Second Edition,* Englewood Cliffs, NJ: Prentice Hall, 1988.

Thomas, Stephen A., *Ipng and the TCP/IP Protocols,* John Wiley & Sons, 1996.

Unisys Corporation, *How to Speak Open Systems,* Copyright 1989, Blue Bell, PA.

Weik, Martin H., *Communications Standard Dictionary,* New York, NY: Van Nostrand Reinhold Publishing, 1983.

All RFCs identified with qualifications throughout this book.

About the Author

Floyd Wilder is the course developer at Advanced Computer Communications (ACC) where he has developed an internetworking class covering many of the topics of this book, including various TCP/IP protocols and Internet interfaces.

Prior to joining ACC, he was an independent consultant based in California, specializing in the interfaces used for data communications, the protocols used in networking (e.g., X.25 and TCP/IP), and network management.

His introduction to data communications began with writing software programs that edited and routed messages defined by ACP127/JANAP128. The programs were used in a military message switching system that replaced a torn-tape center in 1963. This activity was balanced with a part-time college program; he earned his B.A. in mathematics at California State at Los Angeles in 1965.

During the 1970s, he was network development manager at Bell Laboratories for two years and program manager for several major communications systems, including Canadian National Telegraph, Richardson Stock Brokerage, Chrysler Corporation, the Federal Reserve Banks of Atlanta and Philadelphia, Barclays Bank of England, and General Electric. He has written several training handbooks and *Recommendation X.25, Changes for 1988* published by Auerbach Publishers.

Technical comments, suggestions, or questions regarding this book may be directed to Mr. Wilder's e-mail address, fwilder@hotmail.com.

Index

481

Recent Titles in the Artech House Telecommunications Library

Vinton G. Cerf, Senior Series Editor

Computer Telephony Integration, Second Edition, Rob Walters

Convolutional Coding: Fundamentals and Applications,
Charles Lee

Desktop Encyclopedia of the Internet, Nathan J. Muller

*Distributed Multimedia Through Broadband Communications
Services,* Daniel Minoli and Robert Keinath

Electronic Mail, Jacob Palme

*Enterprise Networking: Fractional T1 to SONET, Frame Relay to
BISDN,* Daniel Minoli

FAX: Digital Facsimile Technology and Applications, Second Edition,
Dennis Bodson, Kenneth McConnell, and Richard Schaphorst

Guide to ATM Systems and Technology, Mohammad A. Rahman

Guide to Telecommunications Transmission Systems,
Anton A. Huurdeman

A Guide to the TCP/IP Protocol Suite, Floyd Wilder

Information Superhighways Revisited: The Economics of Multimedia,
Bruce Egan

International Telecommunications Management, Bruce R. Elbert

Internet E-mail: Protocols, Standards, and Implementation,
Lawrence Hughes

Internetworking LANs: Operation, Design, and Management,
Robert Davidson and Nathan Muller

Introduction to Satellite Communication, Second Edition,
Bruce R. Elbert

Introduction to Telecommunications Network Engineering,
Tarmo Anttalainen

Introduction to Telephones and Telephone Systems, Third Edition,
A. Michael Noll

LAN, ATM, and LAN Emulation Technologies, Daniel Minoli and
Anthony Alles

The Law and Regulation of Telecommunications Carriers,
Henk Brands and Evan T. Leo

Videoconferencing and Videotelephony: Technology and Standards,
  Second Edition, Richard Schaphorst

Visual Telephony, Edward A. Daly and Kathleen J. Hansell

World-Class Telecommunications Service Development,
  Ellen P. Ward

For further information on these and other Artech House titles,
including previously considered out-of-print books now available
through our In-Print-Forever® (IPF®) program, contact:

Artech House

685 Canton Street

Norwood, MA 02062

Phone: 781-769-9750

Fax: 781-769-6334

e-mail: artech@artechhouse.com

Artech House

46 Gillingham Street

London SW1V 1AH UK

Phone: +44 (0)171-973-8077

Fax: +44 (0)171-630-0166

e-mail: artech-uk@artechhouse.com

Find us on the World Wide Web at:
www.artechhouse.com